Thrive in Cell Biology

Other titles in the Thrive in Bioscience series

Thrive in Biochemistry and Molecular Biology

Lynne S. Cox, David A. Harris, and Catherine J. Pears

Thrive in Genetics

Alison Thomas

Thrive in Ecology and Evolution

Alan Beeby and Ralph Beeby

Forthcoming: **Thrive in** Human Physiology

Ian Kay and Gethin Evans

Thrive in Immunology

Anne Cunningham

Thrive in Cell Biology

Qiuyu Wang
School of Healthcare Science, Manchester Metropolitan University, UK

Chris Smith
Formerly of Manchester Metropolitan University, UK

Emma Davis
School of Cancer and Enabling Sciences, University of Manchester, UK

Thrive in Cell Biology

OXFORD
UNIVERSITY PRESS

OXFORD
UNIVERSITY PRESS

Great Clarendon Street, Oxford, OX2 6DP,
United Kingdom

Oxford University Press is a department of the University of Oxford.
It furthers the University's objective of excellence in research, scholarship,
and education by publishing worldwide. Oxford is a registered trade mark of
Oxford University Press in the UK and in certain other countries

First Edition published in 2013

Published in the United States of America by Oxford University Press
198 Madison Avenue, New York, NY 10016, United States of America

British Library Cataloguing in Publication Data
Data available

ISBN 978–0–19–969732–8

Contents

Contents

Four steps to exam success

1 *Review the facts*

This book is designed to help your learning be quick and effective:

- Information is set out in bullet points, making it easy to digest.
- Clear, uncluttered illustrations illuminate what is said in the text.
- Key concept panels summarize the essential learning points.

2 *Check your understanding*

- Try the questions throughout each chapter and online multiple-choice questions to reinforce your learning.
- Download the flashcard glossary to master the essential terms and phrases.

3 *Take note of extra advice*

- Look out for revision tips, and hints for getting those precious extra marks in exams.

4 *Go the extra mile*

- Explore the suggestions for further reading listed on the book's web site to take your understanding one step further.

Go to the Online Resource Centre for more resources to support your learning, including:

- Answers to *Check your understanding* questions.
- Suggestions for further reading, and links to relevant materials on other web sites.
- Online quizzes, with feedback.
- A flashcard glossary, to help you master the essential terminology.

www.oxfordtextbooks.co.uk/orc/thrive/

online resource centre

1 Introduction to cell biology

Key features of cells

- All organisms are composed of cells, which can only arise from the division of pre-existing cells.
- Differences in the structures of cells and their molecules allow all organisms to be classified into one of three domains: the Archaea, Bacteria, or the Eucarya.
- The Archaea and Eucarya are the most closely related of all three domains.
- Eucarya possess nuclei, mitochondria, and chloroplasts that are thought to have evolved from bacterial ancestors.
- The fabric of all cells is similar; they contain, for example, carbohydrates, nucleic acid, proteins, and lipids.

1.1 THE DISCOVERY OF CELLS

Living cells (Figure 1.1) were first observed by the Dutch linen merchant Leeuwenhoek who observed unicellular organisms in pond water, and later blood cells, oral bacteria, and spermatozoa.

- His remarkable observations were made using self-made microscopes with almost spherical lenses only about 1 mm in diameter but capable of magnifications of 300–500× (Figure 1.2).

The discovery of cells

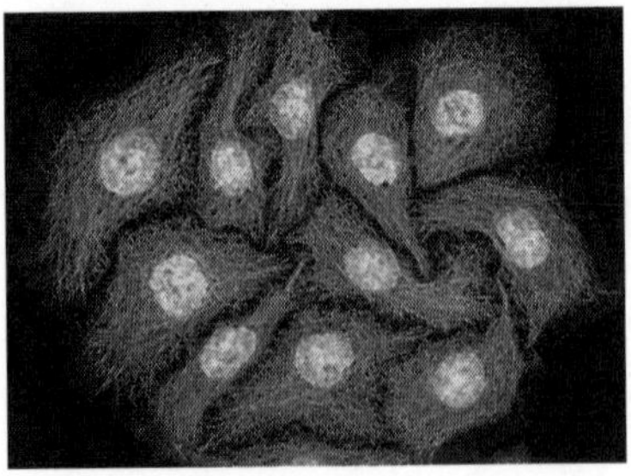

Figure 1.1 Human cervical adenocarcinoma (HeLa) cells viewed by confocal microscopy. Courtesy of Olympus Corporation, UK.

- Leeuwenhoek reported his initial findings to the Royal Society of London in 1674; the term *cell* had been used in 1665 by the Society's curator, Hooke, to describe the empty cell walls of cork (Figure 1.3), and this name became an established term.
- Progress in cell biology was initially slow because of the poor resolution of microscopes, lack of technical equipment, and a suitable paradigm in which new findings could be incorporated.
- In 1831, Brown, using an improved type of microscope, observed the nucleus, which controls the activities and development and is the largest structure in many cells.
- In 1838 and 1839, Schleiden and Schwann working independently proposed that all organisms are composed of cells.
- In 1858, Virchow, using the findings and conclusions of Remak, stated that 'all cells arise from other cells' and firmly established the **cell theory**.

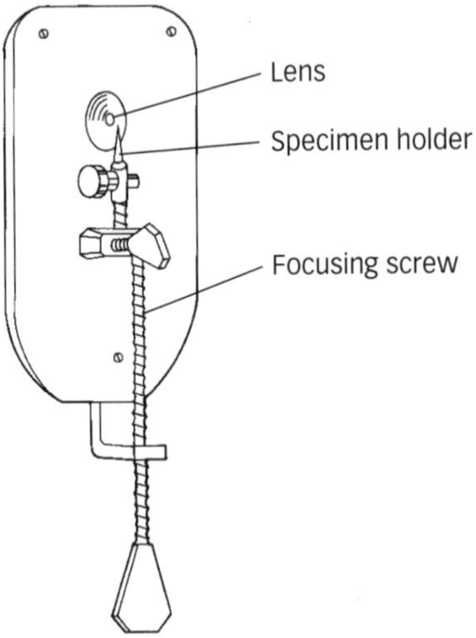

Figure 1.2 A Leeuwenhoek microscope.

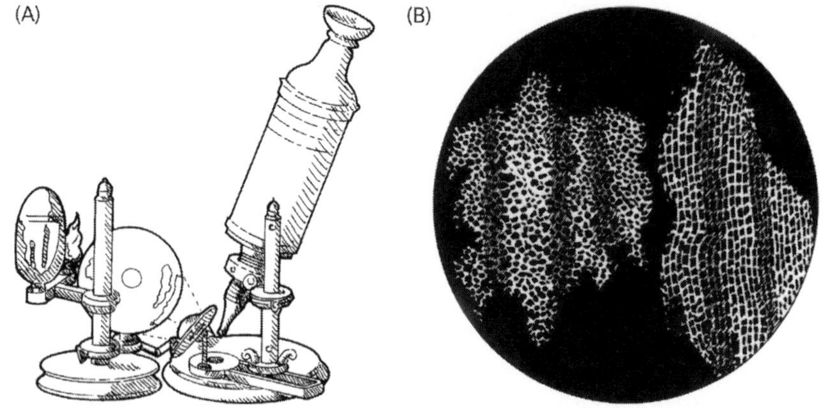

Figure 1.3 (A) Hook's microscope and (B) cork as first observed by Hook in 1665.

- Hence by the beginning of the twentieth century, biologists had established the importance of cells to biological studies.
- In 1937, Chatton used the terms **eukaryotique** to describe those cells with nuclei and **prokaryotique** for those without nuclei, such as bacteria.
- The cell theory, Darwin's theory of evolution by natural selection, and the elucidation of the structure and role of **deoxyribonucleic acid** (**DNA**) by Crick and Watson, form the three great unifying ideas of modern biology.
- Developments in technical equipment, particularly since the 1950s, have allowed cell biology to progress to its present high state of knowledge, although much remains to be discovered.

1.2 CLASSIFICATION OF ORGANISMS

Despite cells being the basic structural and functional unit of all organisms, differences between them are sufficient to classify organisms into different groups.

- Woese and co-workers, for example, examined the sequences of a ribosomal ribonucleic acid (RNA) (the 16 S RNA; see Chapters 5 and 6) from different organisms and concluded they comprised three phylogenetic **domains**:
  1. Archaea
  2. Eucarya
  3. Bacteria.
- Subsequently, the determination of the sequences of nucleic acids from many species has produced a vast amount of data that has been essential in determining the evolutionary relationships between these groups.
- These data suggest that the Bacteria and Eucarya domains are distinct groups that diverged from a **last universal common ancestor** (**LUCA**) and that domain Archaea is more closely related to domain Eucarya and separate from the Bacteria domain (Figure 1.4).

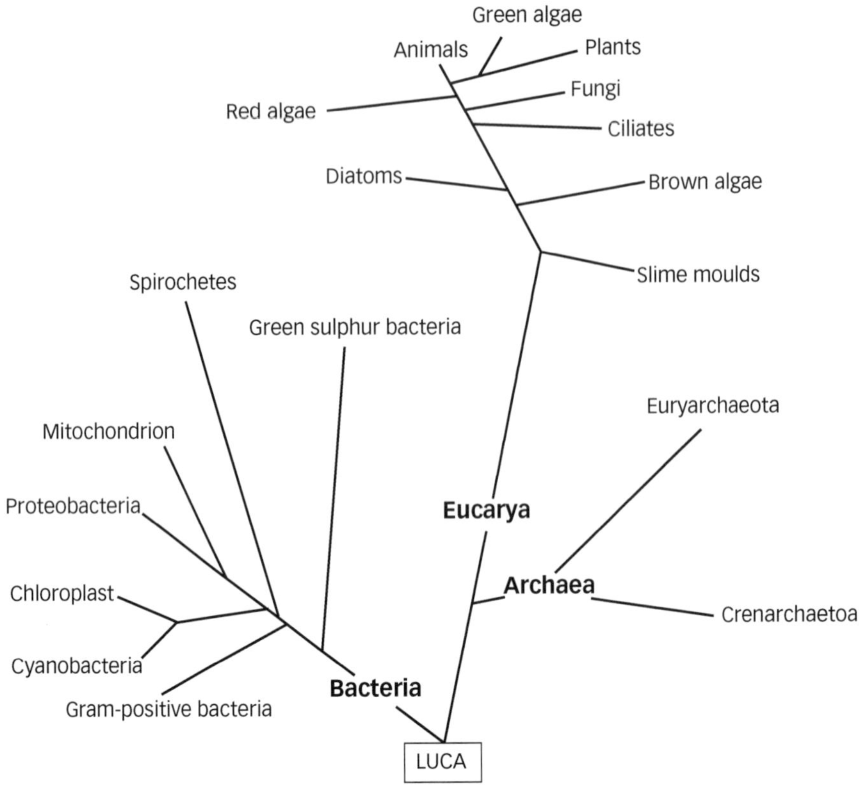

Figure 1.4 Phylogenetic tree showing evolutionary relationships between representative members of the three domains.

Archaea

The domain Archaea consists of prokaryotic organisms.

- Prokaryotic cells are surrounded by a surface membrane and are encased in a cell wall, while the interior is generally undifferentiated **cytoplasm** containing the genetic material as a single, freely suspended **chromosome** (Figure 1.5).

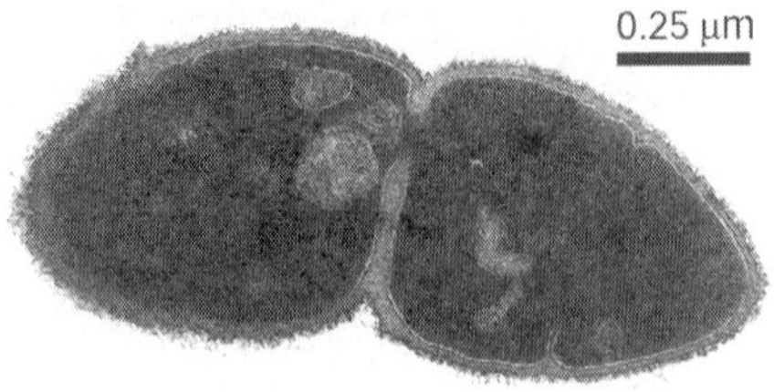

Figure 1.5 Electron micrograph of a dividing archaeon cell, *Methanobacterium ruminantium*. Courtesy of Drs K.A. Schwartz and J.G. Zeikus. Reprinted with permission from Zeikus, J.G and Bowen, V.G. (1976) Ultrastructure of methogenic bacteria. *Canadian Journal of Microbiology*, 21, 121–9.

- Archaea can be distinguished from bacteria, which are also prokaryotic, because of differences in their cell walls and membranes (Chapter 3), and because their molecular biology is more like that of eukaryotes than bacteria (Chapter 5).
- Archaea were originally found occupying ecological niches characterized by extremes: for example, high temperatures (the extreme thermophilic organism), high concentrations of salts (halophiles), or an environment that is extremely chemically reducing in nature (methanogens).
- Archaea consist of two groups:
  1. Crenarchaeota, which contains the extreme thermophiles.
  2. Euryarchaeota, which contains the halophiles and methanogens.
- Archaea are now thought to also occur in more 'normal' conditions.

Eucarya

Eucarya are composed of eukaryotic cells that possess a **nucleus**.

- Their cells are surrounded by a plasma membrane and most types are encased in a cell wall.
- The cytoplasm of eukaryotic cells contains an **endomembrane system** (Chapters 8 and 9) and discrete structures limited by membranes called **organelles** that have specific function(s); cytoplasm surrounding the endomembrane system and organelles is called the **cytosol** (Figure 1.6).
- Eucarya include many of the commonly seen groups of organisms, for example, plants, animals, seaweeds, and fungi as well as diverse groups of microorganisms such as protozoa (Figure 1.4).

Bacteria

Bacteria consist of prokaryotic organisms that superficially appear similar to archaea (Figure 1.7) but are distinguishable from them by differences in their membranes, biochemistry, and molecular biology (Chapters 4 and 5).

- In some cases, the cytoplasm of bacterial cells is divided into microcompartments by infoldings of the surface membrane or shells of proteins.
- Bacteria comprise at least 23 distinct groups of microorganisms that vary in their metabolisms and the ecological niches they occupy.
- Individual species of bacteria were initially classified on the basis of differences in the shapes of their cell, for example, bacilli are rod-shaped, cocci spherical, and spirochetes helical. However, bacterial cells share many common features (Chapter 4).

Major differences between archaea, eucarya, and bacteria

Visually, archaeal and bacterial cells appear similar because both types are composed of prokaryotic cells and are of similar sizes.

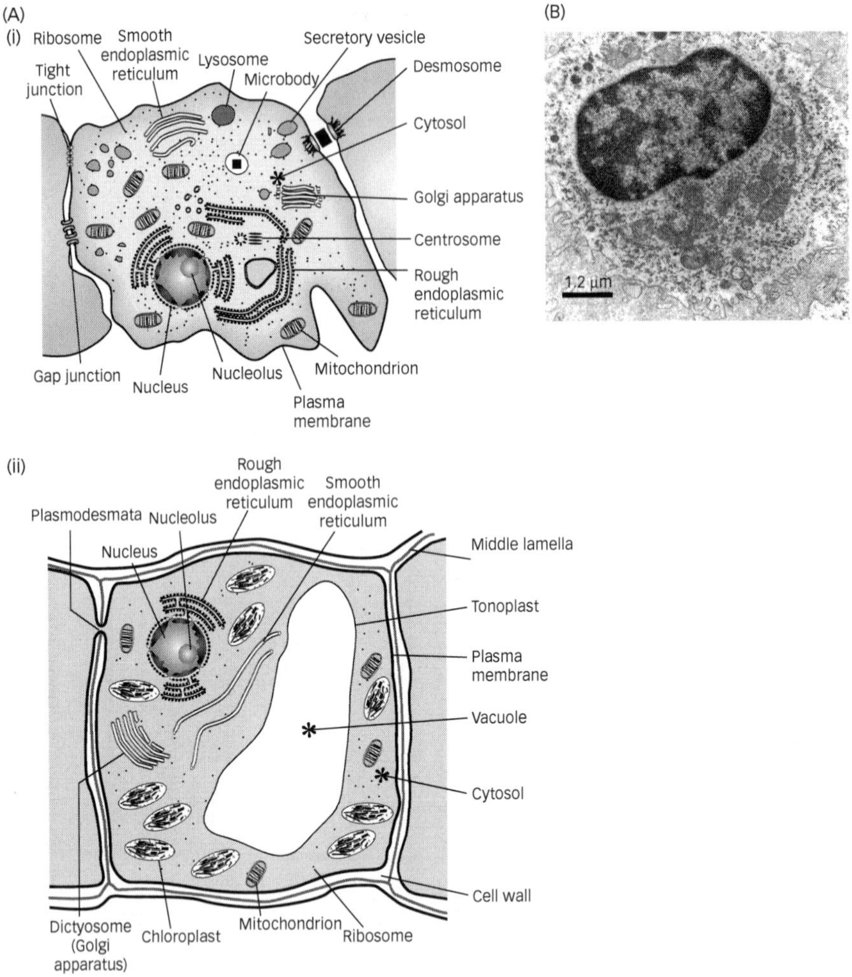

Figure 1.6 (A) Schematic illustrations of a typical eukaryotic (i) animal and (ii) plant cell. (B) Electron micrograph of a human monocyte (a type of lymphocyte). Note the numerous intracellular compartments compared with Figures 1.5 and 1.7. Courtesy of Dr A. Curry, Public Health Laboratory, Withington Hospital, Manchester, UK.

- Archaeal and bacterial cells differ from eukaryotic cells in generally being smaller and not having a nucleus or other organelles, although this is no longer regarded as a strong taxonomic character, and the two groups are thought to be widely different in their evolutionary lineages (Figure 1.4).
- In terms of their biochemistry and molecular biology, archaea and eucarya are the more closely related groups, although bacterial and eukaryotic cells also share a number of features (Table 1.1).

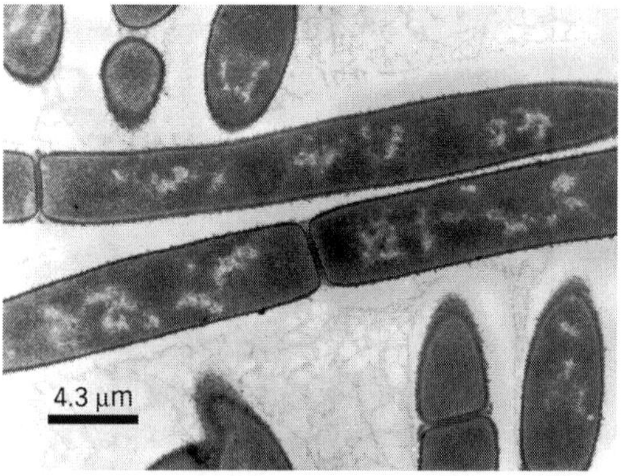

Figure 1.7 Electron micrograph of rod-shaped bacterial cells. Courtesy of Dr I.D.J. Burnett, National Institute for Medical Research, London, UK. Compare with Figures 1.5 and 1.6, and Figure 4.1 in Chapter 4.

Feature	***Archaea***	***Eucarya***	***Bacteria***
Nucleus	Not present	Present (Chapter 6)	Not present
Organelles	Not present	Present	Not present
Cell diameter (μm)	1–10	10–100	1–10
Chromosomes	Circular and associated with histone-like proteins (Chapter 4)	Linear chromosomes associated with histones (Chapter 6)	Circular chromosome associated with bacterial proteins (Chapter 4)
Reproduction	Binary division (Chapter 5)	Mitosis and meiosis (Chapter 16)	Binary division (Chapter 5)
RNA polymerase	Single enzyme with 8–12 subunits (Chapter 5)	Three specific eukaryotic enzymes with 12–14 subunits (Chapter 6)	Single, specific bacterial enzyme with 4 subunits (Chapter 5)
Post-transcriptional activities	Moderate (Chapter 5)	Extensive (Chapter 6)	Some (Chapter 5)
Ribosomes	70 S (Chapter 5)	80 S (Chapter 7)	70 S (Chapter 5)
Translation initiated by	Methionine (Chapter 5)	Methionine (Chapter 7)	Formylmethionine (Chapter 5)
Translation requires	Multiple initiation and elongation factors	Multiple initiation and elongation factors	Two initiation and three elongation factors
Cytoskeleton	Actin- and tubulin-type proteins present	Actin, intermediate filaments, and tubulin proteins present (Chapter 7)	Actin- and tubulin-type proteins present (Chapter 4)
Membrane lipids (Chapter 3)	Branched polyprenyl glycerol esters	Phosphoacylglycerols Sphingomyelins Glycolipids Sterols	Phosphoacylglycerols
Flagella and cilia	Archaeal types (Chapter 4)	Eukaryotic 9 + 2 types (Chapter 7)	Bacterial types (Chapter 4)

Table 1.1 Major differences between archaeal, eukaryotic, and bacterial cells

Differences and similarities between archaea, bacteria, and eukaryotes are highlighted in Chapters 3–7.

Revision tip

Explore the structures of cells using 'The Cell: An Image Library™' of The American Society for Cell Biology at: http://www.cellimagelibrary.org/

Check your understanding

1.1 Which of the following statements is/are true and which is/are false?

a. Leeuwenhoek was the first to observe and use the term *cell*.

b. Virchow was the first to realize all cells are derived from pre-existing cells using the phrase 'all cells arise from other cells'.

c. Differences in the sequences of nucleic acids suggest that bacteria and archaea are more closely related to each other than are the archaea and eucarya.

d. Archaea were first discovered in extreme types of environments.

e. Nuclei are not present in archaea and bacteria.

1.3 EVOLUTION OF CELLS

The earth was formed about 4.5 billion years ago; conditions compatible with life are thought to have occurred within 200 to 300 million years.

- The initial atmosphere was reducing in nature, containing large quantities of dinitrogen (N_2), methane (CH_4), dihydrogen (H_2), and carbon dioxide.
- Various types of energy, such as lightning, ultraviolet solar radiation, or volcanic heat could have caused components of the atmosphere to react and form a variety of organic compounds now associated with organisms.
- It is thought that life began with a group(s) of molecules that were capable of catalysing their self-replication.
- Given that RNA molecules can store genetic information and catalyse a number of biological reactions, it has been suggested that these earliest self-replicating systems consisted of RNA molecules: a so-called *RNA world*.
- Self-replicating RNA molecules could have recruited amino acids and peptides, which would have improved their catalytic powers and been selected for, leading to the present protein-rich biosphere.
- Fossil evidence shows microbial life was present 3.5 billion years ago and had possibly evolved much earlier, with LUCA first appearing approximately 4 billion years ago. This cell is generally thought to have lacked a nucleus and organelles.
- A number of hypotheses have been proposed to explain the origin of eukaryotic cells including the fusion of two or three different types of prokaryotic cells.
- Eukaryotes possess a number of unique characteristics but also some archaeal and bacterial ones.

- For example, eukaryotic genes are partly of the bacterial type and partly of the archaeal type, implying a mixed ancestry.
- Generally, genes with nuclear functions (informational genes) have archaeal characteristics; those whose products have cytoplasmic metabolic and 'housekeeping' functions possess bacterial features.
- One hypothesis to explain this distribution is that the nucleus arose from an archaeon that consumed hydrogen to synthesize methane and which formed an endosymbiotic relationship with a host bacterial cell that produced hydrogen by fermentation. The bacterial genes were eventually transferred into the archaeon, which evolved to become the nucleus.
- Alternatively, the nucleus may have arisen from invaginations of the bacterial surface membrane; to date, no idea has been fully accepted.
- Approximately 2.4 billion years ago, oxygenic photosynthesis began converting the atmosphere from one rich in methane and hydrogen to one containing large amounts of dioxygen (O_2).
- Oxygen-using organelles such as mitochondria (Chapter 11) and peroxisomes (Chapter 12) must have been acquired after that date.
- Although the evolutionary origins of the nucleus are debatable, mitochondria and chloroplasts definitely arose from bacteria.
- The ancestor of mitochondria was an aerobic bacterial endosymbiont of an anaerobic host.
- Plants evolved later when a mitochondrial-containing cell (or at least one with a mitochondrial precursor) acquired a cyanobacterium-like endosymbiont that became the ancestor of chloroplasts.
- Hydrogenosomes and mitosomes (Chapter 11) evolved later from mitochondria as secondary adoptions to anaerobic environments.
- Despite differences in structure and evolutionary origins, all cells have extremely similar compositions.

The evolution of different organelles is revised in Chapters 6 and 10–12.

Check your understanding

1.2 The more primitive a biological structure, the earlier it appeared during evolution. With this in mind, which two of the three domains are considered the most primitive?

1.4 COMPOSITION OF CELLS

Cells are largely composed of water and a variety of inorganic minerals and organic constituents.

- Cells are composed of approximately 60–70% water, most of which is not free but associated with macromolecules.

Composition of cells

- The major minerals present are K^+, Mg^{2+}, and phosphate and sulphate ions (P_i and S_i).
- In multicellular organisms, the extracellular fluid surrounding cells is rich in Na^+, Cl^-, and HCO_3^-.
- The major organic compounds are:
  - carbohydrates
  - nucleotides and nucleic acids
  - amino acids and proteins
  - lipids.
- Some carbohydrates, nucleic acids, and proteins are **macromolecules**: these have a relative molecular mass, M_r, in excess of approximately 5000.
- Biological macromolecules are polymers, which are constructed by covalently linking much smaller units together in condensation reactions (Figure 1.8).
- Polymeric carbohydrates are formed by linking simple sugars together; nucleic acids are polymers of **nucleotides**, and proteins are polymers of **amino acids**.

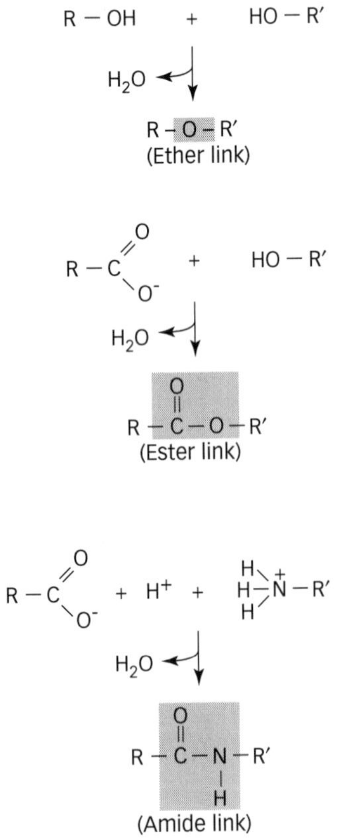

Figure 1.8 Examples of condensation reactions used in forming biological macromolecules.

- The units within the polymer are called *residues*; linking two together gives a dimer, three a trimer, and so on to eventually form a polymer that may contain hundreds or thousands of residues.
- Lipids are not macromolecules but in aqueous environments can aggregate to form large structures such as droplets or lipid bilayers (Figure 3.3).

Revision tip

Explore the structures of macromolecules with the easy to use Jmol molecular modelling program at the Protein Data Base: http://www.rcsb.org/pdb/home/home.do

- Details of the structures and functions of some of these biological molecules can be obtained from the companion volumes of this series and in the 'Further reading' listed on the dedicated website.

➔ *See the companion volumes* Thrive in Biochemistry and Molecular Biology *and* Thrive in Genetics *for a detailed revision of biological molecules.*

Revision tip

Do browse through the further reading references given for this chapter online. Go to http://www.oxfordtextbooks.co.uk/orc/thrive/

Check your understanding

1.3 Which of the following statements is/are true and which is/are false?

a. Archaea and bacteria can be distinguished by differences in their cell walls, ribosomes, and biochemistry.

b. All present-day organisms are thought to have evolved from self-replicating molecules.

c. Mitochondria are thought to have evolved from the domain Bacteria.

d. Within the Eucarya, genes with nuclear functions generally have bacterial characteristics but the products of those with archaeal features normally have cytoplasmic metabolic and 'housekeeping' functions.

e. The major intracellular cations are Na^+ and Mg^{2+}, and the major one in the extracellular fluid of multicellular organisms is K^+.

2 Methods of studying cells

Key points when studying cells

- Studying cells, as in all sciences, is dependent upon the methods and technology available.
- Numerous methods have been devised to culture, identify, and characterize cells and their components, which has resulted in a wealth of advances in cell biology.

2.1 CULTURING CELLS

Culturing prokaryotic and eukaryotic cells means growing them in artificial media in a controlled environment. In addition to providing adequate numbers of specified cells for study, this also provides material for isolating and studying their components.

- Most types of cells can be cultured, including bacteria and archaea, fungi, plant, and animal cells.
- Cells to be cultured can be isolated from the environment or from an already established cell line, or, in the case of multicellular organisms, harvested from tissues or organs.
- The technique chosen to grow cells depends upon the characteristics of the cells concerned and the impact of the technique on the methods to be subsequently used in studying them and their component parts.

Basic requirements for culturing cells

Although the specific requirements for growing different cell types vary, most cell cultures require the following:

- A medium containing nutrients such as amino acids, carbohydrates, vitamins, and minerals.
- An environment regulated for pH, temperature, and osmotic pressure.
- Gases (O_2, CO_2).
- For mammalian cells the environment is commonly kept at 37°C and 5% CO_2, but other cells require different conditions to be cultured successfully.

Culturing prokaryotic cells

Most types of prokaryotic cells, such as *Escherichia coli*, are relatively easy to grow (culture) and can generally survive in less specific conditions; some are much more fastidious and require more complex media containing, for example, blood.

- In most general cases, prokaryotic cells can be grown on plates of solid agar that have been loaded with nutrients, or they are grown in simple nutrient broths containing amino acids, sugars, salts, and, of course, water.
- Archaea have been less studied than bacteria, often because they require more demanding growing conditions.

Growing bacteria on agar

- Cultures can be initiated by streaking a starting culture across the surface of a sterile agar medium using an inoculation loop (Figure 2.1).
- Single cells can be isolated on a plate by dragging the culture in phases across the surface of the plate in a systematic manner.
- After a period of growth, a colony derived from one cell can be isolated from the plate and used to inoculate another agar plate or a bottle of sterile nutrient broth.
- The culture of cyanobacterial and other photosynthetic prokaryotic cells may differ from that of other bacteria because sources of light and nitrogen are essential if they are to be grown as photosynthesizing organisms.

Growing archaeal cells

- Archaea can be split into three main types:
  1. thermophiles and extreme thermophiles that grow at high temperatures
  2. extreme halophiles which grow in high concentrations of salt
  3. methanogens which produce methane.
- Media for growing archaea can be either solid or liquid, although liquid media is usually favoured for thermophiles as solid media are usually unstable at optimum growing conditions.
- The media can contain a source of carbon or sulphur as energy sources, peptone or yeast extracts to supply vitamins, and a complex of salts and minerals.

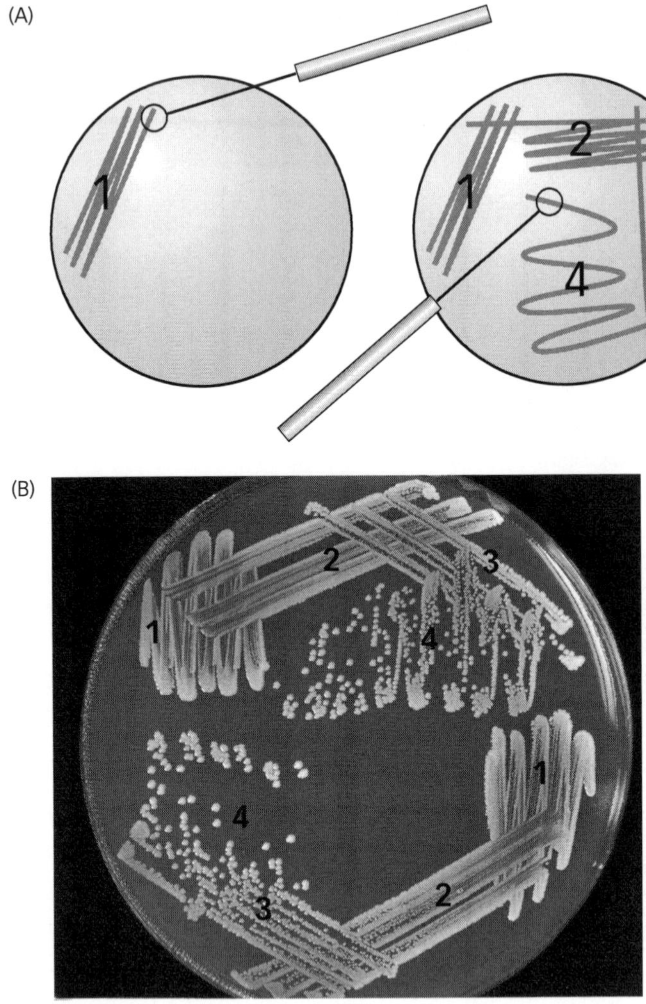

Figure 2.1 (A) Schematic application (streaking) of a culture of bacteria across the surface of nutrient agar in a Petri dish using an inoculation loop. (B) Bacterial colonies growing on the surface of nutrient agar following streaking and incubation at 37°C for 24 hours. The numbers indicate the order in which the cells have been spread over the surface.

Thermophiles and extreme thermophiles

- Thermophiles and extreme thermophilic archaea typically grow at temperatures far higher than those tolerated by bacteria and eukaryotes.
- Ideal growing temperatures for thermophilic archaea are reported to be greater than 75°C, while extreme thermophiles require a very high temperature (80–105°C); culturing these organisms therefore requires specialized ovens or incubators.

- Many thermophiles also require sulphur for growth, either as an electron acceptor in anaerobic or as an oxidizable energy source in aerobic conditions.
- Extreme thermophiles that oxidize sulphur typically grow in acidic conditions partly because they acidify their own environment by oxidizing sulphur to sulphate (SO_4^{2-}).

Extreme halophiles

- Extreme halophiles must be grown in concentrations of NaCl greater than 2 mol dm^{-3}.

Methanogens

- Methanogenic archaea are obligate anaerobes; oxygen is toxic to them.
- Methanogens are cultured in gas-tight containers that have been previously flushed with dinitrogen gas.
- Methanogens must be supplied with CO_2 and H_2 which they metabolize to methane (CH_4).

Culturing eukaryotic cells

Generally, the culture of eukaryotic cells requires a more complex medium than that of most bacteria.

- Eukaryotic cells do require basic nutrients, such as amino acids, carbohydrates, and vitamins, but they may also require the addition of growth factors and hormones to the culture medium.
- Usually these are provided by adding sterile calf serum to the culture.
- The two main methods of culturing animal cells are:
  1. adherent culture methods for anchorage-dependent cells
  2. suspension culture methods for non-anchorage-dependent cells.
- The choice of which type to use therefore depends on the characteristics of the cells being cultured.

Adherent culture

- Cells that must be grown as adherent cultures need to be supplied with a solid or semi-solid surface on which to grow.
- The cells may be grown in specialized culture containers, such as Carrel flasks, or on beads called microcarriers made of materials such as carbohydrates, collagen, acrylamide, or glass (Figure 2.2).
- The growth of adherent-dependent cells is limited by the surface area available for growth, as cell–cell interactions can arrest the cell cycle and prevent further division; a process called **contact inhibition** (Chapter 16).
- This method of culturing allows the cells to be easily visualized using a microscope as the cells are adhered to the walls of culture container or to the microcarrier in a thin layer.
- Most animal cells from solid tissues grow to form a **monolayer** (Figure 2.2) in which the cells adhere to each other.

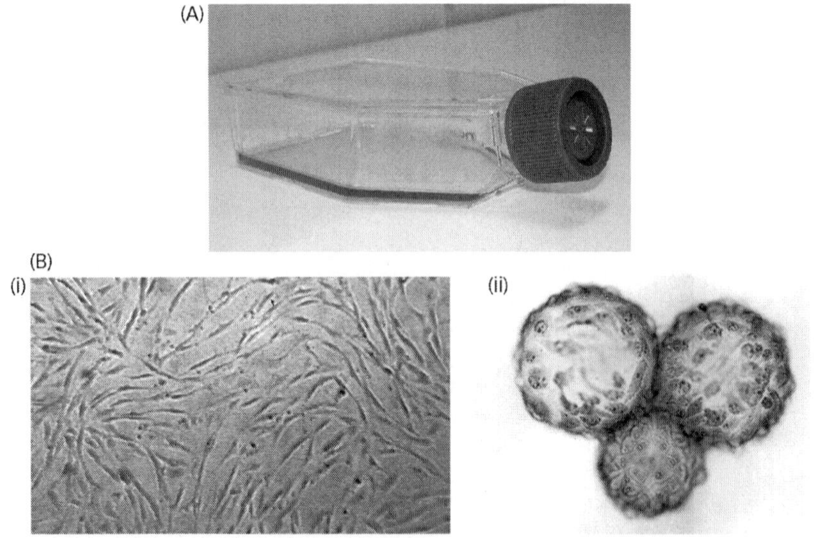

Figure 2.2 (A) A Carrel flask used to culture small quantities of eukaryotic cells. (B) Light micrographs showing monolayers of anchorage-dependent animal cells growing in (i) a Carrel flask and (ii) adhering to microcarrier beads.

Suspension culture

- Non-anchorage-dependent cells can be grown freely suspended in the culture medium.
- Suspension cultures are suitable for cells adapted to living in suspension and hematopoietic cells, which are non-adhesive.
- The number of cells able to grow in the culture is limited by the concentration of the cells in the media; hence frequent cell counts are required to monitor cell growth. Adding more of the growing medium will result in more cell growth.
- Constant shaking is required during culture to maintain an adequate gas exchange.
- A major advantage of growing cells in suspension culture is that cells can easily be removed and subcultured.

Types of cell cultures

- Eukaryotic cell cultures may be **primary**, **secondary**, or **immortalized cultures**.
- Primary cultures are started from cells taken directly from the donor organism and have a finite lifespan.
- **Passaging** or subculturing a primary culture produces a secondary culture; passaging involves removing small quantities of cells from the medium and transferring them into larger volumes of fresh media.
- Passaging is easily performed with suspension cultures, but adherent cells must first be detached from their growing surface by using a protease; a small number of released cells can then be used to make a subculture.
- Secondary cultures are able to undergo a finite number of cell divisions before becoming senescent.

- Immortalized cultures will continue to grow and divide for as long as suitable conditions are maintained.
- HeLa cells, a type commonly encountered in cell biology, are immortalized human epithelial cells that have active telomerase that prevents cell death following shortening of chromosome telomeres (Chapter 6). They are named after the source of their primary culture: a cancer of the cervix from a patient called Henrietta Lacks.

Check your understanding

2.1 Which of the following statements is/are true?

a. Archaea have been less studied than bacteria because they require more demanding growing conditions.

b. Eukaryotic cells are usually easier to culture than prokaryotic ones because they can generally survive in less specific conditions.

c. Mammalian cells are commonly cultured in an environment kept at 37°C and 5% CO_2.

d. The growth of adherent cultures is limited by the available surface area.

e. Primary cultures are started using cells taken directly from the donor organism and most have a finite lifespan.

Maintaining cell cultures

- As cells grow in the culture medium they require monitoring and the replacement of depleted nutrients on a regular basis.
- Techniques for handling and maintaining cultures differ depending on the type of cells being grown, though the following conditions must be monitored to maintain a healthy culture:
  - depletion of nutrients in the growth media
  - excessive cell growth (since cell-to-cell contact can lead to cell cycle arrest and cessation of growth or, in some cases, cell differentiation)
  - accumulation of dead cells.
- Changing the media or subculturing will reduce the effects of all three.
- For adherent cells, the media may be changed by simply aspirating the media out of the culture container and replacing it with fresh media.

Contamination of cell cultures

- The controlled and nutrient-rich environment of cultured cells can leave the cultures susceptible to contamination with bacteria, fungi, and viruses.
- The use of appropriate aseptic techniques is essential to reduce the risk of contamination, though it always remains a possibility.

Bacterial contamination

- Bacterial contamination is easily observed because contaminated cell cultures appear cloudy or turbid.
- It is usually detectable within a few days of infection either by a sudden drop in the pH of the medium or by visual inspection of the cells using a microscope.
- The use of antibiotics is not recommended as a routine additive to cell culture media because they promote the growth of antibiotic resistant strains of bacteria.

Fungal contamination

- The dormant spores of fungi are generally tolerant of harsh conditions and so are difficult to totally eradicate from cell culture areas.
- Cell cultures that are heavily contaminated with fungi have an increased pH and may appear turbid.
- The presence of fungi can be confirmed by a microscopic examination; the thread-like fungal hyphae being diagnostic.

Viral infection

- The initial source of cells or contaminated serum can introduce viruses into cell cultures.
- Unlike bacterial and fungal contamination, viruses are not generally detectable in cultures due to their small size, unless polymerase chain reaction (PCR) techniques or electron microscopy (see 'Electron microscopy' section) are used.
- Virus contamination may result in the destruction of the cell culture.

Chemical contamination

- Chemical contamination of cell cultures can come from a number of sources such as contaminated media or equipment, including deposits of cleaning agents or disinfectants accidentally left on equipment after cleaning.
- Toxins from water or plastics used in the culture process may also contaminate cell cultures.

Checking cultures for viable cells

Checking the viability of the cells in a culture (that they are alive and capable of growth) is essential to estimate how well the culture is growing.

- The stain **trypan blue** is often used to assess viability.
- It readily stains dead or non-viable cells but it cannot pass through the plasma membrane and so does not stain living ones.
- Thus, counting the numbers of stained and non-stained cells in a sample from the culture allows its viability to be estimated.
- This method does not distinguish between apoptotic and necrotic cells.

➔ *Cell death by necrosis and apoptosis are revised in Chapter 17.*

2.2 FLOW CYTOMETRY OR CELL SORTING

Flow cytometry (or cell sorting) is a technique for counting cells, and distinguishing and isolating specific cell types in a population of several different kinds of cells.

- It can also be applied to other microscopic particles 0.5–40 μm in diameter, such as chromosomes.
- Flow cytometry can separate up to approximately 20,000 cells s^{-1} using a number of characteristics, such as their:
  - size
  - shape
  - internal components
  - surface labels.

Flow cytometers

An outline of a flow cytometer is shown in Figure 2.3.

- The cells to be sorted can be tagged with fluorescent-labelled antibodies (fluorophores) to aid detection.
- They are then freely suspended in a stream of fluid that passes through the detection system of the flow cytometer.
- Streaming is achieved by passing the cell suspension through a capillary tube that encloses it within a sheath solution, usually phosphate buffered saline (PBS), which helps to focus the cells into single file.
- The cells and their surrounding salt solution sheath pass through one or more lasers, which excite the fluorophores attached to the cells.
- Incident light is also scattered by the cells; the amount of scattering is proportional to the granularity of the cell: the more granular or complex the internal components of the cell the more light is scattered.
- The emitted fluorescence and scattered light is collected by optical fibres that direct the light to the relevant detectors, which convert the detected light into electrical signals that are then used to direct the separation of the cells.
- Cell separation is achieved at the end of the capillary, which is vibrated to form standard sized droplets of sheath solution each containing a single cell (Figure 2.3).
- The distance between the point of detection and that of droplet formation together with the rate of flow of the sheath solution are used to calculate the time when a cell of interest reaches the centre of a forming droplet.
- As the droplet is formed, a positive or a negative charge is applied to the droplet, which then falls between two oppositely charged plates that deflect it into a collection vessel (Figure 2.3).
- Flow cytometers automatically generate plots or histograms to summarize cell numbers and separations (Figure 2.4).

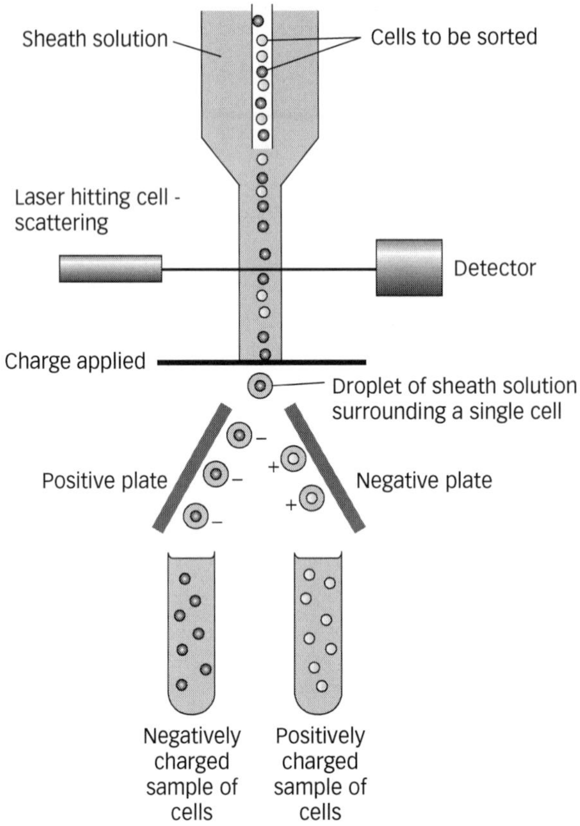

Figure 2.3 Schematic illustration of cell sorting by flow cytometry. The lighter shaded spheres represent cells labelled with fluorescent dye. They are given a positive charge as they move through the flow cytometer. The positive charge on the droplet surrounding the labelled cell ensures it is deflected towards the negatively charged plate and so is separated from the non-labelled darker 'cells'.

Revision tip

Life Technologies™ provides an animation tutorial introducing flow cytometry at: http://probes.invitrogen.com/resources/education/tutorials/4Intro_Flow/player.html

2.3 EXAMINING CELLS BY MICROSCOPY

Microscopy is a technique that uses a microscope to examine objects that are too small to be seen easily with the naked eye.

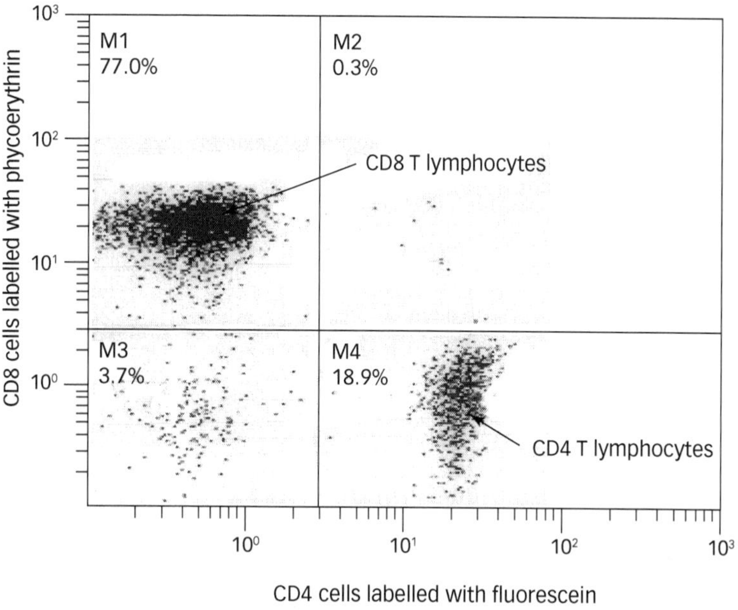

Figure 2.4 Dot plot showing the separation of CD4 and CD8 T lymphocytes by flow cytometry. Reproduced with permission from Glencross, H., Ahmed, N., and Wang, Q. (Eds) (2011) *Biomedical Science Practice*. Oxford: Oxford University Press. Original figure courtesy D. Spradbery, Immunology Department, Hull Royal Infirmary, UK.

A number of different types of microscopes are routinely used to study cells. These include:

- Light microscopes
- Phase contrast microscopes
- Fluorescence microscopes
- Confocal microscopes
- Electron microscopes.

Light microscopy

Light microscopes, also known as optical microscopes, are used to visualize cells. The commonest are compound microscopes, with several lenses (Figure 2.5).

- The light source is usually positioned below the specimen.
- The condenser lens focuses light onto the specimen which is mounted on a glass slide.
- The objective and eyepiece lenses produce the enlarged image that can then be received by the eye or a camera (Figure 2.6).

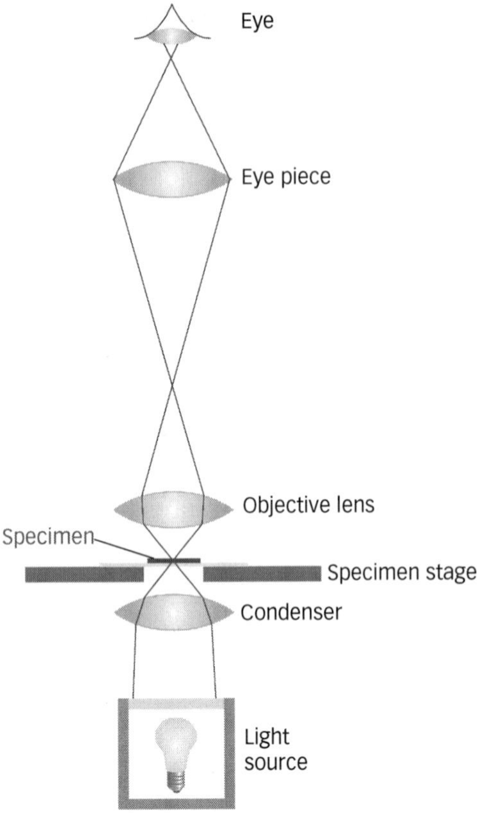

Figure 2.5 Schematic outline of a compound light microscope showing the path light takes through the specimen and lenses.

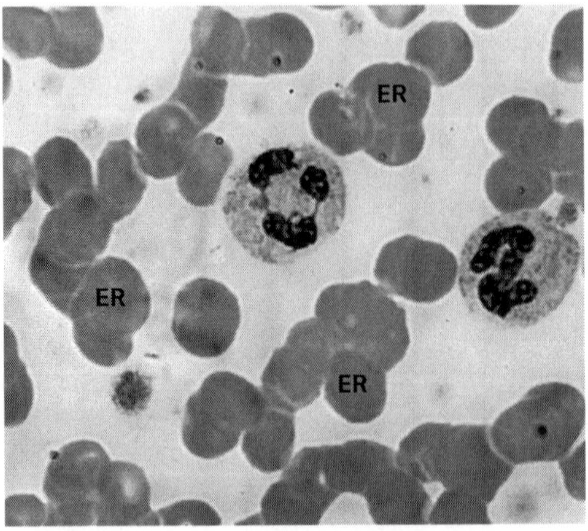

Figure 2.6 Light micrograph of a blood smear. The two central cells are neutrophils. ERs, erythrocytes.

Magnification, resolution, and contrast

- The three major features to consider with any microscope are:
  1. Magnification
  2. Contrast
  3. Resolution.

Magnification

- Magnification is the apparent size of the image formed by the microscope divided by its real size.
- Magnification is expressed in multiples: if a microscope allows an object to be visualized 10 times bigger than its actual size it is said to have a magnification of 10×.
- The magnification of a microscope is estimated by multiplying the magnification power of the objective lens by the power of the eyepiece lens. For example, objective lens power of 20× with an eyepiece power 10× gives a magnification of 200× (20 × 10).
- Light microscopes are limited to magnifications of up to approximately 2000× by the available resolution.

> ***Check your understanding***
>
> 2.2 If a microscope has an objective magnification of 40× and an eyepiece magnification of 5×, what is the total magnification?

Resolution

- Resolution can be defined as the minimum distance separating two points that can still be seen as distinct structures. It is the limit of observable size.
- The better the resolution, the finer the detail visible in the image: if the resolution is too low, objects that are close to each other will appear to blur and will not be distinguishable from each other.
- Resolution limits the magnification available with a microscope; increasing the magnification will not make the image any clearer.
- Lenses with a high resolution may require oil to be placed between the specimen and lens. Oil has a higher refractive index than air and allows more light to be transmitted through the specimen to the lens.
- In light microscopy, as resolution is increased, the brightness of the light source must also be increased.
- The resolution of a light microscope is relatively easy to calculate (see Box 2.1).

Contrast

- Contrast is the difference between the brightness of various details in the object compared to that of the background.

Box 2.1 Calculating the resolution of a microscope

- The resolution (r) of a light microscope can be calculated using:

$$r = \frac{0.61 \times \lambda}{NA}$$

  where λ is the wavelength of the light and NA is the **numerical aperture** of the lens.
- The NA is a numerical value that represents the ability of the condenser lens (Figure 2.5) to gather light.
- The theoretical maximum value for the NA in air is 1.0.
- The best possible value in air is approximately 0.95 although 1.4 can be achieved in oil.
- The limitation of resolution with light microscopy is approximately 300 nm in air and 200 nm in oil.
- The limitation of resolution with electron microscopy (see 'Electron microscopy' section) is roughly 0.2 nm.

Check your understanding

2.3 (a) Calculate the resolution of microscope A in (i) air and (ii) oil, using light with a wavelength of 500 nm. (b) Microscope B has resolutions of 300 and 200 nm in air and oil respectively. Which microscope has the better resolution?

- The contrast of a specimen depends on the interaction between the specimen and light.
- Most biological specimens have poor contrast and appear colourless or grey but their contrast can be improved by staining. Stains for light microscopy are coloured dyes; those for electron microscopy are salts of heavy metals.

Phase contrast microscopy

Phase contrast microscopy uses the phenomenon of interference to produce images of unstained living specimens.

- If two light rays are out of phase then they interfere with one another.
- If the rays are out of phase by half a wavelength, then interference results in the extinction of the ray.
- Figure 2.7 illustrates the operation of a phase contrast microscope.
- The sample is illuminated by a cone of light produced by an annular stop consisting of a ring positioned in the condenser. Light leaving the sample enters the objective lens.

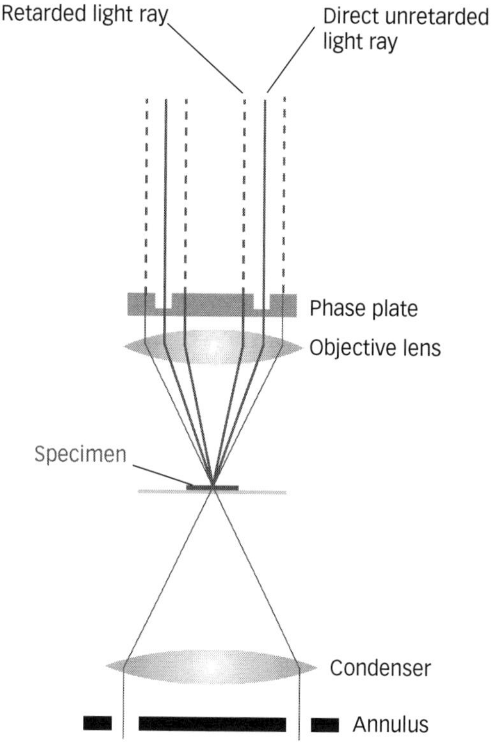

Figure 2.7 Schematic outline of a phase contrast microscope showing the path light takes through the annulus, specimen, and phase plate.

- Phase contrast microscope objective lenses have a glass disc called a phase plate that is exactly the same shape and size as the annulus.
- A ring is etched into the plate of such a depth that light rays passing through the full thickness of the phase plate are retarded by ¼ of a wavelength when compared with light passing through the ring.
- If the lights from the etched and unetched parts of the phase plate are combined they produce an interference pattern but this pattern does not have sufficient contrast to give an image.
- Biological specimens also retard light by approximately ¼ of a wavelength.
- The specimen scatters some of the light rays passing through it and these are gathered by the objective lens and pass through the *non-etched* part of the phase plate and are seen by the observer.
- When these unaltered rays are focused with those retarded by the etched ring they combine to form an image of the specimen.
- Small changes in refractive index within the specimen are seen as varying degrees of brightness in the image against a dark background; hence, phase contrast microscopy produces an image high in contrast despite being unstained (Figure 2.8).

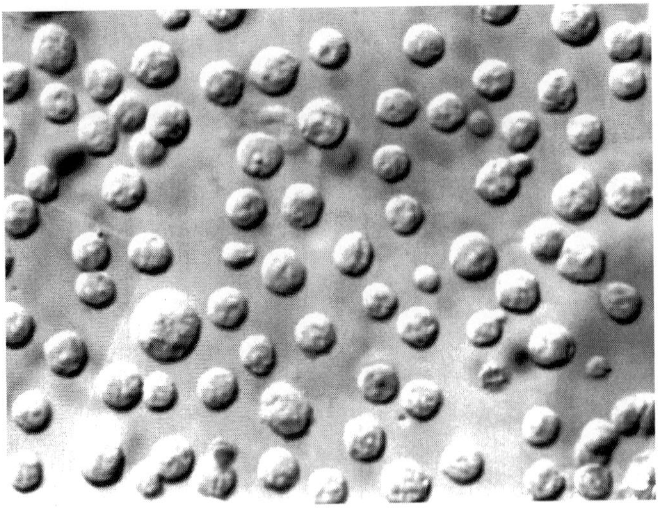

Figure 2.8 Light micrograph showing an interference image of a lawn of tumour cells. Courtesy of Dr J.T. Thornthwaite of the Cancer Research Institute of West Tennessee, USA (http://cancerfoundation.com).

Fluorescence microscopy

Fluorescence microscopy is an optical microscopy technique that uses fluorescence and phosphorescence to visualize the specimen, as opposed to absorption and reflection of light or electrons as in some other techniques.

- To be viewed, specimens must express a fluorescent molecule naturally or be stained with a fluorescent dye (fluorochrome).
- Most fluorescence microscopes have an epifluorescence arrangement with both the detector and the light source positioned above the specimen (Figure 2.9)
- The specimen is excited using light, often in the ultraviolet range, causing the specimen or fluorochrome to fluoresce.
- A filter ensures that only light of suitable wavelengths passes through a dichroic mirror to illuminate the specimen.
- The emitted visible fluorescent light from the specimen is allowed through the same dichroic mirror but in the opposite direction and into the eyepiece lens (Figure 2.9).
- The major advantage of the technique is that the specimen appears bright against a dark background (Figure 2.10).
- Limitations of this technique include photobleaching—where a fluorescent substance loses its ability to fluoresce as it is exposed to the light source.
- The resolution of fluorescence microscopes is 20–30 nm.

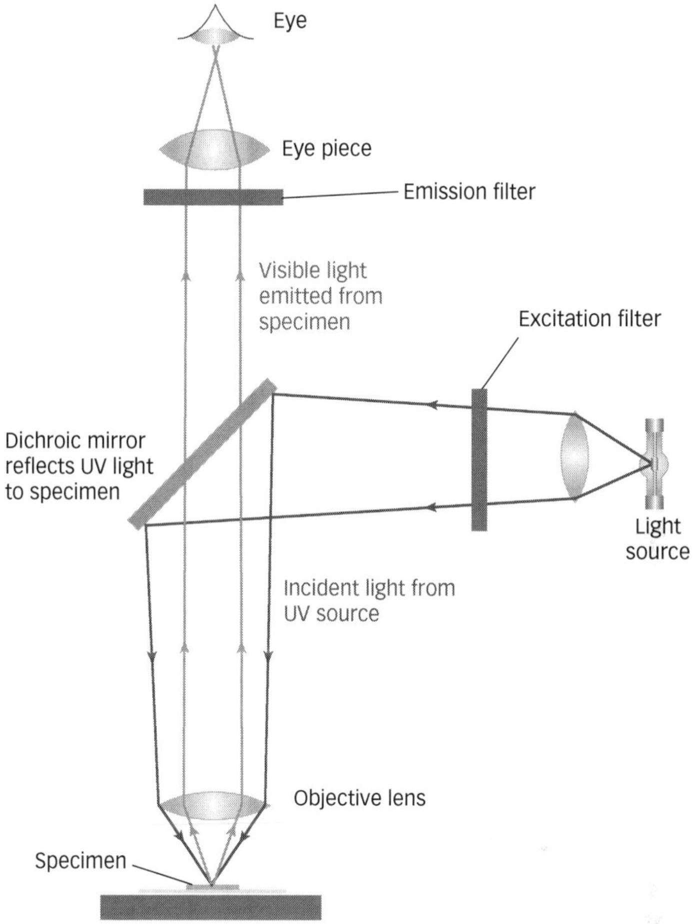

Figure 2.9 Schematic outline of a fluorescence microscope showing the paths of exciting light from its source to specimen and that of emitted light from the specimen after excitation.

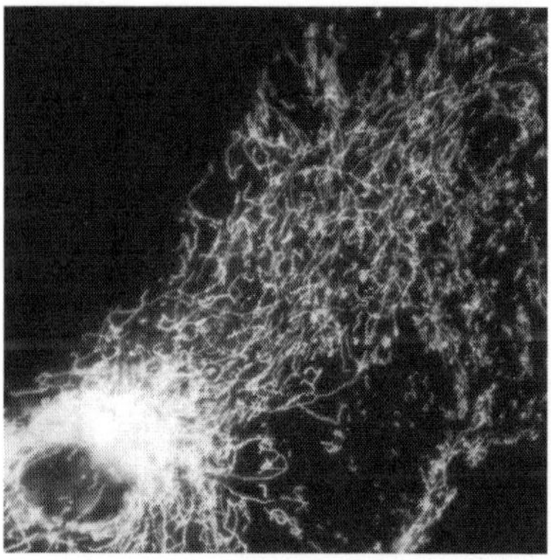

Figure 2.10 Light micrograph of a cell stained with the fluorochrome, rhodamine 123, and viewed using a fluorescence microscope.

Confocal laser scanning microscopy

Confocal laser scanning microscopy (CLSM) is a fluorescence-based microscopy technique.

- Confocal microscopes use a narrow beam of light emitted from a laser to illuminate the specimen (Figure 2.11).
- The laser light is focused through a pin hole so that only one plane of the specimen is illuminated at any one time.
- As the laser light scans the specimen the dyes used to stain it fluoresce.
- The fluorescent light passes through a pinhole that eliminates out-of-focus light from the specimens and only allows light from the plane in focus to fall on a photodetector, which produces a digital electronic signal.
- The signal is interpreted by a computer to form an image (Figure 2.12).
- The specimen can be successively examined at different focal planes to produce a set of images that the computer can store and use to generate a three-dimensional image of the specimen.
- Confocal microscopes have a resolution of 1–2 nm.

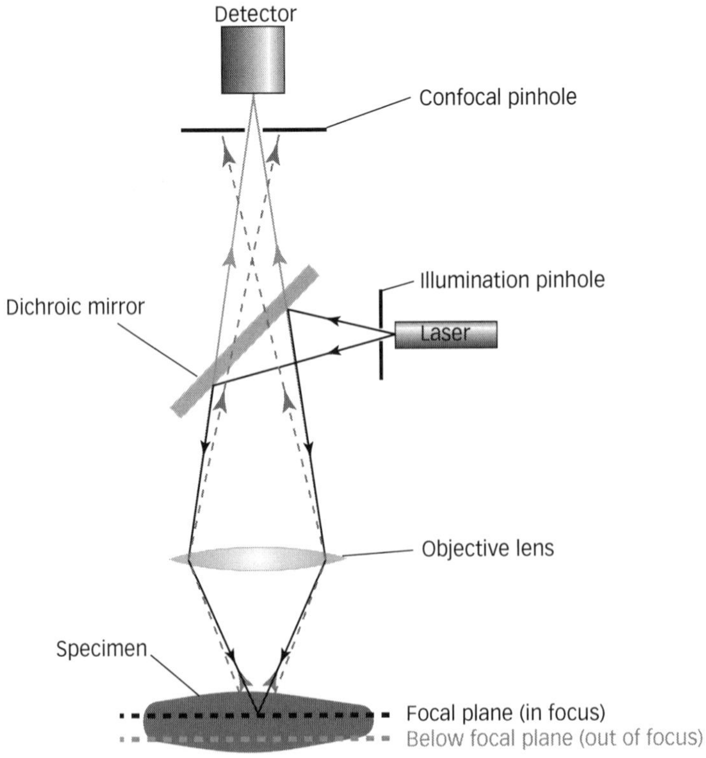

Figure 2.11 Schematic outline of a confocal microscope showing the light path taken and image capture.

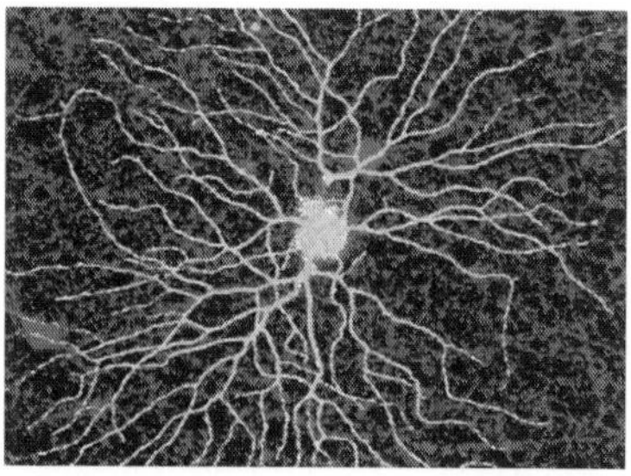

Figure 2.12 Retina ganglion cell (type of neuron found outside the nervous system) viewed by confocal microscopy. Courtesy of Olympus Corporation, UK.

Electron microscopy

Electron microscopes use beams of electrons (e^-) rather than light to visualize the specimen.

- Electrons can be accelerated using a potential difference (V). Moving electrons have wave-like properties such that the faster they are moving, the shorter their apparent wavelength.
- Wavelength (λ) in nm can be calculated using:

$$\lambda = (1.5/V)^{1/2}$$

- Hence the wavelength of a beam of e^- subjected to a potential difference of 100,000 V is approximately 0.004 nm.
- Beams of e^- are focused using electromagnetic lenses, which limits the resolution to less than the theoretical value, though one that is still much better than light microscopes: the development of the electron microscope allowed biologists to view the organelles within a cell in detail for the first time.

Preparing specimens for electron microscopy

- Specimens for electron microscopy require a lengthy and careful preparation.
- They must be:
  - fixed in obnoxious chemicals, such as glutaraldehyde or osmium tetroxide, to stabilize their structure
  - dehydrated
  - embedded in supporting resin
  - sectioned (cut) into thin layers (30–150 nm thick)
  - stained with salts of heavy metals.
- Specimens can only be examined in a vacuum because e^- travel only short distance in air.
- These rigorous preparation procedures and vacuum viewing means only dead biological specimens can be viewed using electron microscopes.

Transmission electron microscopy

- The transmission electron microscope (TEM) is the most commonly used type of electron microscope.
- Its arrangement resembles that of a light microscope because a beam of electrons is transmitted through a thin slice of the specimen (Figure 2.13).
- Electrons are generated by heating a tungsten filament and focused into a beam onto the specimen (Figure 2.13).
- Electrons passing through the specimen are focused to form an image of it on a fluorescent screen or image detection device (Figure 2.13)
- When used with biological specimens, electron microscopes have resolutions of approximately 0.2 nm; about 1000× times better than light microscopy (Figure 2.14).

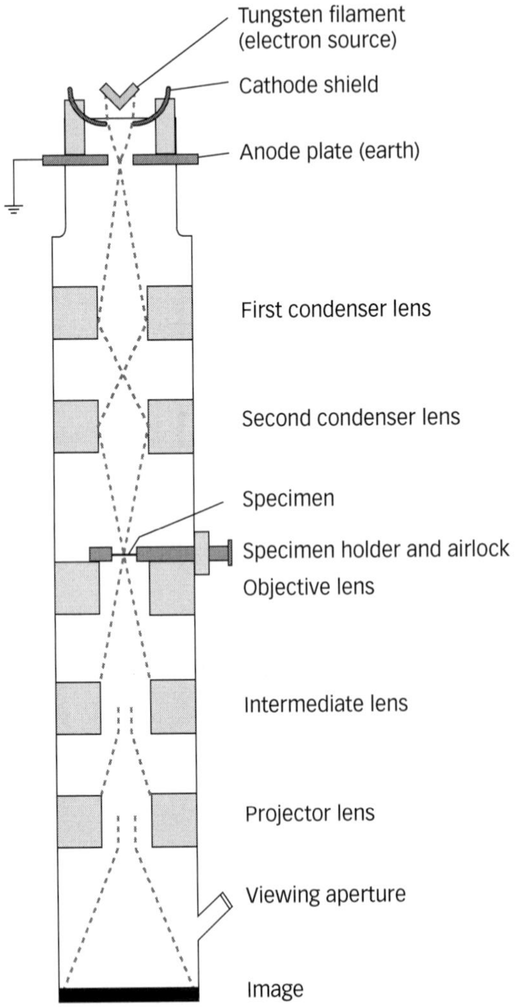

Figure 2.13 Schematic outline of a transmission electron microscope showing the path taken by the beam of electrons to form an image.

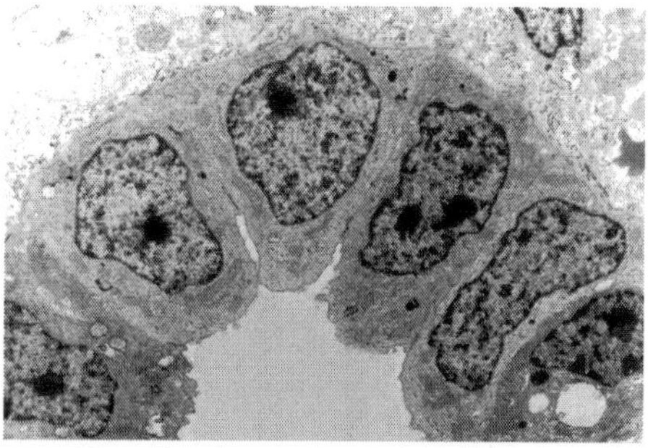

Figure 2.14 Electron micrograph of rat epididymal cells.

Scanning electron microscopy

- The scanning electron microscope (SEM) uses an extremely fine beam of electrons to examine the *surface* of a specimen (Figure 2.15).
- The beam is focused on the specimen surface and systematically scanned across it.
- Electrons scattered by the surface and secondary electrons emitted from it are collected and electronically converted into an output signal.

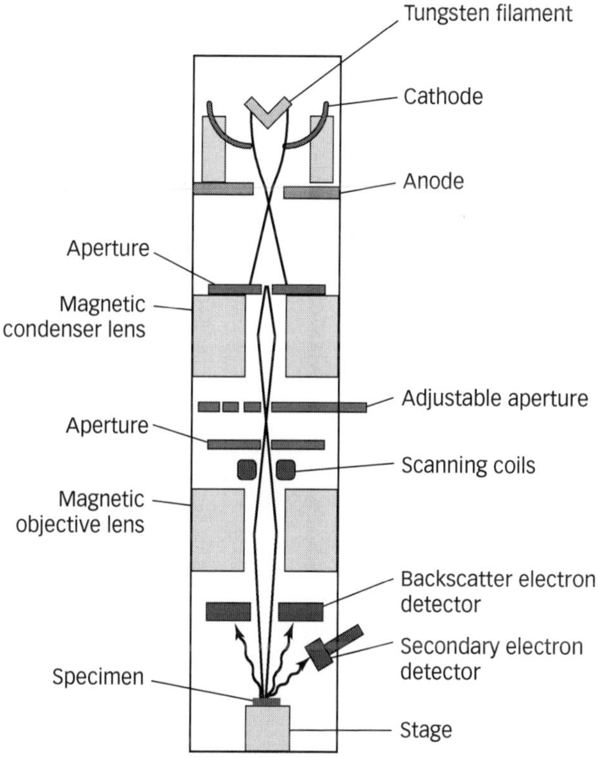

Figure 2.15 Schematic outline of a scanning electron microscope showing the path electrons take through the lenses onto the specimen.

- Movements of the beam are synchronized with the output signal to construct an image of its surface.
- A SEM gives lower resolution than a TEM but the great depth of focus produces images with a three-dimensional appearance (Figure 2.16).

Check your understanding

2.4 Determine the length of the bacterial cell in Figure 2.16 indicated by the arrow head.

Looking for extra marks?

Electron microscopy can be used to determine the distribution of elements in a specimen by electron probe X-ray microanalysis. Interactions between e^- and the specimen produce X-rays whose energies depend on the atomic number of the atom and its electron shell. Thus, the X-rays are generated as a continuous band of radiation that can be analysed to indicate the distribution of the types and amounts of different elements in the specimen.

Revision tip

See an excellent animation on scanning electron microscopy at: http://virtual.itg.uiuc.edu/training/EM_tutorial/

Check your understanding

2.5 Rank the following microscopes in order of resolution (lowest to highest): confocal laser scanning microscope, fluorescence microscope, transmission electron microscope, conventional light microscope, and scanning electron microscope.

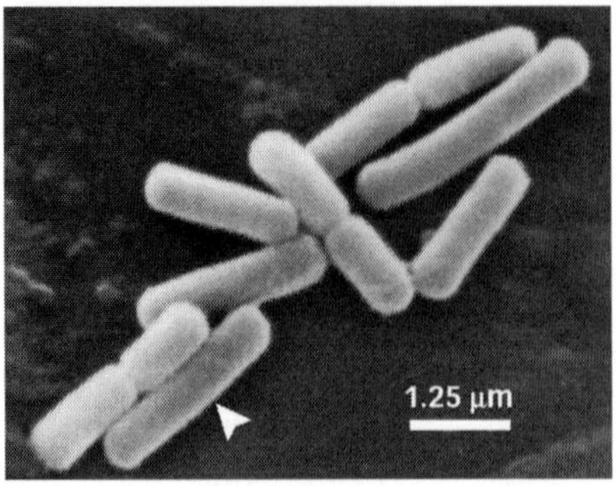

Figure 2.16 Scanning electron micrograph of *Bacillus subtilis*. Courtesy of Professor N.H. Mendelson, University of Arizona, USA.

2.4 CELL FRACTIONATION

Cell fractionation involves separating the cells or tissue into a number of separate fractions which can then be more easily studied separately.

It involves the following steps:

- Suspending the cells or tissue sample in a **homogenizing medium**.
- Disrupting the cell wall (if present) and plasma membrane to release cell contents.
- Separating the homogenate into individual fractions or components.

Homogenizing medium

Suspending the cells or tissue in homogenizing medium is essential to maintain structure and activity of the released cellular fractions.

- A large number of recipes for homogenizing media are available, although they all generally conform to the following rules.
- The medium is buffered to a pH near neutrality to maintain activities of biomolecules.
- They are dilute, with salt concentrations often in the range of 0.025–0.2 mol dm^{-3}, since high ionic strengths denature biomolecules.
- Homogenization media are maintained near osmotic neutrality, often by using 0.25 mol dm^{-3} sucrose, to maintain the morphology of released organelles.
- Specific ions, such as Mg^{2+}, are often added to stabilize enzyme activities.
- Different homogenizing media have specific ingredients as required. For example, the lignin released during the homogenization of plant tissue can denature many proteins; therefore polyvinylpyrrolidone is often included in the medium to precipitate and remove it.

Techniques for homogenizing cells and tissues

- Cell homogenization ruptures the plasma membrane to release the molecules and organelles it contains into the homogenizing medium.
- In all cases, homogenization procedures are imperfect; some cells will remain intact and some organelles will be damaged.
- Homogenization techniques include:
  - pestle and mortar
  - Waring blender
  - Potter homogenizer
  - sonication
  - lysis
  - detergents
  - beadbeating.

Cell fractionation

Pestle and mortar

- Disrupting cell integrity by grinding them with a pestle and mortar is one of the simplest homogenization methods.
- It is most often used to disrupt bacterial and plant cells.
- Sometimes sand or glass microbeads are added to improve disruption.
- The extent of cell disruption and its reproducibility are both difficult to control.

Waring blender

- Waring blenders (Figure 2.17) are kitchen-type blenders.
- Their high-speed spinning blades rapidly rotate the homogenization fluid, which generates shearing (cutting) forces that disrupt plant and animal cells.

Potter homogenizer

- Potter homogenizers consist of a cylindrical Teflon® or glass pestle that is rapidly rotated in a strong glass tube of slightly greater diameter (Figure 2.18).
- The shearing forces are generated in the rapidly rotating fluid in the clearance between the pestle and tube.
- The size of these forces depends upon:
    - speed of rotation
    - clearance distance (0.1–0.15 mm)
    - ratio of cells/tissue to fluid.
- Potter homogenizers are only suitable for use with cultured cells or soft tissues.

Sonication

- Sonication is a popular technique for disrupting cells, especially prokaryotic types, and involves applying an ultrasound probe to the sample, which converts electrical signals into a physical vibration.
- The resulting mechanical action of the high-frequency sound vibrations generated by the probe (typically 20–50 kHz) causes the cells in solution to break apart.

Figure 2.17 A Waring blender.

Figure 2.18 A Potter homogenizer.

- Disadvantages of this technique include:
  - high noise levels
  - variable yields of the desired fractions
  - the large amount of heat that is produced during sonication, which can be localized and difficult to dissipate.

Lysis
- Lysis involves suspending the cells in a low isotonic solution and allowing the osmosis inflow into the cell to cause it to swell and burst.
- This method is usually only successful on cells without cell walls, such as mammalian cells; it is most often used to disrupt cultured cells.
- Cell lysis is usually a gentle process but it may be combined with sonication and a mild surfactant to improve the proportion of ruptured cells.

Detergents
- Detergents lyse cells by disrupting lipid–protein and lipid–lipid interactions in the cell membrane.
- Typically, detergent is used in conjunction with other techniques such as sonication to improve homogenization.
- The type of detergent used depends on the cells being disrupted. For example, cells without walls will lyse with weaker detergent than cells with walls.
- Weaker non-ionic detergents, such as Triton X-100, and zwitterionic detergents, such as 3-[(3-cholamidopropyl)dimethylammonio]-1-propanesulphonate (CHAPS), do not denature proteins (Figure 2.19).
- Strong detergents, like sodium dodecylsulphate (SDS), generally denature biological molecules (Figure 2.19).

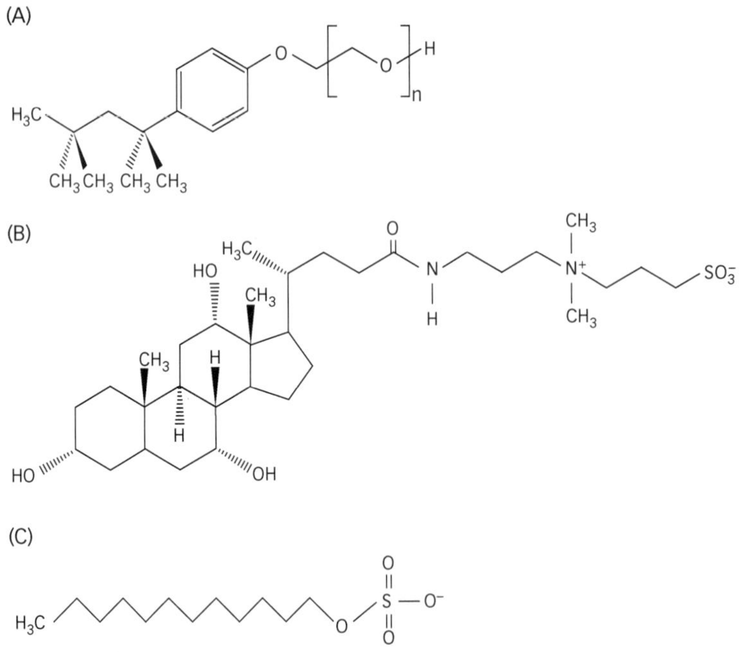

Figure 2.19 (A) The non-ionic detergent Triton X-100, (B) the zwitterionic detergent, 3-[(3-cholamidopropyl)dimethylammonio]-1-propanesulphonate (CHAPS), and (C) the ionic sodium dodecylsulphate (SDS).

Beadbeating

- Beadbeating is commonly used to homogenize small samples of tissues.
- Equal quantities of metal or ceramic beads and tissue are agitated using a laboratory vortex or shaker to disrupt the sample.
- Processing times vary depending on whether manual or specialized automated laboratory equipment is being used.
- Tough tissues, such as skin or muscle, may need to be finely chopped prior to using the bead technique to increase yield.
- Foaming and heating of the suspension can cause difficulties but heating can be reduced by using chilled tubes.

Separating homogenates into subcellular fractions

Separating homogenates into subcellular fractions is virtually always done using centrifugation: the separation of mixtures by applying centrifugal forces.

Centrifugation

- Particles suspended in liquid will eventually sink to the bottom of the container if left for a period of time.

- Larger particles sink to the bottom more quickly than smaller ones, and some minute particles will not fall out of suspension unless the centrifugal force is increased by spinning the suspension in a centrifuge.
- Centrifuges typically involve a central rotating axis, from which containers are suspended. The containers and the rotating axis are referred to as the rotor.
- As the rotor rotates at an increasing speed, the particles in it are subjected to an increasing centrifugal force.
- Both the speed of the rotation and the distance of the particle from the central axis of rotation determine the overall centrifugal force.
- Centrifugal forces are relatively easy to calculate (see Box 2.2).
- Centrifugal forces produced in centrifuges can separate components of a mixture that differ in size, density, or shape.
- In general, centrifugation separates mixtures into two portions:
  - the **pellet** at the bottom of the tube that contains the sedimented material
  - the liquid **supernatant** that contains unsedimented material.
- Centrifugation can therefore be used to separate a homogenate into subcellular fractions.
- The two commonest centrifugation methods used to fractionate homogenates are:
  1. differential centrifugation
  2. density gradients centrifugation.

Box 2.2 Estimating centrifugal fields

- The rate of sedimentation of a particle when spun in a centrifuge depends upon the **centrifugal field** or **acceleration** (G), which is described by the equation:

  $$G = \omega^2 r$$

- The angular velocity (ω) of the rotor is measured in radians per second.
- The radius of the particle from the centre of rotation (r) is measured in cm.
- One revolution equals 2π radians. However, the speeds of centrifuges are always given in revolutions per minute (rev min^{-1}) although this is often expressed as rpm.
- Thus, the angular velocity of a spinning rotor is given by:

  $$\omega = 2\pi\,(\text{speed in rev min}^{-1})/60$$

- Given, $G = \omega^2 r$, then

  $$G = 4\pi^2(\text{speed in rev min}^{-1})^2\ r/60^2$$

- It is usual to express G relative to the earth's gravitational field (g), which is an acceleration of 981 cm s^{-1} per s.
- Hence the relative centrifugal field (RCF) is

continued

$$\text{RCF (in } g) = 4\pi^2 \text{ (speed in rev min}^{-1})^2 r/60^2 \times 981$$

- This simplifies to:

$$\text{RCF (in } g) = 1.12 \times 10^{-5} \text{ (speed in rev min}^{-1})^2 r$$

- In practice, **nomograms** (Figure 2.20) are usually used to determine the speed or RCF.
- These consist of three columns:
  - one is a series of radii expressed in cm
  - the middle one is the RCF/g
  - the third is the rotor speed in rev min^{-1}.
- A straight line across two of the columns gives the corresponding value on the third, and saves doing arithmetic.

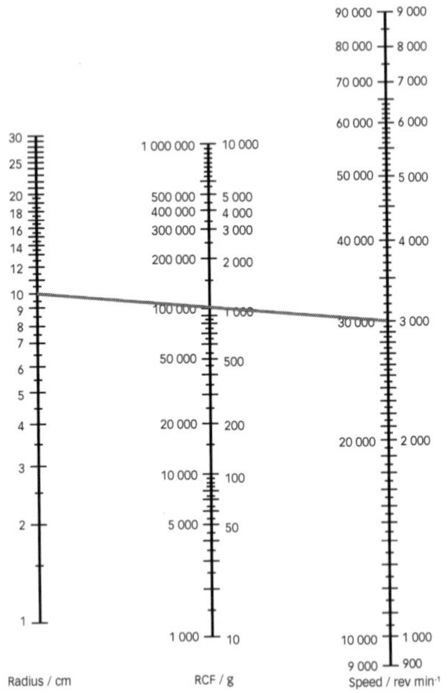

Figure 2.20 Nomogram, greatly reduced in size, used in estimating the speeds of a centrifuge required to produce a given relative centrifugal field (RCF) for a given radius of rotation.

Differential centrifugation

- Differential centrifugation, as the name implies, separates particles mainly on *differences* in the sizes of the particles.
- Cell fractionation of a homogenate involves successive rounds of centrifugation, with each successive centrifugation being at a higher speed and possibly for a longer time.

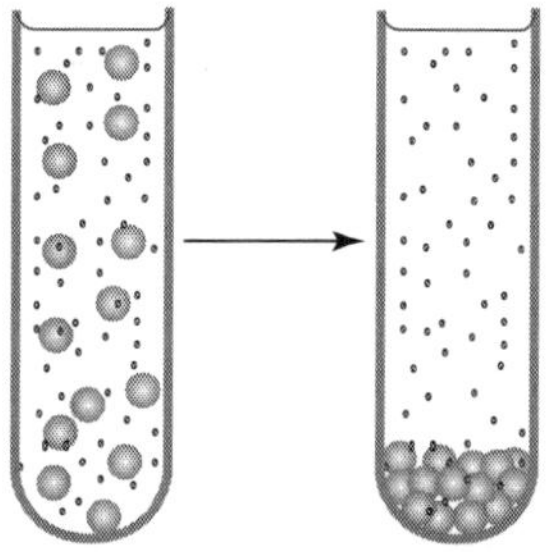

Figure 2.21 Separation of different sized particles using differential centrifugation.

- During centrifugation the largest particles are sedimented and form a pellet, leaving smaller particles suspended in the supernatant (Figure 2.21).
- Once each round of centrifugation is performed the supernatant is removed from the pellet and is placed in a fresh tube for the next round of centrifugation.
- Figure 2.22 shows the RCFs and centrifuge run times required to isolate selected organelles from a homogenate.
- The final volume of supernatant contains the soluble portion of the cell contents, as well as free ribosomes and small structures found in the cytosol.
- Prior to centrifugation, all the components in the mixture are distributed evenly.
- Following centrifugation, a pellet of larger particles will always be contaminated with smaller ones.
- The purification of the pellet can be improved either by carefully decanting off the supernatant, resuspending the pellet in fresh medium, and then recentrifuging it, or by subjecting it to density gradient centrifugation.

Check your understanding

2.6 Which organelle(s) is/are most likely to be found in the pellet if an homogenate prepared from a cell suspension is centrifuged at 12,000 g for 10 minutes?

Density gradient centrifugation

- Density gradient centrifugation separates particles with greater resolution than discontinuous centrifugation because larger particles are not contaminated with smaller ones. It is performed by placing the impure subcellular fraction in a centrifuge tube that contains a column of liquid whose density increases towards the bottom of the tube.
- These types of centrifugation techniques can separate fractions that differ in densities by only one or two parts in 1000.
- Two methods of density gradient centrifugation used routinely are:
  1. rate zonal centrifugation
  2. isopycnic centrifugation.

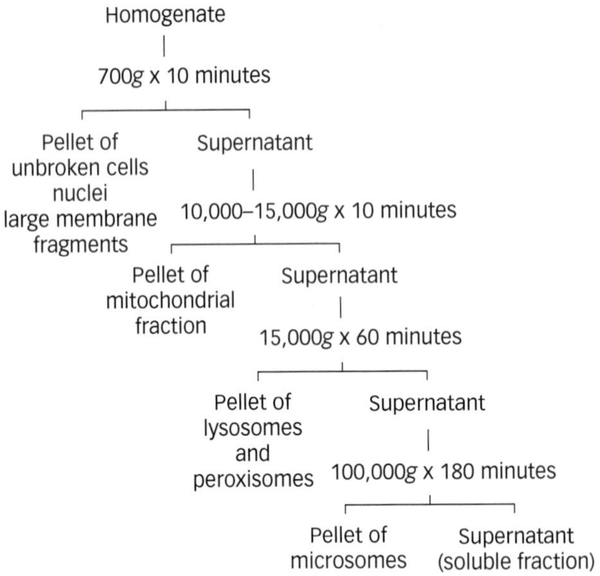

Figure 2.22 Outline of fractionation of a homogenate using differential centrifugation. Note each pellet contains particles of a different size.

Rate zonal centrifugation

- Rate zonal centrifugation separates particles because they sediment through the gradient at different velocities and form discrete zones, each consisting of one type of particle.
- The zones of separated particles are stabilized because the gradient slows diffusion and convection currents.
- The length of the gradient must be sufficient to allow the zones to separate.
- Given that the densities of the particles in the sample exceed that of the densest part of the gradient, the centrifuge run must be stopped before any or all of the zones reach the bottom of the tube.

Isopycnic density gradient separation

- Isopycnic density gradient separation uses a density gradient that exceeds the densities of all the particles in the sample.
- Particles therefore sediment in the gradient to a position that equals their own density.
- Once all the particles have reached these positions, they will remain there and further centrifugation will not increase their migration.

Preparing density gradients

Density gradients are commonly made using sucrose solutions.

- They may be discontinuous or continuous.

- In discontinuous gradients, the increase in density occur as discrete steps, whereas in continuous gradients the density increases without steps towards the bottom of the tube (Figure 2.23).
- Both types of gradients are suitable for harvesting cells from culture media (see earlier), purifying subcellular organelles from homogenates and isolating some types of viruses (Chapter 18).
- In a discontinuous gradient, particles within a suspension collect as discrete zones or bands at the interfaces of each gradient step (Figure 2.24a).
- In a continuous gradient, particles separate into bands in the gradient depending upon their densities (Figure 2.24b).
- Once the purified fractions have formed within the gradient, they can be removed using a syringe.

Assessing purity of isolated fractions

Organelles and biological membranes differ in their sizes and densities (Table 2.1).

- Suitable combinations of differential and density gradient centrifugations should be able to produce purified fractions from a cell homogenate.

Check your understanding

2.7 Can (a) differential and (b) density gradient centrifugation be used to separate peroxisomes from mitochondria?

Yield and purity of cell fractions

- Once a homogenate has been fractionated, it is essential to check that the required organelle is present in the fraction and to determine the yield of the organelle.

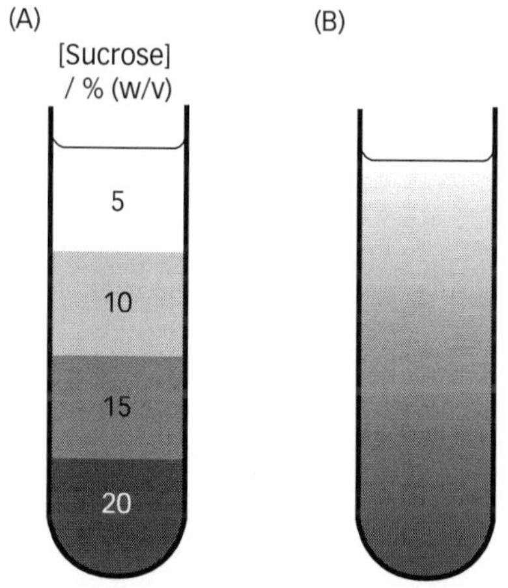

Figure 2.23 Schematics showing (A) discontinuous gradient and (B) continuous gradient.

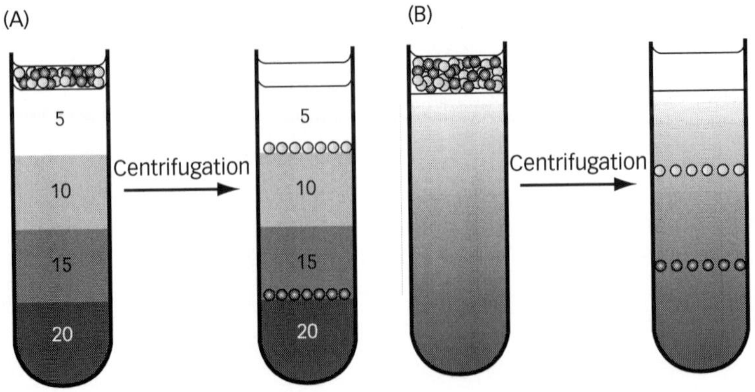

Figure 2.24 Schematic examples of purified components following density gradient centrifugation using (A) discontinuous and (B) continuous gradients.

Organelle or membrane	***Diameter (μm)***	***Density (mg cm⁻³)***
Golgi apparatus		1.06–1.10
Lysosomes	1.05–1.0	1.21
Mitochondria	0.50–1.0	1.19
Peroxisomes	0.10–1.5	1.23
Plasma membrane		1.16
Ribosomes	0.025	1.60–1.75
Smooth ER membrane		1.16
Tonoplast		1.11

Table 2.1 Approximate diameters and densities of some cell organelles and biological membranes

- The two major ways of assessing the purity of a cell fraction are to examine it by electron microscopy (see 'Electron microscopy' section) and to determine the activities of marker enzymes, whose distribution in the cell is restricted to specific sites, such as the cytosol, mitochondria, and so on (Table 2.2).
- Determining the specific activity of the marker enzyme in the fraction compared to its value in the homogenate allows both the purity and yield to be assessed.

Fraction	***Marker enzyme***
Cytosol	Lactate dehydrogenase
Golgi apparatus	Galactosyltransferase
Lysosomes	Acid phosphatase
Microsomes	Glucose-6-phosphatase
Mitochondria	Succinate dehydrogenase
Peroxisomes	Catalase

Table 2.2 Marker enzymes used in estimating both the purity and yield of subcellular preparations

Enzymes and specific activities are revised in the companion volume Thrive in Biochemistry and Molecular Biology.

Check your understanding

2.8 Which of the following statements is/are false?

a. Sonication is a popular technique used to disrupt bacterial cells.
b. Cells with cell walls require milder detergents for their disruption than those without walls.
c. During centrifugation, both the speed of rotation and the radius of rotation affect the overall separation.
d. Lower rotation speeds and shorter spin times are used to isolate smaller particles from a mixture.
e. Succinate dehydrogenase is the common marker enzyme used to detect the presence of microsomes in a cell fraction.

2.5 STUDYING BIOLOGICAL MOLECULES

Numerous physical, chemical, biochemical, and molecular biological techniques are available to isolate, purify, and study biological molecules.

- Biological molecules may be purified using, for example, a number of chromatography techniques.
- Their sizes and some properties can be analysed using one of a number of electrophoretic methods.
- Their properties can be investigated by spectrophotometric methods including ultraviolet/visible and infra-red spectroscopy.
- The structures of biological molecules can be determined using nuclear magnetic resonance spectroscopy, X-ray crystallography, and electron microscopy.
- The ability of macromolecules to bind to complementary substances is the basis of many now routine molecular biological investigative techniques, including various blotting methods, the sequencing of nucleic acids and proteins, recombinant DNA technologies, fluorescence *in situ* hybridization (FISH) and enzyme linked immunosorbent assays (ELISAs), PCR-based methods, and microarray analyses.

See the companion volumes, Thrive in Biochemistry and Molecular Biology *and* Thrive in Genetics, *in particular to revise the investigative methods.*

Revision tip

Do browse through the further reading references given for this chapter online. Go to http://www.oxfordtextbooks.co.uk/orc/thrive/

Check your understanding

2.9 Which of the following statements is/are true?

a. Suspension culture is a suitable technique for growing cells taken directly from body tissues.

b. Trypan blue is able to stain live or viable cells.

c. Use of antibiotics is recommended as a routine additive to eukaryotic cultures as they promote the growth of antibiotic resistant strains of bacteria.

d. Flow cytometry or cell sorting is restricted to separating and counting populations of cells.

e. Streaming of cells in flow cytometry is achieved by tagging them with fluorescent-labelled antibodies called fluorophores.

f. Increasing the magnification can help overcome the limits imposed by the resolution of a microscope.

g. Specimens for light and electron microscopy require staining to improve their contrast.

h. Electron microscopes have resolutions of approximately 1000× better than that of light microscopes.

i. Differential centrifugation is a technique used in fractionating cells.

j. During centrifugation the smallest particles are sedimented first.

3 Biological membranes

Key features of membranes

- Biological membranes are essential features of cells: life is not possible without membranes.
- Biological membranes form the boundary of the cell, separating it from its external environment. In eukaryotes, and to some extent prokaryotes, they divide the protoplasm into separate compartments.
- Eukaryotic plasma membranes are composed of phospholipids, glycolipids, and steroids, together with proteins and glycoproteins.
- Bacterial membranes resemble those of eukaryotes in being composed of lipids and proteins but they generally have simpler compositions.
- Archaeal surface membranes are based upon different types of lipids to eukaryotes and bacteria.
- The fluid mosaic model describes the accepted structure of biological membranes.
- The lipids are organized into a bilayer to which membrane proteins are either attached or embedded.
- Although archaeal membranes differ from eukaryotic and bacterial membranes in composition their overall structure is the same and they perform similar biological roles.

continued

- Biological membranes are fluid structures although organisms can adjust their lipid composition and so adjust the fluidity of their membranes to changing circumstances and environments.
- The relatively large surface areas of membranes provide attachment sites, for example, for structural proteins, enzymes, and ribosomes.
- Membranes are selectively permeable and control the flow of ions and molecules between the cell and its exterior, and between separate compartments within the cell.
- The fluid nature of biological membranes is essential for activities such as the transmembrane transport of particles and solutes, and for endocytosis and exocytosis.
- Membranes are the sites of most energy transduction activities, which, for example produce most of the cell's ATP.
- Membranes also mediate signal transduction and cell–cell interactions.

3.1 COMPONENTS OF MEMBRANES

Membranes are composed mainly of:

  - lipids
  - proteins.
- The eukaryotic plasma membrane and some internal membranes also contain carbohydrate.
- The carbohydrate components are covalently bound to lipids and proteins in glycolipids and glycoproteins (Chapter 1) respectively.
- Carbohydrate appears absent from many eukaryotic internal membranes and the surface membranes of bacteria.
- Many membranes are composed of approximately 40–50% lipid, 40–50% protein, and 2–10% carbohydrate, although many also have very different compositions (Table 3.1).
- Archaeal membranes are distinctively different in composition to those of eukaryotes and bacteria.

Membrane	*Protein (%)*	*Lipid (%)*	*[b]CBH (%)*
Liver plasma membrane	58	42	5–10
Myelin sheath (nerve cell membrane)	18	79	3
Chloroplast lamella membrane	70	30	0
Mitochondrial outer membrane	55	45	0
Mitochondrial inner membrane	76	24	0
Nuclear envelope	66	32	2
Gram-negative bacterial surface membranes	60	40	0
Gram-positive bacterial surface membranes	75	25	0

Table 3.1 Approximate compositions of selected membranes[a]

[a]Data taken from a number of sources.
[b]CBH, carbohydrate present in glycoproteins and glycolipids.

3.2 STRUCTURE OF MEMBRANES

The accepted structure of biological membranes is described as the **fluid mosaic model**, which was first proposed by Singer and co-workers in 1972 but has since been increasingly refined.

- The *general* structure of the eukaryotic plasma membrane and its modes of function also apply to other types of membranes, including the bacterial cell surface membrane.
- The basic structure of the eukaryotic and bacterial membranes is a lipid bilayer.
- Membrane proteins are either attached to or embedded within this bilayer.

Membrane lipids

Three major types of lipids are found in eukaryotic plasma membranes:

- phospholipids (based on the alcohols glycerol or sphingosine)
- glycolipids (based on the alcohol sphingosine)
- steroids (often cholesterol in animal membranes, sitosterol and stigmasterol in plant membranes, and ergosterol in fungi).

➔ *The structures of lipids are revised in* Thrive in Biochemistry and Molecular Biology.

- Bacterial membranes lack glycoproteins and cholesterol but some bacterial membranes contain steroid-like molecules called **hopanoids** (Figure 3.1).
- The surfaces of membrane lipid molecules are a mixture of **hydrophilic** (water loving) and **hydrophobic** (water hating) regions: the lipids are said to be **amphipathic** (Figure 3.2).
- The plasma membrane of eukaryotic cells often contains significant quantities of glycolipids and steroids; although bacterial surface membranes lack glycoproteins and steroids, some contain hopanoids (Table 3.2).
- When membrane lipids are agitated in aqueous environments and then allowed to stabilize they spontaneously form a bilayer in which their hydrophilic portions are in contact with the water but their hydrophobic regions are associated together and protected from the aqueous environments.
- This structure is often loosely referred to as the **phospholipid bilayer** (Figure 3.3).
- Bilayers (Figure 3.3) can be extensive in size and form the structural basis of biological membranes.

Membrane fluidity

- Lipid bilayers are extremely dynamic structures with the molecules in each layer or **leaflet** in constant motion. However, movement from layer to layer, so-called **flip flop movements**, are rare.
- The dynamic nature of the bilayer means biological membranes are also **fluid** structures.

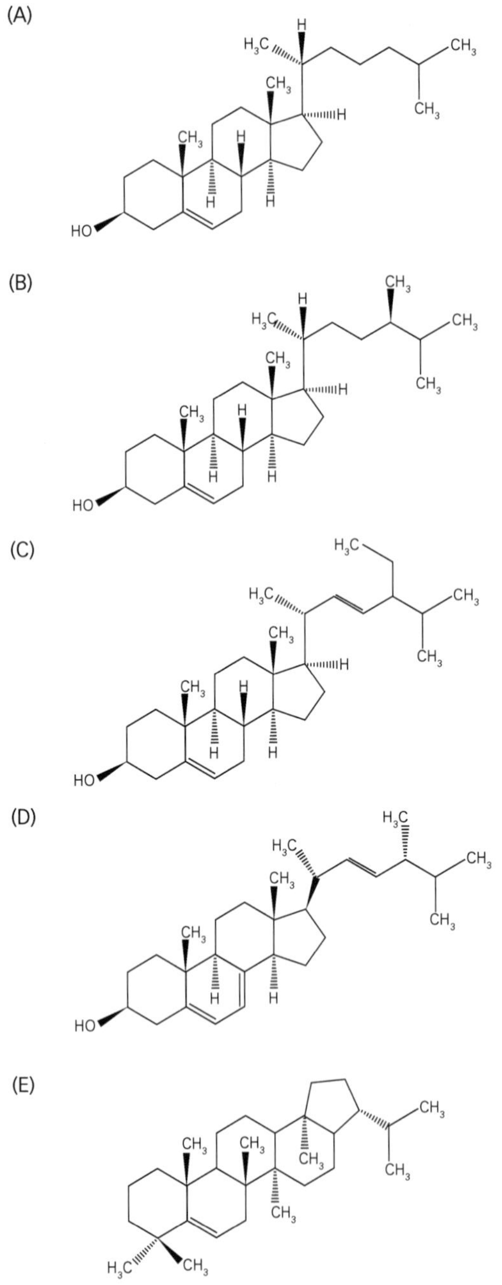

Figure 3.1 Structures of (A) cholesterol, (B) sitosterol, (C) stigmasterol, (D) ergosterol, and (E) the hopanoid, diploptene.

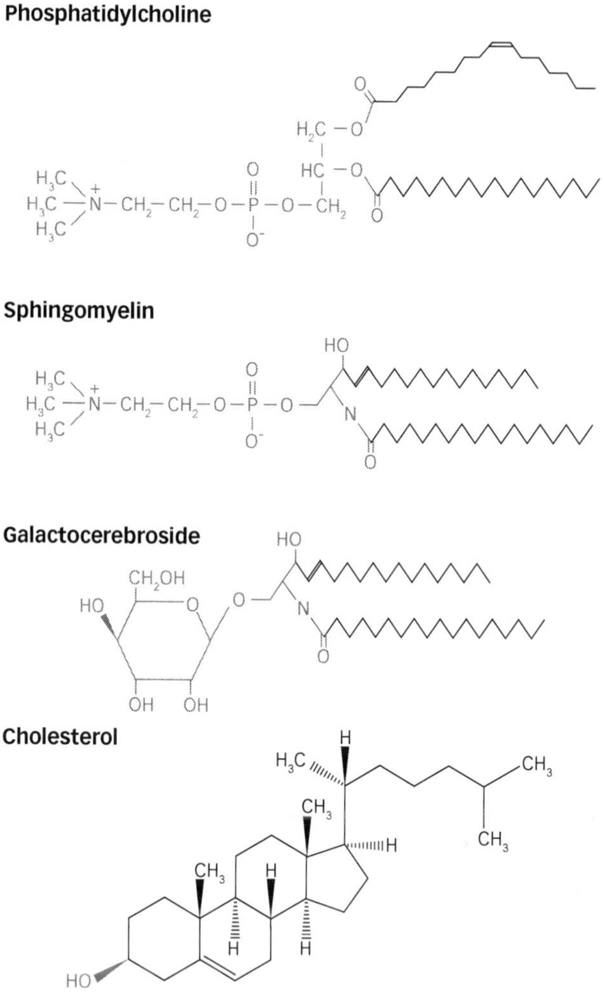

Figure 3.2 Amphipathic nature of representative membrane lipids. Phosphatidylcholine and sphingomyelin are phospholipids, galactocerebroside is a glycolipid, and cholesterol is a steroid. Note how each lipid molecule has hydrophilic (lighter shade) and hydrophobic (darker shade) regions.

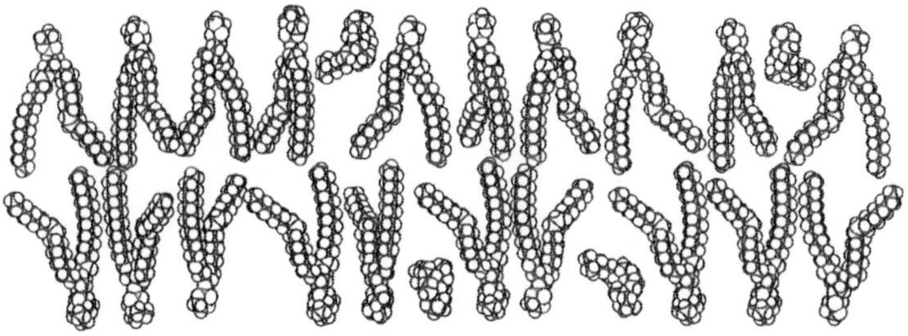

Figure 3.3 Molecular models illustrating the structure of a lipid bilayer in an aqueous environment.

Lipid		*Membrane type*		
		Liver plasma (%)	*Erythrocyte plasma (%)*	*Escherichia coli surface (%)*
	phosphatidate		1.5	12
	Phosphatidylcholine (PC)	24	19	
	phosphatidylethanolamine (PE)	7	18	65
	phosphatidylglycerol (PG)		0	18
	phosphatidylinositol (PI)		1	
	phosphatidylserine (PS)	4	8.5	
	sphingomyelins (SMs)	19	17.5	
Glycolipids	___	7	10	
Cholesterol	___	17	26	

Table 3.2 Approximate lipid composition of three membranes[a]

[a] Data from several sources.

- **Unsaturation**, the presence of carbon–carbon double bonds, is common in the fatty acyl chains of membrane lipids.
- Double bonds always occur in the *cis* configuration (rather than the *trans*) as shown in Figure 3.4.
- *Cis* double bonds increase the packing volume of the lipids in the bilayer and so increase the fluidity of the membranes.
- In contrast, *trans* arrangements reduce the volume of the membrane bilayer and reduce its fluidity.
- Steroids and hopanoids are rigid molecules: their presence in a membrane greatly decreases its fluidity.
- Organisms are able to adjust the lipid composition of their membranes and so regulate its fluidity to suit changing environmental conditions.

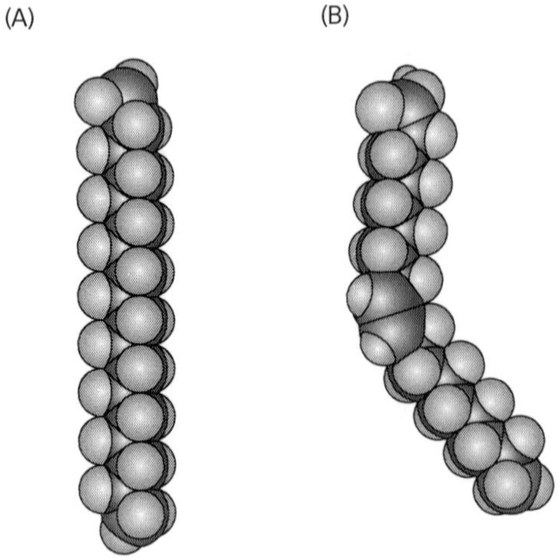

Figure 3.4 Molecular models indicating the effects of *cis*/*trans* isomerization on single double bond on the conformation of a C_{18}-long fatty acid. (A) Octadecenoic (stearic) acid and (B) octadecenoic (oleic) acid (*cis*).

Check your understanding

3.1 Which of the following statements is/are false?

a. All membrane lipids are amphipathic.

b. Triacylglycerols are one class of membrane lipids.

c. The rigid molecules of steroids and hopanoids in membranes decrease their fluidity.

d. The molecules of membrane lipids have hydrophilic (water hating) and hydrophobic (water loving) regions making them amphipathic.

e. The carbohydrate components of membranes are found as free glycans in the bilayer.

Membrane proteins

Membrane proteins appear as discrete, suspended particles within membranes examined by electron microscopy of freeze-fracture preparations.

- The pattern of proteins within the lipid bilayer gives rise to the term *mosaic*, in the fluid mosaic model of membrane structure.
- Freeze-fracturing of membranes entails freezing the cells in a block of ice in liquid nitrogen at −196°C and then fracturing the block along its plane of least resistance, which is usually between the two leaflets of the membrane.
- A platinum replica of the exposed plane(s) is then prepared which can be viewed by electron microscopy (Figure 3.5).

Different types of membrane proteins

- The different ways in which membrane proteins are associated or attached to a membrane bilayer allows them to be divided into three major groups (Figure 3.6):
  1. Peripheral proteins (PPs)
  2. Integral proteins (IPs)
  3. Lipid-anchored proteins (LAPs).

Peripheral membrane proteins

- Peripheral proteins are (relatively) loosely associated with the surface of the membrane and are held in place by ionic interactions.
- Examples of PPs include:
  - cytochrome *c* (Figure 3.7) of the inner mitochondrial membrane
  - enzymes such as phospholipases, cholesterol oxidases, and cholinesterases
  - G proteins, for example, the α and γ subunits of transducing, which is involved in vision (Chapter 15).
- The study of membrane PPs can often be made easier by releasing them from the membrane into solution—a process called solubilization.

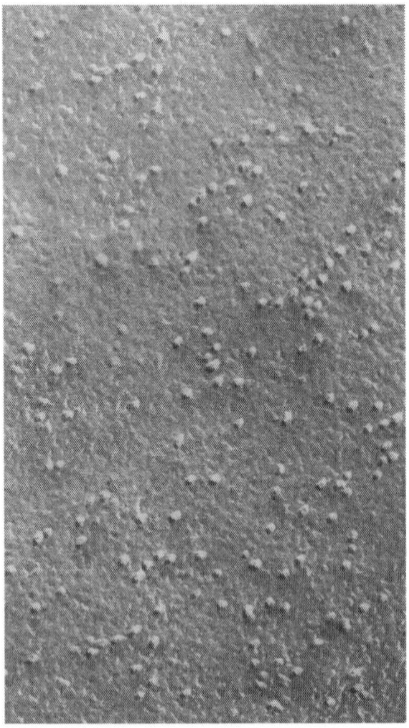

Figure 3.5 Electron micrograph of a chloroplast thylakoid membrane viewed following freeze fracture. Note how the membrane proteins appear as discrete particles ('bumps') embedded within the lipid bilayer; hence the use of the term 'mosaic'.

- Solubilization is achieved by washing the membrane with solutions of very low or very high ionic strengths or perturbing the pH of the suspending liquid to disrupt the ionic interactions holding the PP in place.
- The liquid can then be centrifuged (Chapter 2) to remove the remains of the membrane, leaving the protein in solution.

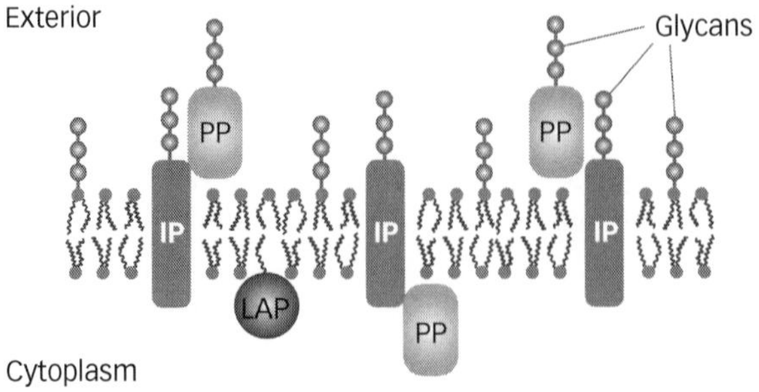

Figure 3.6 Schematic representation of the fluid mosaic model of a biological membrane. Glycolipids and glycoproteins have carbohydrates (glycans) covalently attached to them. IP, integral membrane proteins; LAP, lipid-anchored protein; PP, peripheral protein.

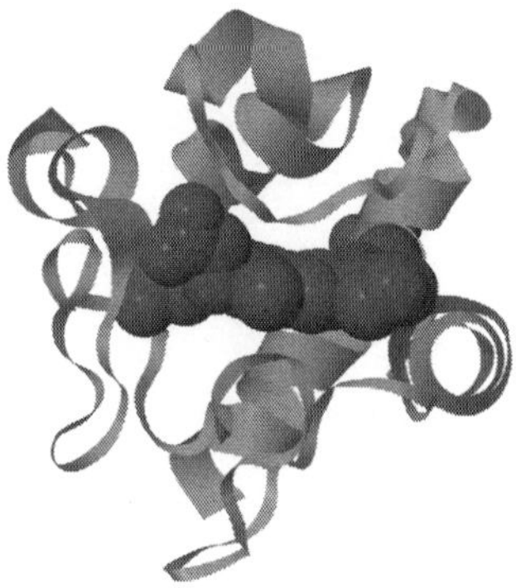

Figure 3.7 Molecular model of horse (*Equus caballus*) heart cytochrome *c*, a peripheral protein of the inner mitochondrial membrane. The solid space-filled part represents the rather flat haem group that carries electrons. The model was constructed using PDB file 1HRC.

Integral membrane proteins

- Integral membrane proteins have hydrophobic surfaces that penetrate the hydrophobic core of the lipid bilayer and hydrophilic portions projecting from the membrane. Integral proteins are therefore amphipathic.
- Integral proteins are classified into four groups:
  1. monotopic
  2. bitopic
  3. oligotopic
  4. polytopic.
- Monotopic IPs are only associated with one leaflet of the bilayer; bitopic ones cross the bilayer once; oligotopic IPs cross it twice; and polytopic proteins more than twice as illustrated in Figure 3.8.
- The intramembrane domains of IP usually consist of membrane-spanning α helices (Figure 3.9).
- Porin proteins form pores in membranes through which material can cross the membrane.
- Eukaryotic and archaeal porins are typical in having membrane spanning α helices (Figures 3.9 and 3.10a) However, bacterial porins differ in having membrane-spanning regions consisting of a rolled up β sheet called a **β barrel** (Figure 3.10b).
- Membrane IPs can only be released by disrupting the integrity of the membrane itself. They are normally solubilized for study using detergents or organic solvents, which disrupt hydrophobic interactions.

Looking for extra marks?

Not all β barrels are transport porins. In some smaller bacterial ones with fewer β strands, the interior is completely occluded with amino acid side chains. These β barrels function as enzymes or receptor sites.

For example, the outer membrane phospholipase A (OMPLA) of *Escherichia coli* is, as its name implies, an enzyme, and the outer membrane protein A (ompA) of the same organism acts as a receptor for a number of T-even bacteriophages (Chapter 18).

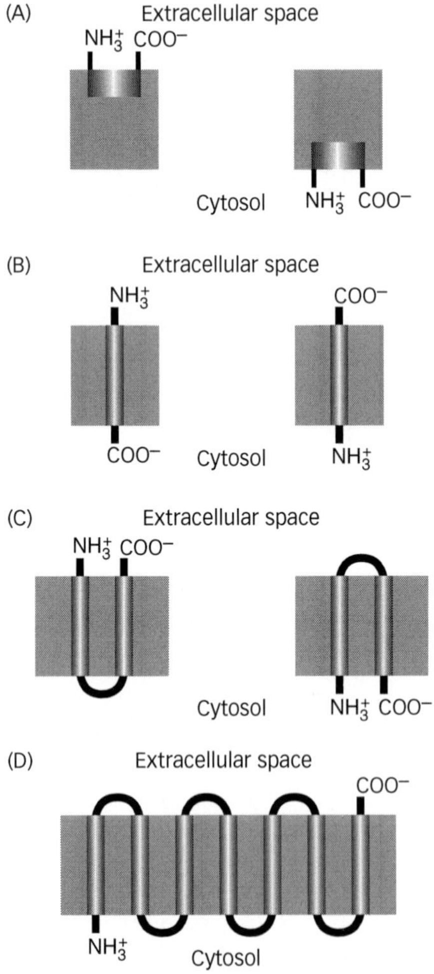

Figure 3.8 Topological arrangements of (A) monotopic, (B) type I and II bitopic, (C) oligotopic, and (D) polytopic integral membrane proteins.

Lipid-anchored proteins

- Lipid-anchored proteins are covalently linked to lipid molecules that are part of the bilayer and so are found on the membrane surface (Figure 3.6). Some IPs are also linked to the membrane by lipid-based anchors.

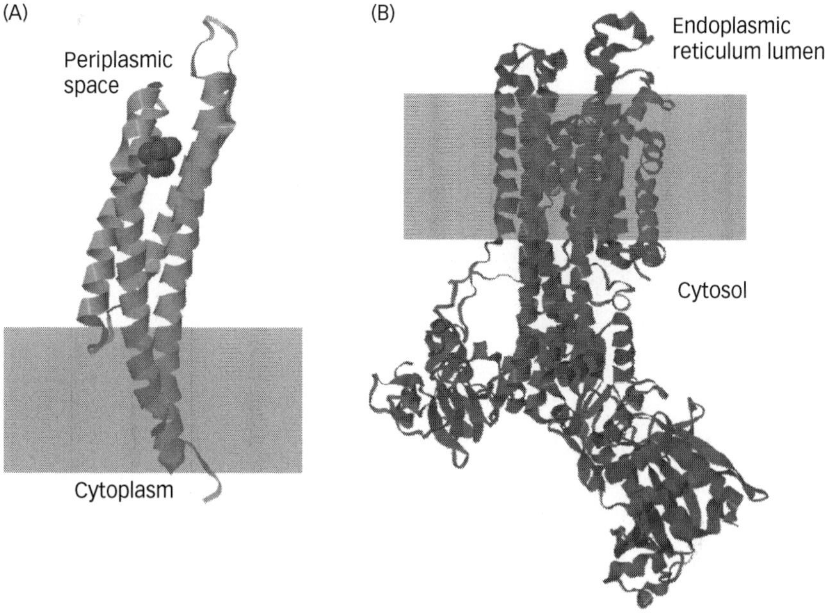

Figure 3.9 Molecular models showing the α helical membrane spanning domains of two integral proteins. (A) The aspartate receptor of *Salmonella typhimurium*. (B) Ca^{2+}-ATPase of the endoplasmic reticulum. The two lighter shaded spheres represent Ca^{2+}. Models constructed using PDB files 2LIG and 1SU4 respectively.

- They show features of both peripheral and integral proteins.
- Examples of lipid-anchored proteins include:
  - guanine nucleotide binding proteins (G proteins)
  - membrane-associated tyrosine kinases of the Src family, which are believed to function in signal transduction (Chapter 15)
  - a number of membrane-associated proteins of the immune system.
- Lipid-anchored proteins can be released from the membrane by using detergents or by hydrolysing the protein-lipid link using lipases.

Carbohydrates

Membrane carbohydrates are covalently attached to proteins and lipids of the outer leaflet of plasma membrane as glycoproteins and glycolipids respectively; free carbohydrates are not found in biological membranes.

- The carbohydrates of membrane glycoproteins and glycolipids are called **glycans**.
- They are often branched oligosaccharides of up to 15 or so monosaccharide residues.
- Glycans form a layer called the **glycocalyx** that covers the surface of the cell (Figure 3.11).

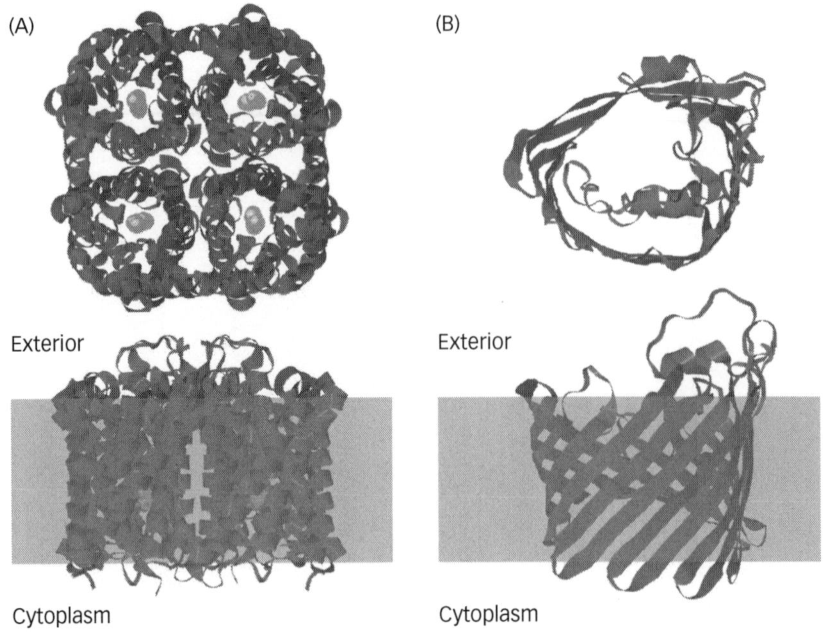

Figure 3.10 Molecular models of aquaporins. (A) From the archaeon *Methanothermobacter marburgensis* (aquaporin, AqpM). Water molecules in transit are represented by the lighter shaded structures. (B) A bacterial aquaporin from *Rhodobacter capsulatus*. Note the α helix-rich nature of the archaeal porin whereas the bacterial one is a β barrel. Models constructed using PDB files 2F2B and 2POR respectively.

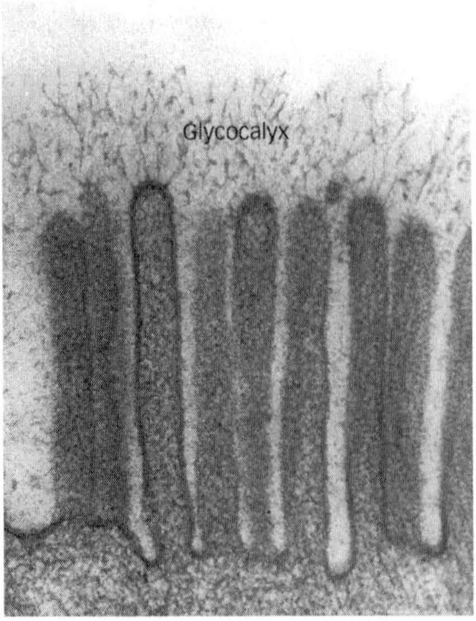

Figure 3.11 Electron micrograph showing a well-developed glycocalyx extending from the surfaces of microvilli of bat intestinal cells. Courtesy of Professor D.W. Fawcett.

Check your understanding

3.2 Which of the following statements is/are false?

a. Peripheral proteins are (relatively) loosely associated with the surface of the membrane and are held in place by hydrophobic interactions.

b. The distribution of integral membrane proteins can be investigated by freeze-fracturing the membrane and then viewing it by electron microscopy.

c. All integral membrane proteins are amphipathic.

d. The intramembrane domains of IP proteins usually consist of membrane-spanning α helices.

e. Membrane lipids may be covalently associated with some types of proteins.

3.3 STRUCTURE OF ARCHAEAL MEMBRANES

Archaeal membranes differ from those of eucarya and bacteria in lacking fatty acid residues.

- Archaeal membranes contain side chains of four or eight repeating isoprene units, giving chains lengths of C_{20} and C_{40}, which are bound to glycerol units by ether links (Figure 3.12).
- The lipid layer can be a bilayer or a monolayer, depending upon whether four or eight isoprene units are present in the archaeal lipids as shown in Figure 3.13.
- Regions of bilayers and monolayers may be present in the same membrane.
- In many cases, archaeal lipids contain hydrocarbon rings (Figure 3.12), which presumably stiffens the membrane and reduces its fluidity.
- The presence of monolayers and hydrocarbon rings means archaeal membranes are stable in the harsh environments they inhabit, for example extreme temperatures or high salt concentrations (Chapter 1).
- Despite their chemical differences, archaeal membranes have a similar overall arrangement to those of eucarya and bacteria: their surfaces are hydrophilic and enclose a hydrophobic interior (compare Figures 3.6 and 3.13).
- The proteins of archaeal membranes are arranged in a similar manner to those of bacteria and eukaryotes.

3.4 MEMBRANE ASYMMETRY

Membranes are asymmetric structures.

- For example, all the carbohydrate of plasma membranes is found only attached in glycoproteins and glycolipids of the outer leaflet.

Membrane asymmetry

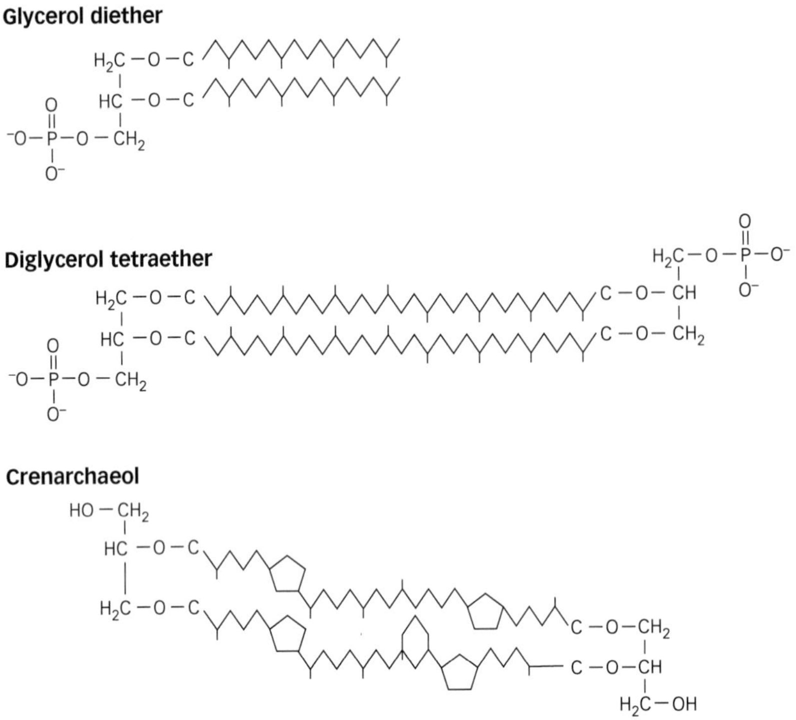

Figure 3.12 General structures of archaeal membrane lipids. Compare with Figure 3.2.

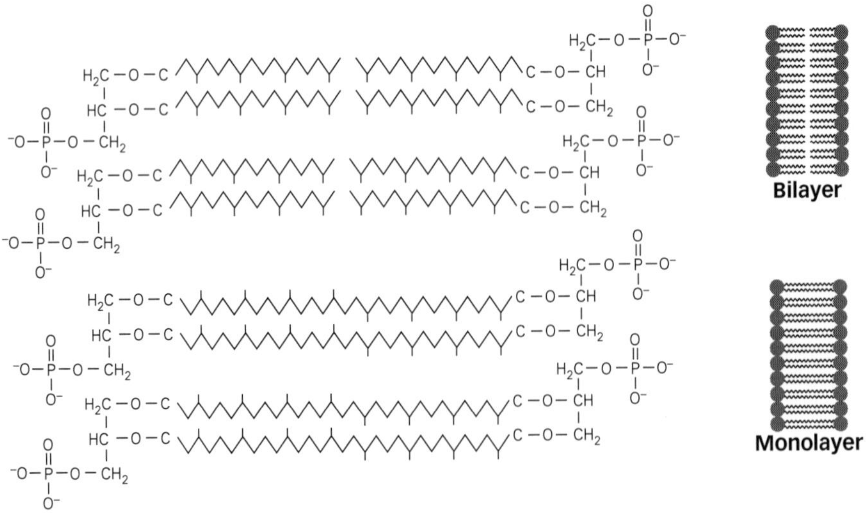

Figure 3.13 Archaeal membrane lipids can form bilayers or monolayers.

- Membrane proteins are also unequally distributed, both within each leaflet and between the leaflets, as implied by the manner of their attachments to the lipid bilayer (described previously).
- Lipids are also distributed in the membrane in an asymmetric manner—both *between* leaflets and *within* a leaflet.

Lipid asymmetry between leaflets

Phospholipids are synthesized on the cytosolic side of the endoplasmic reticulum (ER) membrane.

Glycolipid synthesis also begins on the ER membrane and is completed in the lumens of the ER and Golgi apparatus.

➔ *Membrane lipid synthesis is outlined in Chapter 8 and revised more thoroughly in the companion book,* Thrive in Biochemistry and Molecular Biology.

- The dimensions of the inner and outer leaflets of the ER membrane are maintained by a **scramblase** that catalyses the equilibration of the newly formed membrane lipids between both leaflets.
- Both phospholipids and glycolipids are eventually incorporated into the plasma membrane, but the two leaflets of the plasma membrane have very different compositions (Figure 3.14).
- Flippases use the free energy of ATP hydrolysis to transfer phosphatidylserine and phosphatidylethanolamine across the membrane to the inner leaflet producing a membrane with an asymmetric distribution of lipids between the outer and inner layers.

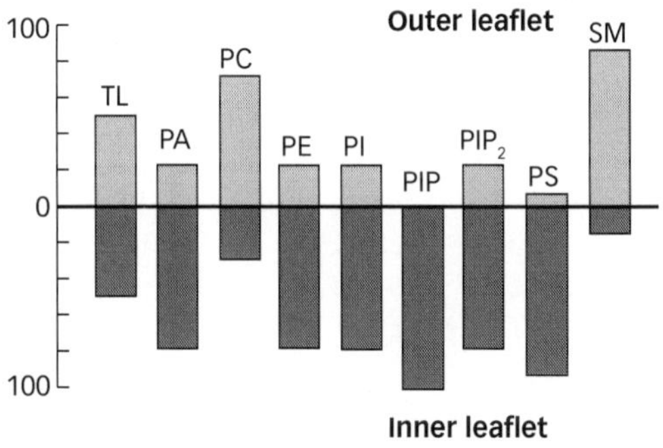

Figure 3.14 Relative amounts of some of the lipids in the plasma membrane of human erythrocytes. PA, phosphatidate; PC, phosphatidylcholine; PE, phosphatidylethanolamine; PI, phosphatidylinositol; PIP, phosphatidylinositol 4-phosphate; PIP_2, phosphatidylinositol 4,5 bisphosphate; PS, phosphatidylserine; SM, sphingomyelin; TL, total lipids.

Lipid asymmetry within a leaflet

Lipid rafts or **microdomains** are regions of the leaflets that are enriched in specific types of lipids.

- Some long-chain saturated sphingolipids and cholesterol are found predominantly in rafts of the outer leaflet of eukaryotic plasma membranes making them thicker and less mobile than the surrounding membrane.
- The compositions of lipid rafts are not fixed and change as lipids enter and leave them.
- Rafts attract raft-associated proteins, which are thought to be involved in a variety of functions, including:
  - detecting and responding to extracellular chemical signals (Chapter 15)
  - the recognition and binding of immune cells to pathogens
  - transport of materials into the cell (see later in this chapter).

Looking for extra marks?

Leaflets of biological membranes do not consist merely of *rafts* and *non-raft* regions. Other microdomains consist of many types of specialized lipid domains that are too small to be easily observed by microscopy, and caveolae, which are small invaginations of the plasma membrane, and clathrin coated pits, which are both thought to function in endocytosis (Chapter 9).

3.5 FUNCTIONS OF THE MEMBRANE LIPID BILAYER

A major general function of the lipid bilayer is to form a permeability barrier around cells and, in the case of eukaryotic cells, around organelles.

- Many membrane lipids are also essential for the appropriate orientation and correct folding and functioning of membrane proteins.
- For example:
  - The lac permease, which allows lactose to cross the cytoplasmic membrane of bacterial cells (see Figure 3.18 and associated text), requires the presence of phosphatidylethanolamine to be appropriately assembled in the membrane.
  - The ADP/ATP exchanger protein of the inner mitochondrial membrane (Chapter 11) requires the presence of cardiolipin for activity.
  - The Na^+/K^+-ATPase of the muscle plasma membrane and the Ca^{2+}-ATPase of the sarcoplasmic reticulum both function optimally in the presence of phosphatidylserine or cardiolipin.

Check your understanding

3.3 Which of the following statements is/are false?

a. The two leaflets of the plasma membrane have different lipid compositions.

b. The activities of scramblase and flippase are responsible for the uneven lateral distribution of phospholipids in the leaflets of biological membranes.
c. Lipid rafts, which are also called macrodomains, are regions of a membrane leaflet that are enriched in specific types of lipids.
d. Lipid bilayers form the major permeability barriers around cells and eukaryotic organelles.
e. Some membrane proteins are only fully functional in the presence of certain membrane lipids.

3.6 ROLES OF BIOLOGICAL MEMBRANES

Membranes are essential for life. For example they:

- provide numerous binding sites
- mediate solute transport across the membrane
- are essential for **endocytosis** (and **exocytosis** or secretion (Chapter 9))
- are essential for energy transduction
- are involved in many signal transduction reactions
- are necessary for cell–cell adhesion
- are required for the structures and activities of specialized cell–cell junctions.

- Many of these roles are mediated by proteins of the membrane, although many membrane proteins only function in an appropriate lipid environment.

Membrane binding sites

The large surface area of membranes provides many binding sites for other biological molecules and structures. These include the following:

- Enzymes of the smooth endoplasmic reticulum membrane concerned with the synthesis of phospholipids.
- Ribosomes involved in producing membrane or secretory proteins, which bind temporarily to the inner face of bacterial surface membranes and in eukaryotes, the cytosolic side of the rough endoplasmic reticulum membrane.
- A number of structural proteins, of which perhaps the best known example are the spectrins.
- Spectrins have extended structures that are attached to the inner face of the erythrocyte plasma membrane and help stabilize their shape as they are propelled through blood vessels.

Transport of solutes across membranes

Biological membranes are **selectively permeable**; they allow the passage of some solutes but restrict or prevent the movements of others.

Roles of biological membranes

- Solutes can cross membranes by one of two methods:
  - **diffusion**
  - **active transport.**
- The diffusion of solutes across a membrane may be **simple** (also called **passive**) or **facilitated.**

Diffusion

- Diffusion is the movement of molecules from a region of *high* concentration to one of *low* that is down its **concentration gradient.**
- Diffusion will continue until the concentrations in both regions are equal and equilibrium is reached.
- If the solute is charged, then the ionic gradient most also be taken into account.
- Movement of an ion across a membrane carrying a potential is in accordance with the **electrochemical gradient.**
- If an ion is travelling to a compartment of opposite charge, its movement will be aided. Conversely, transport into one of the same charge will be opposed.
- An unequal concentration of ions on different sides of a membrane produces a potential difference across it called the **membrane potential.**

Simple diffusion

- Only uncharged small hydrophobic molecules cross membranes by simple diffusion.
- The rate of diffusion through the membrane is described by the equation:

  $$v = P \times \Delta[S]$$

  where v is the rate of diffusion, P is the permeability coefficient of the membrane to the solute and $\Delta[S]$ is the difference in concentration between each side of the membrane that is the concentration gradient.
- The rate of diffusion of hydrophobic gases, such as dioxygen (O_2) or steroid hormones across biological membranes is directly proportional to the difference in concentrations or concentration gradient as shown in Figure 3.15.

Facilitated diffusion

- Facilitated diffusion is the movement of a solute down its chemical (or electrochemical gradient) using integral membrane proteins, rather than diffusing directly through the lipid bilayer. Movement will occur until equilibrium is established.
- Lipid bilayers are impermeable to hydrophilic molecules and ions; thus, facilitated diffusion is essential if hydrophilic molecules, such as sugars and ions, are to cross the membrane.
- Movement across the membrane may be achieved by changes in the conformation of the protein transporter or through pores in the membrane.

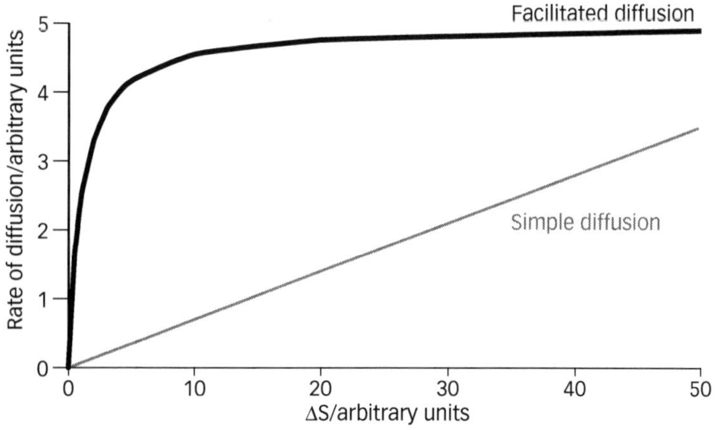

Figure 3.15 Comparison of the movement of solutes across a biological membrane by simple (passive) and facilitated diffusion.

- In the **alternating conformation model** of facilitated diffusion, the solute binds to the membrane transport protein, which then changes its conformation to release the solute on the other side of the membrane.
- Once the solute is released, the transporter resumes its previous conformation and can transport another solute molecule.
- In other cases, transport may be facilitated by pores through the membrane formed by integral transmembrane proteins.
- The pores are lined with hydrophilic amino acid side chains, which can substitute for the water molecules binding to the hydrophilic solute and allow it to cross the membrane.
- Transport proteins have the following nomenclature:
  - membrane proteins that transport a single substance are called **uniporters**
  - those that transport two substances simultaneously in the *same* direction are called **symporters**
  - those transporting two substances in *opposite* directions are **antiporters**.
- Examples of solutes which cross membranes by facilitated diffusion include the uptake of glucose by erythrocytes, ions by nerve cells, and water by a variety of tissues.

➔ *Chemiosmosis is revised in Chapters 4, 10, and 11.*

Facilitated diffusion of glucose and erythrocytes

- Erythrocytes are bathed in glucose-rich blood plasma.
- Thus, glucose always enters the erythrocyte down its concentration gradient.
- A **glucose uniporter** is thought to facilitate its movement through the membrane using the alternating conformation model described earlier.
- The movement of glucose across the membrane shows **saturation kinetics**.

- Unlike simple diffusion, the rate of diffusion does not increase in proportion to concentration.
- At low concentrations glucose crosses the membrane faster than would be expected for simple diffusion but reaches a maximum rate and increases in concentration have no effect (Figure 3.15).
- Further, the transporter shows specificity only for L-glucose and structurally related sugars, for example L-galactose and L-mannose.
- The saturation curve for facilitated diffusion is a rectangular hyperbola; which is described by the equation:

$$v = \frac{V_{max} \times [S]}{K_a + [S]}$$

where v is the rate of transport at a concentration of S, V_{max} is the maximum rate of transport and K_a is the affinity constant of the transporter for the particular sugar (in this case glucose) and is the concentration required to give half the maximum rate.

- Note how this equation is identical in form to the famous Michaelis–Menten equation that describes the kinetics of many enzymes activities.

➔ *The Michaelis–Menten equation is revised in the companion volume,* Thrive in Biochemistry and Molecular Biology.

Check your understanding

3.4 Estimate (a) the V_{max} of transport and (b) the K_a of the hypothetical facilitated diffusion transporter in Figure 3.15.

Facilitated diffusion of ions

- Ions cross membranes through transmembrane ion channels.
- Surprisingly, ion channels are selective; one type of channel allows only the passage of one type of ion such as K^+, Na^+, Ca^{2+}, and Cl^-.
- Selectivity is achieved by a constriction in the centre of the channel that functions as a size filter (Figure 3.16) and binding sites, which are formed of amino acid side chains and atoms of the polypeptide backbone that are ion-specific.
- Most ion channels open and close in response to specific stimuli; this is often termed **gating**.
- The three most common types of gated channels are:
  1. ligand-gated channels
  2. mechanosensitive channels
  3. voltage-gated channels.
- Ligand-gated channels open or close in response to a specific substance binding to the channels or to a protein that regulates the channel (Chapter 15).

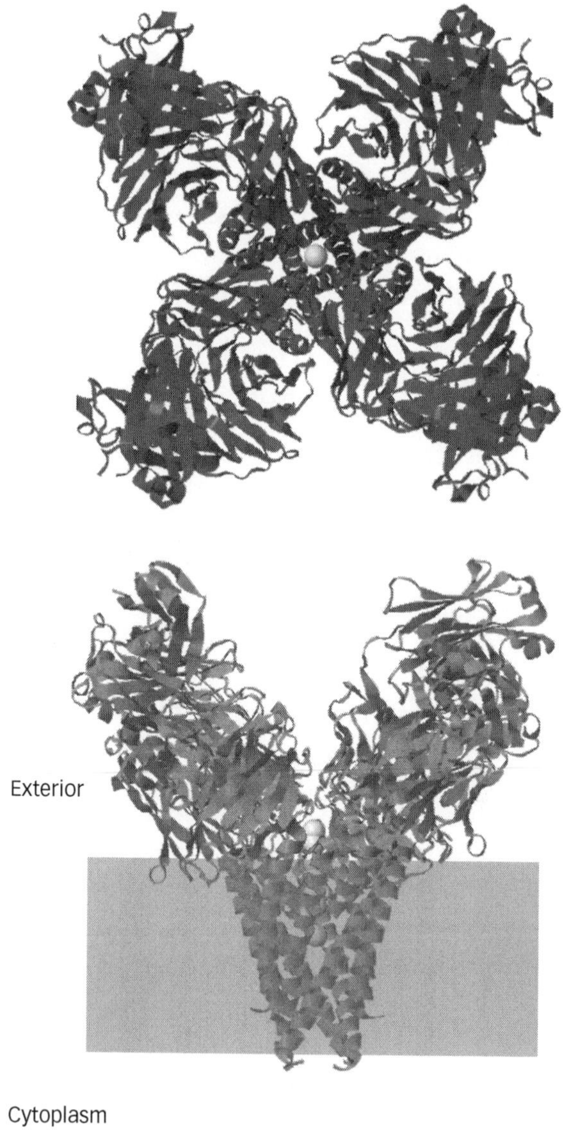

Figure 3.16 Molecular model of a K^+ channel. The lighter coloured spheres represent K^+ moving through the central channel. Model constructed using PDB file 1K4C.

- Mechanosensitive channels are triggered by mechanical stresses acting on the membrane.
- Voltage-gated channels open and close in response to changes in the membrane potential, that is the potential difference that occurs across the membrane (described earlier).

Facilitated diffusion of water

- Water, although hydrophilic, crosses membranes by simple diffusion although this process is slow.
- Its diffusion is increased in all cells by pore-forming transmembrane channels called aquaporins, which facilitate its movement (Figure 3.10).
- Water moves through aquaporins in accordance with the prevailing **osmotic gradient** across the membrane.
- The osmotic gradient is the concentration gradient of water.
- If the solute is at a lower concentration on one side of the membrane, the concentration of water will be higher and so water will tend to diffuse over to the other side.

Revision tip

Simple animations showing passive and facilitated diffusion may be seen at: http://bcs.whfreeman.com/thelifewire/content/chp05/0502001.html

Active transport

- Active transport is the movement of solutes *against* their concentration or electrochemical gradient.
- Active transport can be achieved only by coupling the movement of the solute to the expenditure of metabolic energy.
- Active transport is energy intensive: up to 50% of the cell's available energy can be used in these processes.
- The energy may be supplied by one of two methods:
  1. directly coupling transport to an **exergonic (energy-releasing) reaction** such as the hydrolysis of ATP
  2. linking the active transport to the facilitated transport of an ion.
- The first are called direct or primary active transport mechanisms, the second, naturally, indirect, or secondary active transport mechanisms.

Group translocation

- Group translocation mechanisms chemically modify the solute *during* its transfer across the membrane. Thus, the unmodified substance does not accumulate in the cell and the diffusion gradient is maintained—but the modified substance can accumulate.
- Among the best studied group translocation systems are the uptake of the hexose sugars, glucose, mannose, and fructose, by *E. coli* cells.
- The passage of these sugars across the membrane involves modification to form their 6-phosphate derivatives. The addition of the phosphate group imparts a negative charge that prevents the sugar 6-phosphate diffusing out of the cell.

- The transport mechanism involves five proteins—enzyme I, H protein (HPr, so-named because it has an active histidine residue), enzyme II_a, enzyme II_b, and enzyme II_c—which function as a phosphotransferase system.
- Enzyme I, HPr, and enzyme II_a are cytosolic proteins; enzyme II_b is found on the cytosolic side of the bacterial surface membrane, while enzyme II_c is an integral transmembrane protein.
- Enzyme I catalyses the phosphorylation of HPr using phosphoenolpyruvate (PEP) as the phosphate donor (Figure 3.17).
- The phosphate is transferred from HPr to each of the remaining proteins in turn (enzyme II_a to enzyme II_b to enzyme II_c) until the transporter protein, enzyme II_c, catalyses the phosphorylation of the sugar during its passage across the membrane (Figure 3.17).
- Enzyme I and HPr are non-specific and participate in the uptake of several different types of sugars.
- However, several different forms of enzyme II are found, each specific for a different sugar.

Lac permease or lactose transporter

- Cells of *E. coli* are able to use active transport to concentrate lactose in their cytoplasm to values greater than in the extracellular space.
- The protein responsible for this transport is the lac permease or lactose transporter, which functions as a symport using an indirect active transport mechanism.

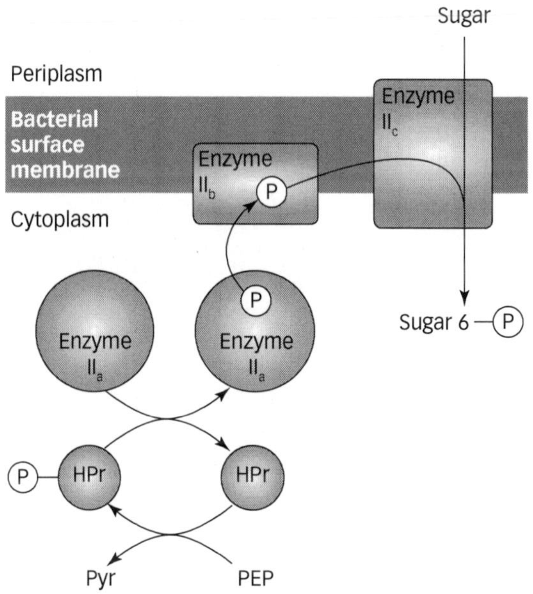

Figure 3.17 Group translocation mechanism for hexose uptake in *Escherichia coli*. See text for details. HPr, H protein; PEP, phosphoenolpyruvate; Pyr, pyruvate.

- Electron transport in the surface membrane is linked to the active expulsion of H^+ from the cytoplasm into the periplasmic space, as shown in Figure 3.18.

➔ *Electron transport and the generation of transmembrane H^+ gradients are revised in more detail in Chapters 4, 10, and 11.*

- This activity generates a transmembrane H^+ gradient across the membrane; the periplasmic space being more concentrated and more positively charged than the cytoplasm.
- Transmembrane gradients of ions are temporary stores of energy, which can be used to drive active transport mechanisms.
- The lac permease allows H^+ to return to the cytoplasm by facilitated diffusion down their electrochemical gradient.
- However, an H^+ can only pass through the permease if a lactose molecule is transported at the same time (Figure 3.18).
- Thus, lactose enters the cell by active transport, against its concentration gradient.

Transport ATPases

- Four major types of transport ATPases are involved in active transport:
  1. **ATP-binding cassette (ABC) ATPases**
  2. **factor (F) ATPases** or **ATP synthases**
  3. **phosphorylation (P) ATPases**
  4. **vacuole (V) ATPases.**
- All employ direct active transport mechanisms using the free energy of ATP hydrolysis to drive the transport mechanism.

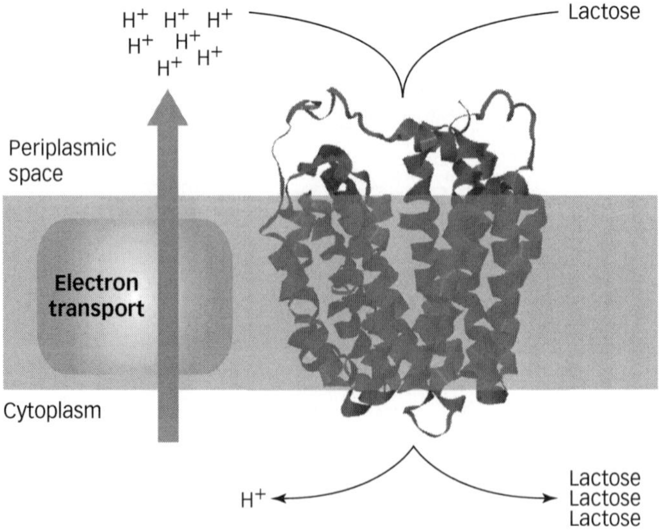

Figure 3.18 Role of the lac permease of *Escherichia coli* in the active uptake of lactose. See text for details. The molecular model of the lac permease was constructed using PDB file 2CFP.

1. ATP-binding cassette ATPases
    - ATP-binding cassette ATPases share a common structure consisting of two membrane-spanning domains and two peripheral domains that bind ATP and associated with the cytoplasmic side of the membrane (Figure 3.19).
    - Over 200 different types of ABC-ATPases have been described in the plasma membranes of most cell types.
    - In bacteria and archaea, ABC-ATPases function largely as importers of nutrients.
    - In eukaryotes, their roles are more varied and they are exporters as well as importers.
    - ATP-binding cassette ATPases are highly specific; each type transports only one type of solute or a group of closely related compounds.

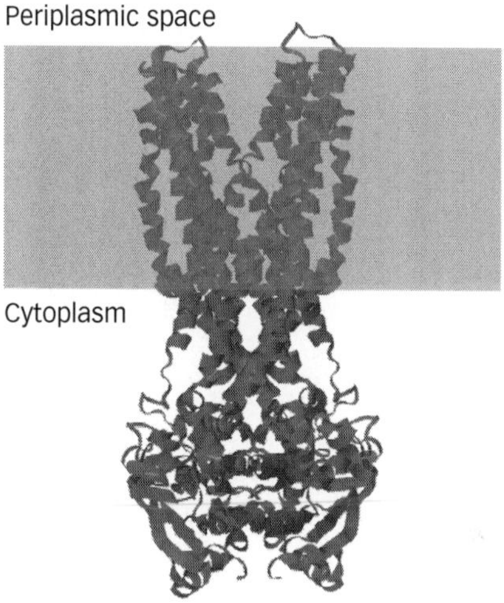

Figure 3.19 Molecular model of ATP-binding cassette (ABC) ATPase from *Staphylococcus aureus*. Note the two upper α helical membrane-spanning domains and two lower ATP binding domains. Model constructed using PDB file 2ONJ.

Looking for extra marks?

The **multidrug resistance (MDR) transport protein** was the first ABC-ATPase found in humans. It differs from other types in having a *broad* specificity for small relatively hydrophobic compounds. During cancer chemotherapy, the MDR transport protein actively pumps anticancer drugs out of the cells and so reduces the effectiveness of treatment.

2. Factor ATPases or ATP synthases
    - Factor ATPases or ATP synthases are found in the surface membranes of prokaryotic cells, the inner mitochondrial membrane and the thylakoid membrane of chloroplasts.

- The transmembrane H^+ gradients generated using electron transport can be used as an energy source to synthesize ATP, as well as driving active transport mechanisms.
- Hydrogen ions can cross the membrane back to their original compartment through a pore. When doing so they activate the ATP synthase to form ATP from ADP and P_i.
- Thus, F-ATPases link the transmembrane H^+ gradients to the synthesis of ATP.
- The actions of ATP synthases during oxidative phosphorylation provide most of the cells ATP in aerobic conditions.

➔ *Oxidative phosphorylation is revised in Chapters 4 and 11.*

3. Phosphorylation (P) ATPases
   - Phosphorylation ATPases are a large family of transport proteins, which are phosphorylated using ATP at specific aspartate residues.
   - Most are found in the plasma membranes of both eukaryotes and prokaryotes and are mainly concerned with transporting cations across the membrane. An exception is the flippase of plasma membranes, which distributes phospholipids between the membrane leaflets (see earlier).
   - Some P-ATPases are found in intracellular membranes. For example, the Ca^{2+}-ATPase of the sarcoplasmic reticulum (Figure 3.9 (b)) pumps Ca^{2+} out of the sarcosol of muscle cells (Chapter 7).
   - Perhaps the best characterized P-ATPase is the **Na^+/K^+-ATPase** of plasma membranes.
     - Most animal cells maintain intracellular concentrations of K^+ and Na^+ that are respectively higher and lower than their extracellular fluid (Table 3.3).
     - These differences in concentration are maintained by the Na^+/K^+-ATPase of plasma membranes.
     - This enzymic activity is essential for maintaining the appropriate ionic balance in the cytoplasm and generating the membrane potential.
     - Membrane potentials are necessary for driving indirect active transport mechanisms, the transmission of nerve impulses, the contractile events of muscle tissues and regulating cell volume.
     - The Na^+/K^+-ATPase is a tetrameric protein, consisting of two α and two β subunits (Figure 3.20).

Cell	*[K^+]/mmol dm^{-3}*		*[Na^+]/mmol dm^{-3}*	
	I	*O*	*I*	*O*
Frog muscle	400	20	50	440
Mammalian motor neuron	140	5	10	145
Human skeletal muscle	150	5	14	150
Squid axon	400	20	50	440

Table 3.3 Intra- and extracellular concentrations of K^+ and Na^+

I and O, intracellular and extracellular concentrations respectively.

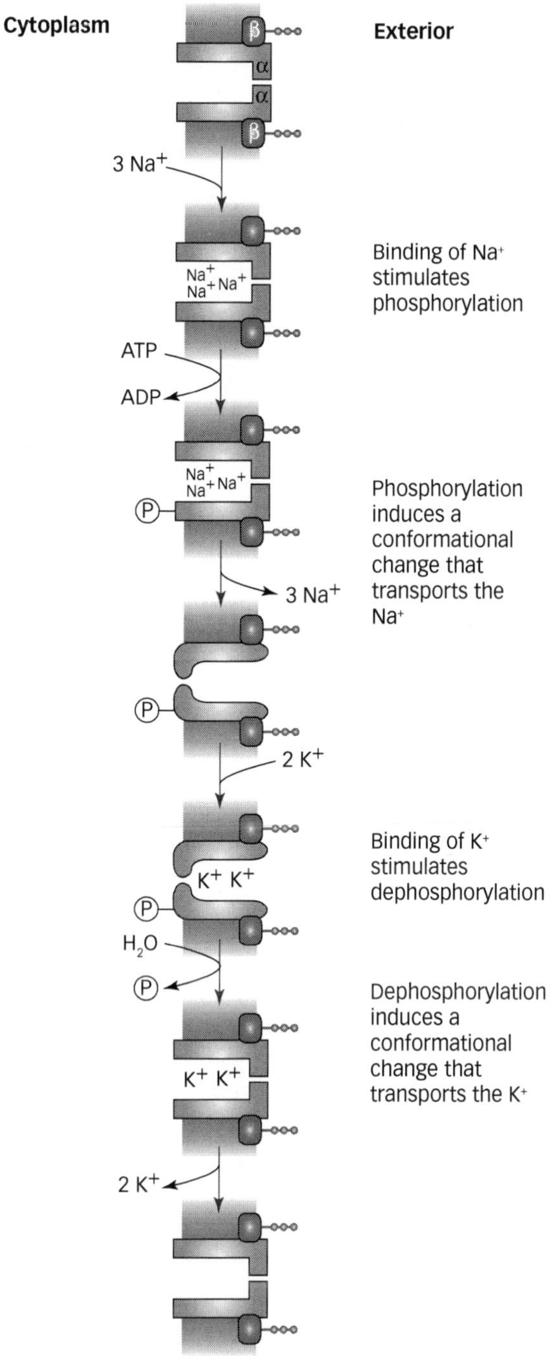

Figure 3.20 Schematic mechanism of the Na^+/K^+-ATPase in transporting K^+ into the cell and Na^+ out of it.

  - Three Na^+ are transported out of the cell and two K^+ imported for every ATP hydrolysed using an alternating conformation model (Figure 3.20).
  - This 3:2 stoichiometry means the Na^+/K^+-ATPase is **electrogenic**: it maintains the cytoplasm at a negative charge compared to the exterior, generating a potential difference (the membrane potential) across the plasma membrane.
  - In many mammalian cells, the resting membrane potential is approximately –60 mV.
  - In addition to its roles in nerve transmission and indirect transmembrane transport, the Na^+/K^+-ATPase has a role in maintaining cell volume.
  - The osmolarity of the cytoplasm is regulated by pumping out any Na^+ that leaks into the cell with the membrane potential preventing the entry of excess Cl^-.
  - Thus, the amount of water entering the cell through aquaporins (Figure 3.10), and consequently the volume of the cell, is controlled.

4. Vacuole (V) ATPases
   - Vacuole ATPases are found in the membranes of the endosomal-lysosomal system of animal cells and the vacuole of plant cells (Chapter 9).
   - These ATPase pump H^+ into the organelles at the expense of ATP.
   - The resulting acidification of the contents is essential for the acid optimal enzymes they contain.

Glucose uptake by the gastrointestinal tract

- The membrane potential generated by the Na^+/K^+-ATPase is used to drive numerous indirect active transport systems.
- Both amino acids and monosaccharides are absorbed by **enterocytes** lining the gastrointestinal tract (GIT) in Na^+-dependent fashion.
- Both glucose and galactose are transported from the lumen of the GIT into an enterocyte by the same **Na^+/glucose symporter**.
- A single molecule of either of these sugars can only move through the symporter into the cell if an Na^+ moves through it at the same time (Figure 3.21).
- The concentrations of the sugars can build up within the cytoplasm, such that they are able to leave the cell through the basolateral membrane by facilitated diffusion.

Energy changes associated with transport across membranes

- The energy driving the diffusion of an electrically neutral substance is the difference in its concentrations on either side of the membrane.
- The free energy change (ΔG) associated with the movement of one mole of the substance from the exterior (concentration C_O) to the interior of the cell (concentration C_I) is:

  $$\Delta G = RT \ln ([C_I]/[C_O])$$

  where R is the gas constant (8.314 J K^{-1} per mol) and T is the absolute temperature.

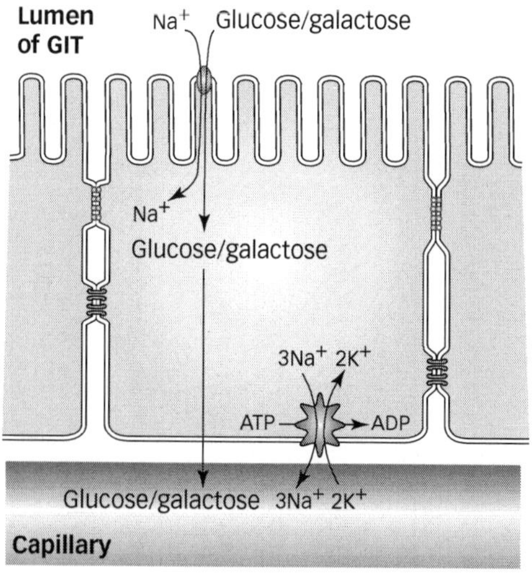

Figure 3.21 Schematic import of glucose or galactose by an enterocyte of the gastrointestinal tract.

- However for ions, the nature and size of their charge, the ionic nature of the compartment they are entering and the membrane potential must all be taken into account.
- The change in energy associated with transporting ions is described as:

$$\Delta G = RT \ln ([C_I]/[C_O]) + zF\Delta\psi$$

where z is the valence number of the ion, *F* the Faraday constant (96,500 J V^{-1} per mol) and $\Delta\psi$ the membrane potential.
- A negative value for ΔG indicates that uptake is energetically favourable.

Revision tip

A simple animation of active transport is available at: http://bcs.whfreeman.com/thelifewire/content/chp05/0502002.html

Check your understanding

3.5 (a) Determine the free energy associated with pumping one mole of glucose into a mammalian cell if its concentration in the cell is twice that outside. (b) Determine the energy expenditure involved in transporting one μmole of K^+ across a membrane with a potential of –60 mV if the concentrations inside and outside the cell are 145 and 5 mmol dm^{-3} respectively.

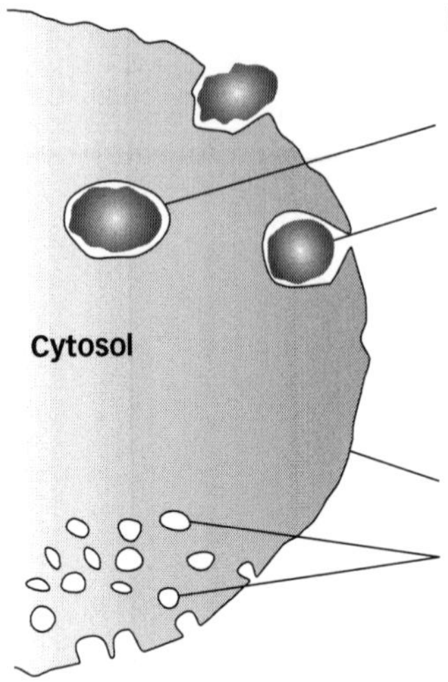

Figure 3.22 Overview of phagocytosis and pinocytosis. See text for details.

Bulk uptake of material by cells

Macromolecules and suspended particles (other cells, parts of cells, or macromolecular aggregates) can be taken up by cells using endocytosis.

- The two commonest methods of endocytosis are **phagocytosis/pinocytosis** and **receptor-mediated endocytosis.**

Phagocytosis and pinocytosis

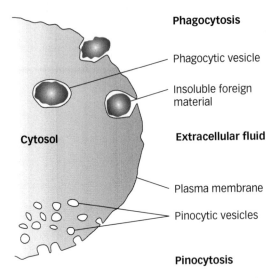

- Phagocytosis, meaning *cell eating*, is the uptake of suspended particles larger than about 0.5 μm in diameter.
- In phagocytosis, the plasma membrane flows forward to surround the particles, which are ingested into a phagocytic vesicle or vacuole (Figure 3.22).
- The vacuole then fuses with vesicles of the endosomal-lysosomal system, which results in the digestion of the particle.

➔ *The endosomal-lysosomal system is reviewed in Chapter 9.*

- Phagocytosis is the routine method by which a variety of protozoa, including amoebae, catch and digest their prey.
- It is also the method used by neutrophils and macrophages of the immune system to attack parasitic microorganisms.
- Pinocytosis, or 'cell drinking', involves the continual infolding of parts of the plasma membrane (Figure 3.22).
- These infoldings form small pinocytic vesicles containing extracellular fluid, which may contain dissolved or suspended nutrients.

Receptor-mediated endocytosis

- Receptor-mediated endocytosis is the uptake of materials (ligands) by endocytosis following their binding to specific cell surface receptors.
- Following binding to receptors, the receptor-ligand complexes move laterally in the membrane and become concentrated in regions called **coated pits** (Figure 3.23a).
- The concentration of complexes at the coated pit leads to the accumulation of additional proteins called **adaptor proteins** (**dynamin** and **clathrin**) on its *cytoplasmic* surface.
- The presence of the proteins allows the membrane to tightly curve inwards and form a **clathrin-coated vesicle** (Figure 3.23b–d).
- The clathrin coat of proteins is then shed, releasing an uncoated vesicle.
- The receptors and coat proteins are then returned to the cell surface for re-use.
- The vesicle then fuses with those of the endosomal-lysosomal system as outlined in Chapter 9.
- Depending upon the nature of the ligands, it may then be delivered intact to intracellular sites or digested.
- The use of receptor-mediated endocytosis allows a ligand to be absorbed with little extracellular fluid.
- Coated pits can form within a minute and occupy 20% or more of the surface of a cell.
- Receptor-mediated endocytosis is the main way in which cells internalize macromolecules.
- Depending upon the type of cell and receptor involved, hormones, growth factors, proteins, including enzymes and antibodies, and lipoprotein particles may all be taken up by receptor-mediated endocytosis.
- Thus, the fate of the vesicle depends upon the nature of the ligand and the type of cell involved.

Looking for extra marks?

Receptor-mediated endocytosis was discovered by Brown and Goldstein who received the Nobel Prize in Physiology or Medicine in 1985 for their studies on the regulation of cholesterol metabolism.

Energy and signal transductions

Transduction is the conversion of energy or a signal from one form to another.

- Membranes are essential for both types of transduction.
- For example, the energy of transmembrane H^+ gradients (chemiosmotic energy) can be used by ATP synthases to make ATP.

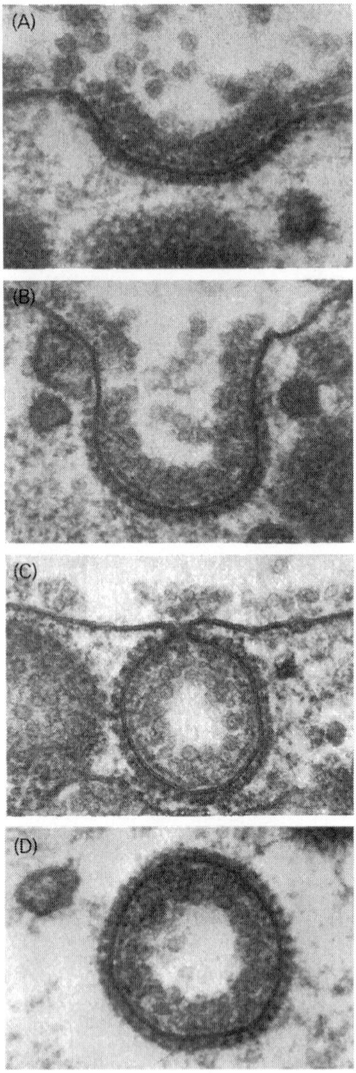

Figure 3.23 Electron micrographs illustrating receptor-mediated endocytosis of yolk protein in chicken (*Gallus domesticus*) oocytes. (A) Coated pit. (B) to (D) Stages in the formation of a coated vesicle. See text for explanation of events. Reproduced with permission from Perry, M.M. and Gilbert, A.B. (1979) Yolk transport in the ovarian follicle of the hen (*Gallus domesticus*): lipoprotein-like particles at the periphery of the oocyte in the rapid growth phase. *Journal of Cell Science*, 39, 257–72.

- **Cell signalling** is the communication networks and pathways used by cells following receipt of chemical or electrical signal from other cells or following external stimuli, for example light and sound waves.
- Many signals are received by integral membrane protein receptors on the surface of the cell and if they are weak must be amplified and communicated to the interior of the cell by **signal transduction** processes.

- These promote *intracellular* biochemical changes associated with receipt of the initial signal.

  ➔ *ATP synthesis and cell signalling are revised in more detail in Chapters 4, 10 and 11, and 15 respectively.*

Cell–cell adhesion

Within multicellular organisms, cells are organized into tissues. In most organisms, cell–cell interactions involve contacts between cell walls. However, in animals, it involves the use of cell–cell adhesion macromolecules.

- In animal tissues, cells may be separated by an extracellular matrix (ECM).

  ➔ *The extracellular matrix is revised in Chapter 14.*

- In other instances, the cells are closely packed and so must recognize, bind, and communicate with one another.
- Binding of cells in tissues is mediated by integral membrane proteins, whose extracellular domains recognize and bind to specific ligands or identical types of molecules on neighbouring cells.
- If an adhesion protein on one cell binds to the *same* type of molecule on the other cell, the interaction is said to be **homophilic**. If, however, it involves the adhesion protein interacting with a *different* type of molecule on the surface of the neighbouring cell, then the binding is **heterophilic**.
- Four major types of proteins are involved in cell–cell adhesion as shown in Figure 3.24; they are:
  1. **cadherins**
  2. **cell adhesion molecules** (CAMs)
  3. **integrins**
  4. **selectins**.
- In many cases, the intracellular domains of proteins involved in cell–cell adhesion are linked to elements of the cytoskeleton (see Chapters 7 and 14).
- This allows cell–cell binding to be signalled to the interior of the cells and stimulate appropriate responses.

Check your understanding

3.6 What is the nature of binding shown by (a) cadherins and cell adhesion molecules, and (b) integrins and selectins?

Cadherins

- Cadherins are a family of proteins that includes E-, N-, P-, and VE-cadherins.
- Different family members are expressed in different tissues: E-cadherins are found in epithelial cells, N in neural tissues, P was first found in the placenta, and VE in vascular endothelial cells.

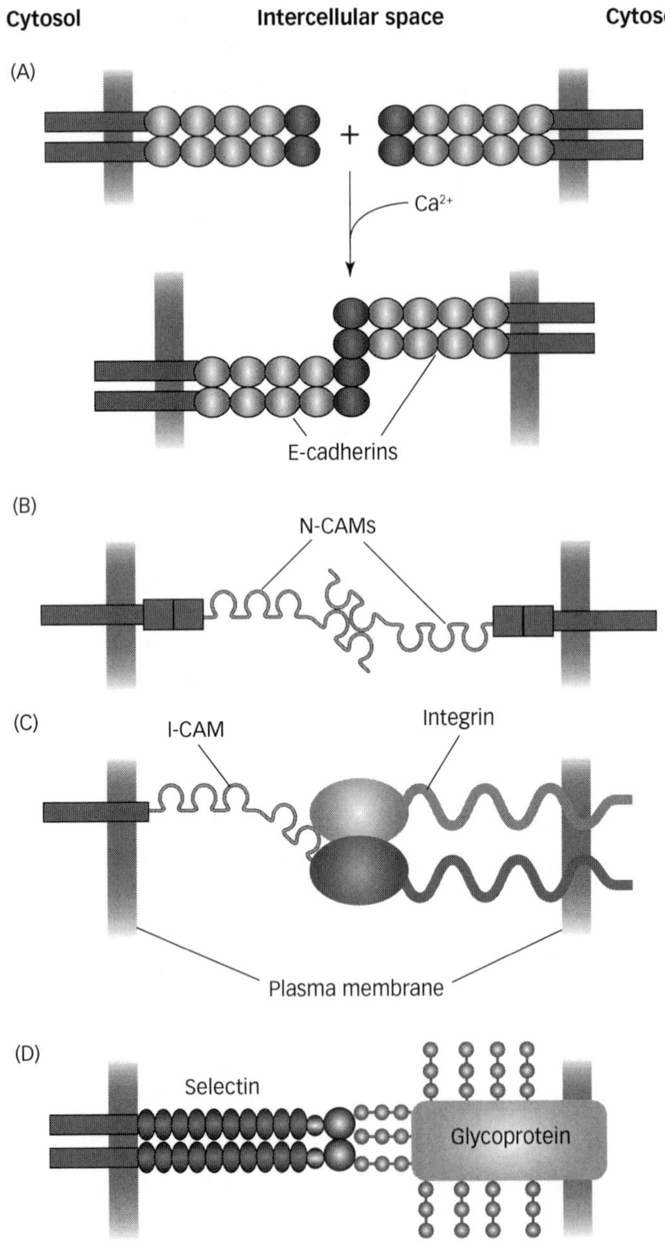

Figure 3.24 Schematic interactions of the four main types of cell adhesion molecules. (A) Dimers of cadherins linking adjacent cells together. (B) Two cells linked together by N-cell adhesion molecules (N-CAMs). (C) Mutual binding of an intercellular adhesion molecule (I-CAM) and an integrin. (D) The binding of a selectin on one cell to the glycan of a glycoprotein on another.

- All are involved in cell–cell adhesion, with cadherin dimers binding to identical cadherin molecules in a Ca^{2+}-dependent manner as shown in Figure 3.24a.
- E-cadherins are particularly concentrated in cell–cell junctions called **adherens junctions** and **desmosomes** (see later in chapter).
- Cadherin expression is tightly regulated, particularly during embryonic development where the assembly and disassembly of cadherin-based adhesion sites allows different cells to bind and separate in an appropriate manner during tissue development.

Cell adhesion molecules

- Cell adhesion molecules (Figure 3.24b) are members of the **immunoglobulin superfamily (IgSF)**.
- Members of the CAM family include **neural** and **intercellular adhesion molecules (N-CAMs** and **I-CAMs**).
- Both groups are involved in adhesion events during embryonic development.

Integrins

- Integrins are dimeric glycoproteins that function in cell–cell and cell–ECM interactions (see Chapter 14).
- However, integrins on the surfaces of cells are able to bind to specific receptors such as I-CAMs (Figure 3.24c) on target cells and facilitate binding.

Selectins

- Selectins are a family of structural glycoproteins.
- Different members of the family are expressed on the surfaces of different types of cells—E-selectins on endothelial cells of blood vessels, L-selectins on leucocytes, and P-selectins on blood platelets.
- All selectins are **lectins**—glycoproteins which are able to recognize and bind to carbohydrates.
- As indicated in Figure 3.24d, selectins recognize and bind to glycans on the surfaces of cells and so mediate a *weak* adhesive interaction between the cells.
- For example, during inflammation, initial selectin-based interactions allow leucocytes to bind to the inner wall of a blood vessel. However, these weak interactions readily break and reform, allowing the leucocytes to move along the wall to positions near to the site of inflammation, whereby stronger specific integrin-ICAM-based interactions (Figure 3.24b) stabilize the positions of the leucocytes and allow them to migrate through the vessel wall to the inflammatory site.

Cell–cell junctions

Four major types of specialized cell–cell junctions are found in cells:

1. **adhesive junctions**
2. **tight junctions**

3. **gap junctions**
4. **plasmodesmata**.

- The first three types are associated with animals; plasmodesmata are found in plants (see Chapter 13).

Adhesive junctions

- Adhesive junctions link cells strongly to one another or link cells with the ECM.
- Adhesive junctions of animal cells include:
  - adherens junctions
  - desmosomes and **hemidesmosomes**.

Adherens junctions

- Adherens junctions consist of a large patch of cadherin molecules assembled side-by-side in the *extracellular* space (Figure 3.25).
- The *intracellular* domains of the cadherin molecules are linked to actin filaments of the cytoskeleton (Chapter 7) by **catenin** proteins.
- The cells are joined together by the cadherin molecule, which also ensures the bundles of actin filaments in adjacent cells are also linked.
- Thus, adherens junctions convert the actin filaments in the tissue into an organized 'supracellular' network.
- Myosin motor proteins can allow the actin bundles to contract, as explained in Chapter 7, so the tissue is able to move during development—such as the folding of sheets of epithelial cells into tubes.

➔ *The cytoskeleton and development are revised in Chapter 17.*

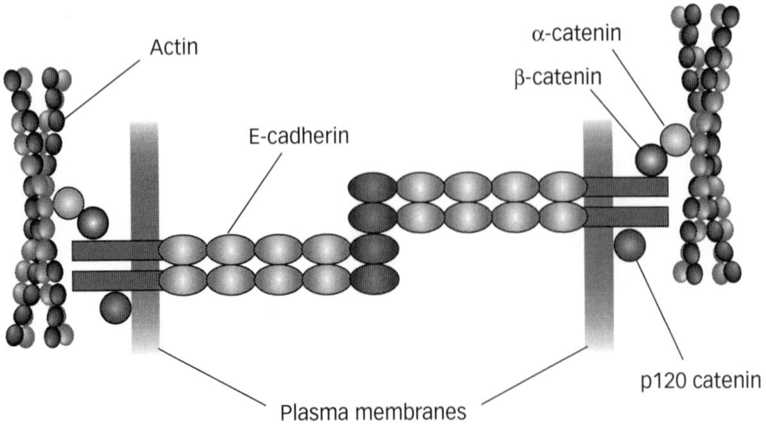

Figure 3.25 Schematic partial structure of an adherens junction. Only one cadherin–cadherin link of the many present is shown.

Desmosomes and hemidesmosomes

- The structure of desmosomes (Figure 3.26) resembles that of adherens junctions but differs in that:
  - their adhesive proteins are dimers of the specialized cadherins, desmoglein and desmocollin
  - the cytosolic tails of these cadherins are linked to intermediate or tonofilaments (IFs), not actin filaments of the cytoskeletons
  - dense plaque of proteins, including desmoplakins and plakoglobin, in the cytoplasm links the cadherins to the IFs.
- Intermediate filaments are strong fibres able to resist mechanical stress.
- Thus, desmosomes act rather like rivets and strongly anchor neighbouring cells together.
- Further, the interlinked network of IFs is able to distribute any stress throughout the tissue, rather than allowing it to act at any one point.
- Desmosomes are particularly common in tissues subject to stress such as epithelia and heart muscle.
- Hemidesmosomes resemble half a desmosome in appearance, but like the focal adhesions discussed in Chapter 14, link a cell to its basement membrane.

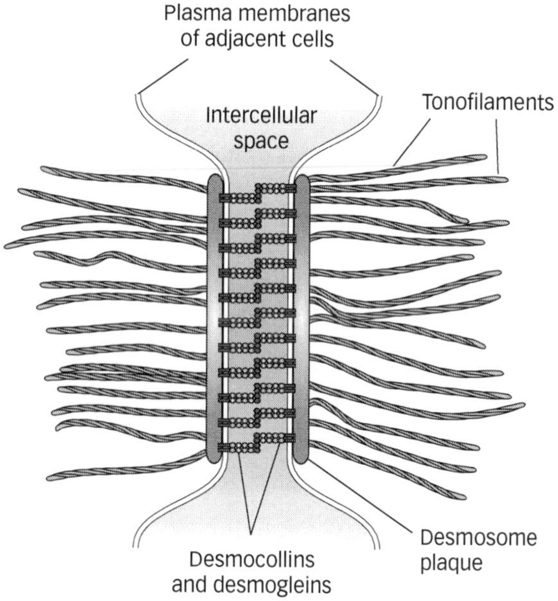

Figure 3.26 Schematic structure of a desmosome.

Looking for extra marks?

Defective desmosomes are associated with clinical conditions. The potentially fatal skin condition, pemphigus is caused by the production of autoantibodies that attack desmoglein or desmoplakin, resulting in severely blistered skin and leakages of body fluids. Mutations in genes encoding desmocollins can lead to defective cardiac muscle tissue.

Tight junctions

- Tight junctions are areas where the plasma membranes of two animal cells are so closely apposed that the transport of materials through spaces *between* the cells cannot occur.
- The membranes at tight junctions are held in close contact by a continuous row(s) of transmembrane proteins as shown in Figure 3.27.
- Proteins associated with tight junctions include occludin, claudins, and IgSF members called junctional adhesion proteins (JAMs).
- Tight junctions vary in their 'tightness'; the more rows of transmembrane proteins, the more leak-proof the junction.
- Tight junctions are prevalent in epithelial tissues that isolate particular compartments, such as the lumen of the GIT.
- Tight junctions also prevent the lateral movements of lipids in the outer leaflet of the membrane and of integral membrane proteins completely and so can divide the membrane into macrodomains.

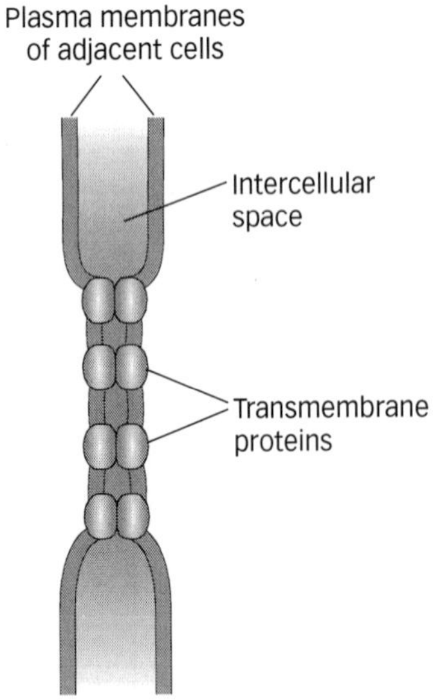

Figure 3.27 Schematic showing how rows of transmembrane proteins form a tight junction.

Looking for extra marks?

Defects in the gene encoding claudin 16 results in the autosomal recessive condition, familial hypomagnesemia with hypercalciuria and nephrocalcinosis (FHHNC), which as its name implies is characterized by extreme imbalances in the homeostasis of Mg^{2+} and Ca^{2+}.

Gap junctions

- Gap junctions consist of opposing protein-lined channels in the plasma membranes of two adjacent animal cells that form an almost continuous pore connecting the cytoplasms of both cells.
- Each pore is formed of a hexamer of **connexin** proteins forming a **connexon**. Opposing connexons in adjacent cells form a transmembrane pore as shown in Figure 3.28.
- Connexons have overall diameters of 7 nm and contain hydrophilic pores of approximately 3 nm diameter.
- Given the distance separating the cells is only about 2 nm, ions and molecules of M_r 1000–1500 are able to be directly transferred between them through the pores.
- Thus, gap junctions allow animal cells to directly communicate using ionic and chemical signals.

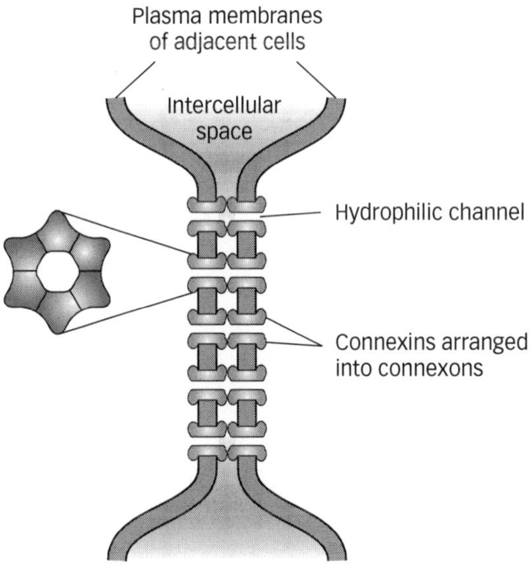

Figure 3.28 Schematic structure of gap junctions.

- The analogous plasmodesmata of plant cells, discussed in Chapter 13, perform a similar role.
- A gap junction may consist of relatively few to several thousand connexons.
- They are particularly numerous in nerve and muscle cells, which require rapid electrical (ionic) communication systems.

Looking for extra marks?

Mutations in genes encoding connexins are associated with numerous medical complaints, including a demyelinating disease of peripheral nerves and congenital deafness that arises from the death of cells in the developing organ of Corti.

Junction complexes of intestinal epithelial cells

- Epithelial cells of the small intestine are specialized for the absorption of nutrients from digested food.
- Their apical surfaces face inwards, towards the lumen of the GIT, while their basal surfaces rest on a basement membrane (Figure 3.29).
- Tight junctions are found at the extreme apical area of the plasma membrane, ensuring that nutrients can only be absorbed through the microvilli at the apical surfaces.
- Gap junctions are found towards the basal ends of the cells and allow the cells to communicate with one another by transferring ions and molecules of small M_r.

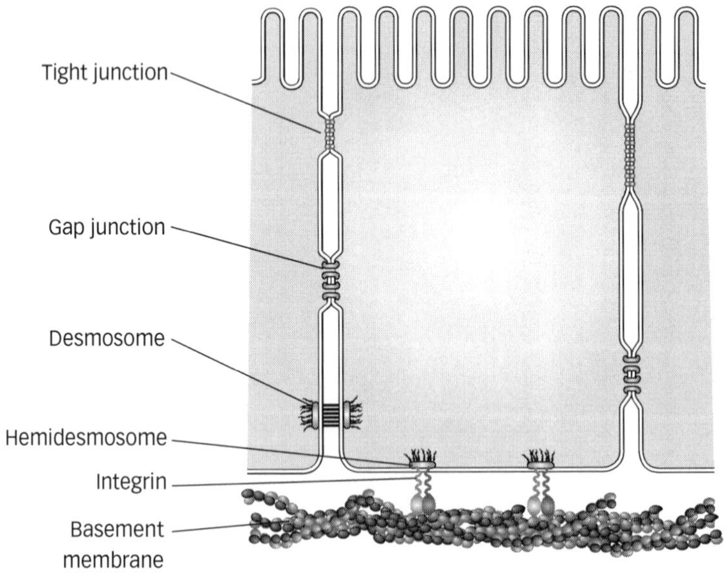

Figure 3.29 Schematic illustration of junctional complexes of vertebrate epithelial cells.

- Adherens junctions and desmosomes are found more basally and give mechanical strength to the tissues.
- Hemidesmosomes and focal adhesions (Chapter 14) anchor the tissue to its basement membrane.

Revision tip

Do browse through the further reading references given for this chapter online. Go to http://www.oxfordtextbooks.co.uk/orc/thrive/

Check your understanding

3.7 Which of the following statements is/are false?

a. The three major types of lipids found in eukaryotic plasma membranes are phospholipids, glycolipids, and steroids such as hopanoids.

b. Unsaturation is the presence of carbon–carbon double and triple bonds and is a common feature of membrane lipids.

c. Bitopic integral proteins have two membrane-spanning regions.

d. Unequal concentration of ions on different sides of a membrane produces a membrane potential.

e. Group translocation is a direct or primary active transport mechanism.

f. The Na^+/K^+-ATPase is responsible for transporting Na^+ into the cell and K^+ out of it.

g. Adherens junctions contain of a large patch of cell adhesion molecules assembled in a side-by-side fashion in the extracellular space.

h. Desmosomes resemble adherens junctions in containing dimers of cadherin-type proteins but the cytosolic tails of these proteins are linked to actin filaments.

i. Membranes connected by tight junctions are held in close contact by continuous row(s) of transmembrane proteins that include occludin, claudins, and junctional adhesion proteins.

j. Gap junctions connect epithelial cells to their basal lamina.

4 Structure and activities of prokaryotic cells

Key features of prokaryotic cell structure

- Figures 1.5 and 4.1 show the structure of typical prokaryotic cells.
- Prokaryotic cell walls give shape to the cell and help protect it from osmotic lysis.
- The surface (cytoplasmic) membrane of bacterial and archaeal cells regulates the composition of the cytoplasm; it also synthesizes ATP and is the site of prokaryotic photosynthesis.
- Cytoplasm consists largely of water containing a variety of inorganic ions, small organic molecules, and macromolecules.
- It has a gel-like consistency due to its high concentrations of macromolecules, particularly of proteins, but nucleic acids and polysaccharides are also present.
- Electron microscopy shows cytoplasm consists of two prominent regions; one is granular because of its numerous ribosomes, the other is the nucleoid or prokaryotic chromosome.
- A variety of different microcompartments and inclusions are also embedded within the cytoplasm.
- Surface feature of prokaryotes include capsules or slime layers, fimbriae, and pili.
- Flagella are the major means of locomotion in both bacteria and archaea.

Looking for extra marks?

Emphasize that eukaryotic and prokaryotic are not the names of taxonomic groupings but descriptions of types of cells.

4.1 PROKARYOTIC CELL WALLS

Almost all bacterial and archaeal cells have tough cell walls (Figures 1.5 and 4.1). Exceptions include *Mycoplasma* (bacteria) and *Thermoplasma* (archaea).

- Functions of the cell wall include:
  - protection against osmotic lysis and mechanical damage
  - conferring shape to the cell.
- Many groups of eukaryotic cells also produce cell walls (Chapter 13).

Bacterial cell walls

Peptidoglycans or **mureins** are major constituents of bacterial cell walls. These are typically polymers of *N*-acetylglucosamine (β,1$\rightarrow$4) *N*-acetylmuramic acid residues that are cross-linked to short peptides (Figure 4.2).

- Bacteria form two major types of cell walls that can be distinguished using the **Gram stain**.

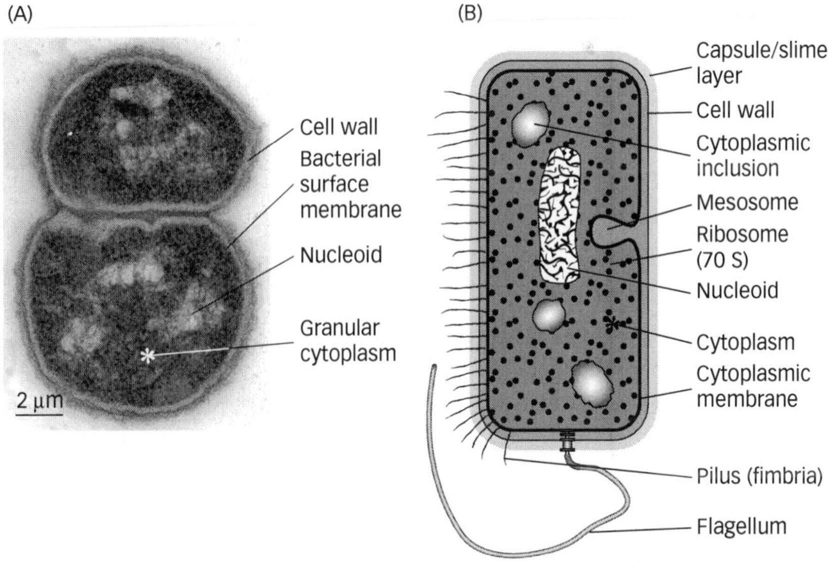

Figure 4.1 (A) Electron micrograph of a bacterial cell (*Neisseria gonorrhoea*). Courtesy of Dr A. Curry, Manchester Royal Infirmary, UK. (B) Schematic structure of a typical bacterial cell.

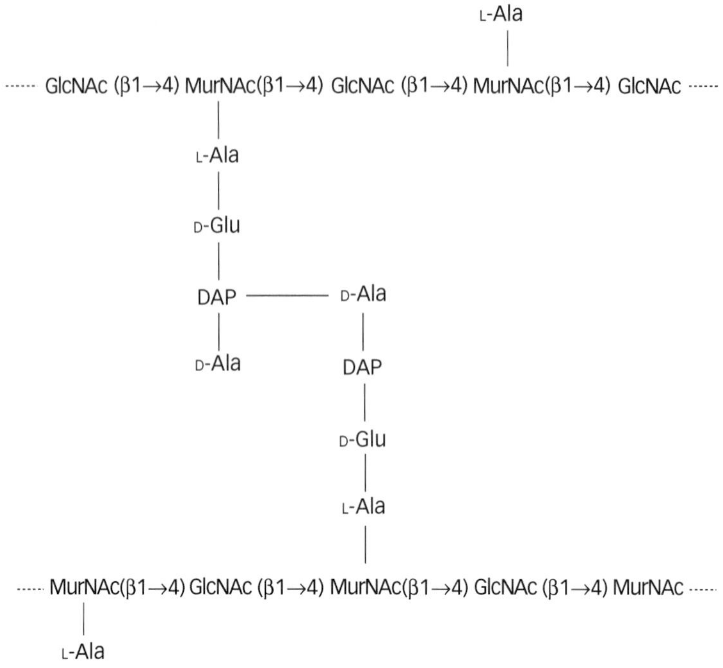

Figure 4.2 Schematic structure of a portion of a peptidoglycan molecule from *Escherichia coli*. Note the presence of D- and L-amino acids. DAP, diaminopimelic acid; GlcNAc, *N*-acetylglucosamine; MurNAc, *N*-acetylmuramic acid.

- Gram-positive bacterial cell walls retain the Gram stain and are dyed blue-black.
- Gram-negative walls do not retain Gram stain and would appear colourless but are normally counterstained pink or red to allow them to be visualized by light microscopy (Chapter 2).
- Figure 4.3 illustrates the structures of Gram-positive and Gram-negative cell walls.

Gram-positive cell walls

- Gram-positive cell walls are constructed largely of peptidoglycan and teichoic acids.
- Several molecules of peptidoglycan are cross-linked to form single fibres approximately 50 nm in diameter.
- Each fibre surrounds the bacterial cell.
- Eventually the fibres are also cross-linked to give a cell wall of stacked sheets of peptidoglycans, which form the bulk of the wall.
- Teichoic acids are polymers of glycerol or ribitol linked by phosphodiester bonds.
- Teichoic acids can be covalently bonded to *N*-acetylmuramic acid residues of the wall and to phosphate heads of the surface membrane and so anchor the wall to cell surface membrane.

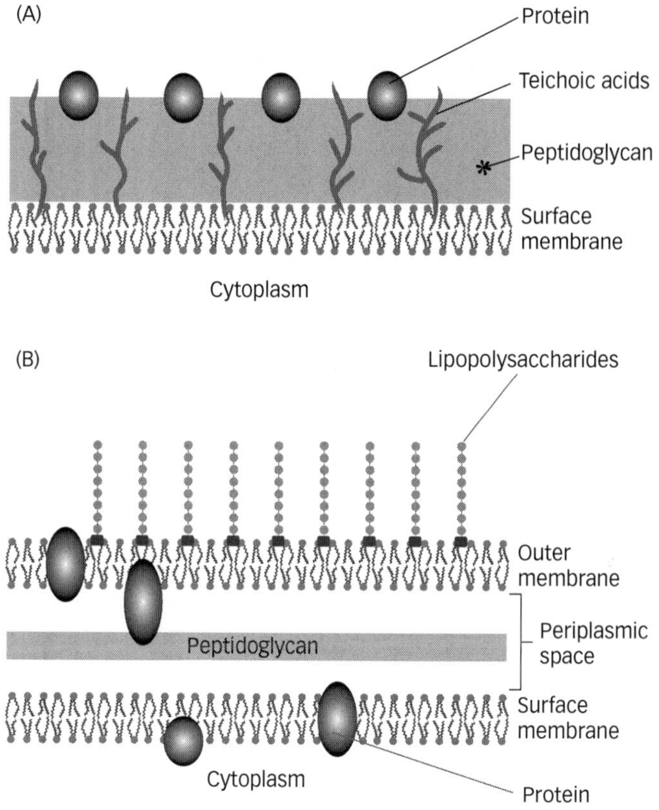

Figure 4.3 Schematic structures of (A) Gram-positive cell wall and (B) Gram-negative cell wall. See text for explanations of structure.

Gram-negative cell walls

- Gram-negative cell walls contain much less peptidoglycan than Gram-positive types but have a more complex structure (Figure 4.3).
- They are multilayered with an outer membrane rich in **lipopolysaccharides** (**LPS**) and contain pore-forming transmembrane proteins called **porins** (Figure 4.4).
- Porins allow water-soluble ions and molecules of small M_r to diffuse into the periplasmic space.
- Some porins are non-specific but others have specific binding sites and allow only one or a restricted range of molecules to cross the membrane.
- The outer membrane of the wall encloses a **periplasmic space** about 15 nm deep that contains the peptidoglycan layer.
- The periplasm of Gram-negative bacteria has a high concentration of proteins that have major roles in hydrolysing nutrient molecules, environmental sensing, and specifically binding ligands for transport into the cell.
- A major role of the outer membrane is to maintain a high concentration of these proteins within the periplasmic space.

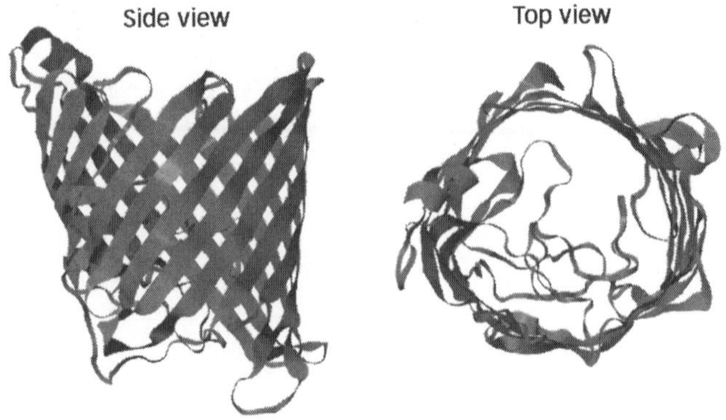

Figure 4.4 Molecular models showing two views of a porin from *Pseudomonas aeruginosa.* Constructed using PDB file 3T24.

Looking for extra marks?

Penicillin, the first antibiotic used clinically, is based on a β lactam ring structure. β lactams are effective agents against bacteria because they interfere with cell wall synthesis. See an animation of their effects at: http://www.youtube.com/watch?v=qBdYnRhdWcQ

Archaeal cell walls

A variety of types of cell walls are found in archaeal cells.

- All lack peptidoglycan and none have an outer wall of the type found in Gram-negative cells.
- The commonest type consists of an extremely strong, almost crystalline, array of interlocking proteins or glycoproteins called an **S-layer**.
- Other types consist of **pseudopeptidoglycan** (**pseudomurein**), which has a similar overall structure to bacterial peptidoglycans but is formed of alternating *N*-acetylglucosamine (β,1→3) *N*-acetyltalosaminuronic acid repeats linked by peptides (Figure 4.5).
- Some methanogens have thick walls of complex polysaccharides containing glucose, galactose, and acid sugars.
- Extreme halophiles have similar but sulphated polysaccharides.

Check your understanding

4.1 Which of the following statements is/are false?

a. Sugar residues found in bacterial cell walls include *N*-acetylglucosamine, *N*-acetylmuramic acid, and *N*-acetyltalosaminuronic acid.

b. Bacteria cell walls can be divided into one of two types on the basis of their response to the Gram stain.
c. Gram-negative cell walls have more peptidoglycan than Gram-positive types and are multilayered with an outer membrane rich in lipopolysaccharides.
d. The periplasm of Gram-negative bacteria is rich in proteins.
e. The commonest type of archaeal cell walls are largely composed of proteins or glycoproteins and are called an S-layer.

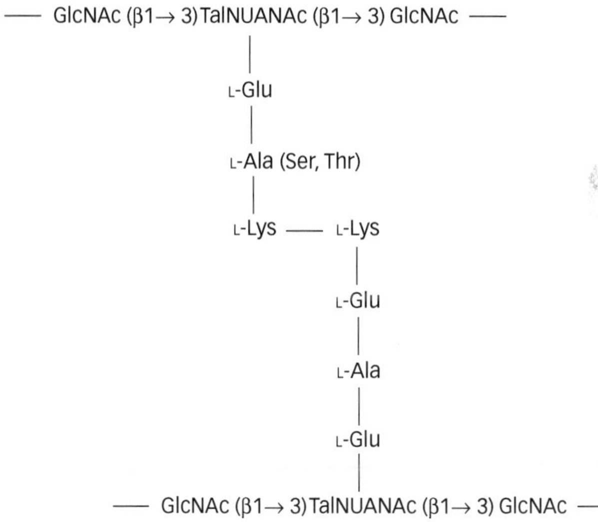

Figure 4.5 Structure of pseudopeptidoglycan (pseudomurein). GlcNAc, *N*-acetylglucosamine; TalNUANAc, *N*-acetyltalosaminuronic acid.

4.2 BACTERIAL CYTOPLASMIC MEMBRANE

The bacterial cytoplasmic membrane is largely a phospholipid bilayer containing proteins (Chapter 3).

- Proteins mediate many of its roles, which includes selective permeability, and the export of proteins such as the subunits of fimbriae and flagellar filaments.
- Materials pass through the membrane by simple or facilitated diffusion in accordance with their osmotic, concentration, or potential gradients, or against their gradient by active transport (Chapter 3).
- However, prokaryotic cytoplasmic membranes are also involved in **energy transduction** reactions, for example, **oxidative phosphorylation** and **photosynthesis.**
- These processes occur in eukaryotic cells in mitochondria (Chapter 11) and chloroplasts (Chapter 10) respectively.

Bacterial cytoplasmic membrane

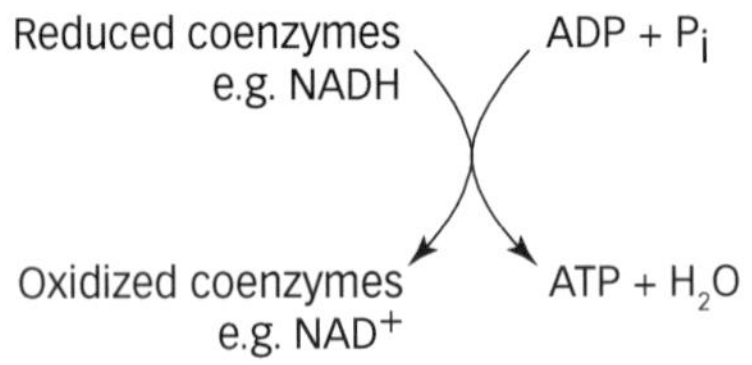

Figure 4.6 Overview of oxidative phosphorylation. The 'kissing' arrows indicate that both processes are tightly coupled.

Oxidative phosphorylation

Oxidative phosphorylation is the coupled oxidation of reduced coenzymes to the phosphorylation of ADP (Figure 4.6).

- Coupling means neither process can occur in isolation; they must happen together.
- Coupling is effected by biological membranes: in bacteria and archaea by the surface membrane; in eukaryotic cells by the inner mitochondrial membrane (Chapter 11).
- Coupling in oxidative phosphorylation occurs by a mechanism called **chemiosmosis.**

Chemiosmosis

- A highly schematic version of chemiosmosis in prokaryotes is shown in Figure 4.7; more detailed explanations are given in Chapters 10 and 11.
- Electrons (e^-) from reduced compounds, in this instance the reduced coenzymes NADH and $FADH_2$, are passed along an electron transport chain (ETC) to a final acceptor.
- In aerobic respiration the final acceptor is dioxygen (O_2).
- The ETC are mainly components of the surface membrane.

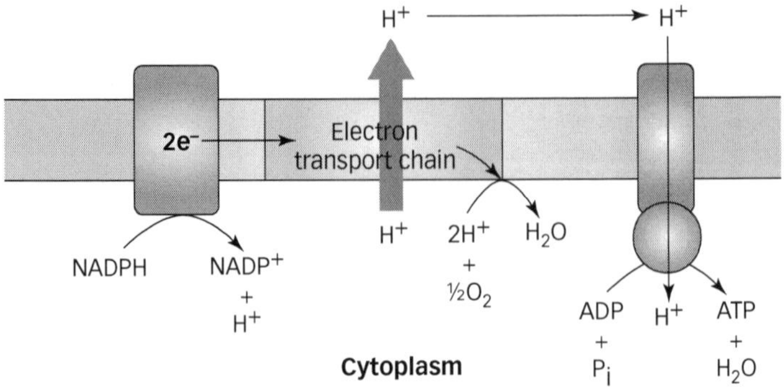

Figure 4.7 Schematic showing the synthesis of ATP by chemiosmosis (oxidative phosphorylation) in prokaryotic cells. See text for explanations of the events.

- The transport of e^- through the ETC drives an efflux of H^+ across the membrane; the concentration of H^+ outside the membrane is higher than in the cytoplasm, so the exterior pH is lower and it is more positively charged.
- The differences in pH (ΔpH) and charge ($\Delta\mu_H{}^+$) across the membrane form an electrochemical gradient called a **proton motive force** (**pmf**), which is a form of free energy.
- Cytoplasmic membranes are impermeable to H^+ but if they are returned to the cytoplasm through an ATP synthase, as shown in Figure 4.7, the energy of the pmf can be used to form ATP.

Revision tip

Look at an animation showing the H^+-driven actions of an ATP synthase at: http://www.sigmaaldrich.com/life-science/metabolomics/learning-center/metabolic-pathways/atp-synthase.html

Anaerobic respiration

- Many prokaryotes are able to perform anaerobic respiration.
- The pathway of anaerobic respiration follows that for aerobic chemiosmosis but any one of a number of inorganic compounds may be used as the final e^- acceptor depending on the conditions and/or species concerned.
- These include the reduction of NO_3^- to NO_2^- or NO or N_2, that of SO_4^{2-} to H_2S, and CO_2 to CH_4 respectively but not dioxygen (Figure 4.8).
- These e^- acceptors have lower redox potential than dioxygen and so less ATP is produced in anaerobic conditions compared to aerobic ones.

➔ *Redox potential is revised in the companion text,* Thrive in Biochemistry and Molecular Biology.

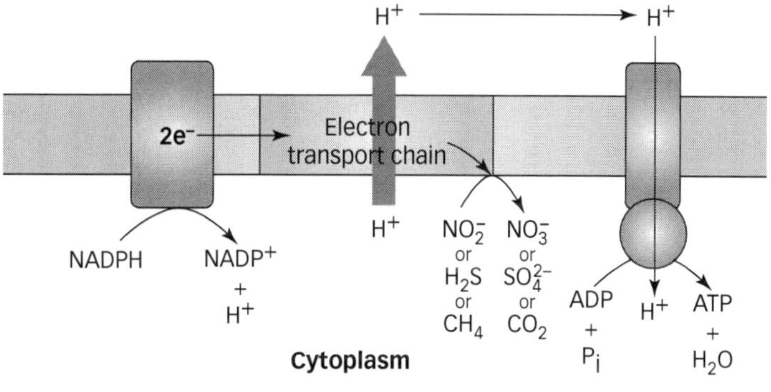

Figure 4.8 Schematic showing the anaerobic synthesis of ATP in prokaryotic cells. See text for explanations of the events.

Looking for extra marks?

Point out the similarities between aerobic and anaerobic respirations in bacteria; only the final electron acceptors differ.

Bacterial photosynthesis

Photosynthesis is the use of light energy to drive the formation of carbohydrate from CO_2.

- The reduction of CO_2 to carbohydrate requires ATP and a source of reducing power such as NADPH.
- Photosynthesis in purple and green sulphur bacteria is called **anoxygenic** photosynthesis because O_2 is not formed.
- In cyanobacteria (and plants and algae (Chapter 10)), O_2 is generated by **oxygenic** photosynthesis.

Anoxygenic photosynthesis

- Anoxygenic photosynthetic bacteria have specialized membranes that are rich in bacteriochlorophyll pigments which capture light energy.
- These membranous regions are called **chromatophores** in purple bacteria and **chlorosomes** in green sulphur and green non-sulphur bacteria.
- Chromatophores are intracytoplasmic invaginations of the bacterial cytoplasmic membrane or of its inner leaflet.
- Chlorosomes are flattened, ellipsoidal-shaped membranous structures attached to the inner face of the bacterial surface membrane. They are the largest light harvesting systems found in photosynthetic bacteria and are associated with organisms inhabiting environments of low light intensity.
- Both chromatophores and chlorosomes greatly increase the surface area available for light absorption.

ATP formation

- Light energy is trapped by numerous pigment molecules and funnelled to protein complexes in the membrane containing special pairs of bacteriochlorophylls called **reaction centres** (**RCs**).
- The energy oxidizes the RC which delivers e^- to an ETC in the membrane.
- The ETC, in turn, transfers e^- to the small water-soluble protein, cytochrome *c2* in the periplasmic space that transfers them back to the oxidized RC and reduces it.
- The movement of e^- through the ETC is used to drive H^+ from the cytoplasm into the periplasmic space, forming a pmf across the membrane (see 'Chemiosmosis' section).

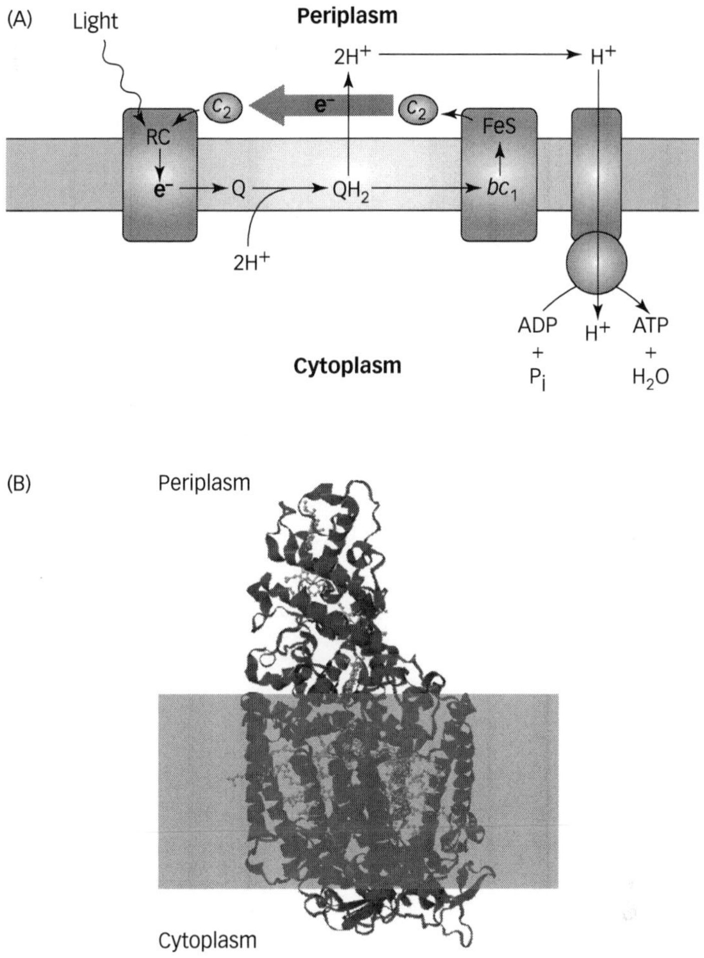

Figure 4.9 (A) Formation of ATP by photophosphorylation in anoxygenic photosynthesis. *bc*$_1$, cytochrome *bc*$_1$ complex; *c*$_2$, cytochrome *c*$_2$; FeS, iron-sulphur protein; RC, reaction centre; Q, quinone. The actions of the ATPase in forming ATP are revised in Chapters 10 and 11. (B) The membrane protein complex that constitutes the photosynthetic reaction centre of *Rhodopseudomonas viridis*. The three membrane proteins are shown in the darker tone. The photosynthetic pigments bound in the reaction centre are denoted in the lighter tone. Model constructed using PDB file1PRC.

- The return of H^+ to the cytoplasm through the ATP synthase generates ATP.
- Hence, ATP is produced by the light-driven cyclic flow of e^- from and back to the RCs. However, NADPH is also required to biosynthesize carbohydrate.
- In all groups, light energy is trapped by photosynthetic pigments and used to form ATP by photophosphorylation, as illustrated in Figure 4.9.

NADPH formation

- In purple sulphur bacteria H_2S is typically used to give reducing power by reducing NAD^+ or $NADP^+$.
- The e^- released in these bacteria are insufficiently electronegative to reduce NAD^+ or $NADP^+$.
- In this case, the free energy of ATP hydrolysis by the ATP synthase is used to generate a pmf across the membrane that drives **reverse electron transport.**
- The e^- are eventually used to reduce $NADP^+$ to NADPH.
- When green sulphur bacteria require reducing power they produce it using a variety of e^- donors including hydrogen, H_2S, sulphur, but never H_2O.
- In this case, e^- from energized RCs enter the ETC and are sufficiently energetic to be used to reduce another small protein called ferredoxin (Fd).

$$Fd_{reduced} + NADP^+ + H^+ \xrightarrow{\text{Ferredoxin-NADP}^+\text{ reductase}} Fd_{oxidized} + NADPH$$

- Reduced ferredoxin can directly reduce CO_2.
- The ATP and NADPH are then used in the reduction of CO_2 to carbohydrate.

Oxygenic photosynthesis by cyanobacteria

- Oxygenic photosynthesis by cyanobacteria occurs essentially as described for green plants in Chapter 10.
- In cyanobacteria, the light reactions occur in flattened layers of membrane that form stacks called **thylakoids** (Figure 4.10).
- Chlorophyll *a* within thylakoids is able to absorb sufficiently energetic light that it is able to split water to release e^-, H^+, and O_2.

$$2\,H_2O \rightarrow 4\,H^+ + 4\,e^- + O_2$$

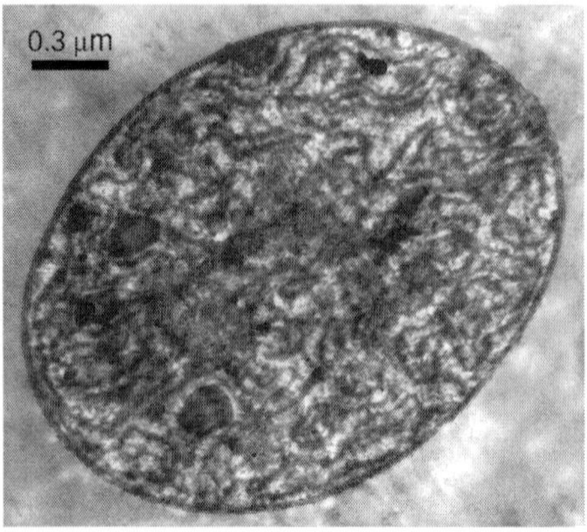

Figure 4.10 Electron micrograph of the cyanobacteria, *Anabaena azollae*.

- The e^- enter an ETC chain to generate a proton motive force and so promote ATP formation, and are also used to reduce $NADP^+$. The O_2 is excreted.
- The evolution of cyanobacteria and the use of chlorophyll *a* was responsible for changing the atmosphere of the earth from a reducing one (rich in H_2 and CH_4) to its present dioxygen-rich composition (Chapter 1).

Looking for extra marks?

Unlike other types of photosynthetic bacteria, cyanobacteria can perform non-cyclic electron transport during photosynthesis, as do chloroplasts which evolved from them.

Carboxysomes

- Carboxysomes are **microcompartments** (see 'Cytoplasmic microcompartments and inclusions' section) found in cyanobacteria and some chemoautotrophs.
- Carboxysomes of cyanobacteria are limited by a proteinaceous shell or envelope that encloses the enzymes ribulose 1,5-bisphosphate carboxylase/oxygenase (rubisco) and carbonic anhydrase.
- The former is the key enzyme for fixing CO_2 (see Chapter 10) in its conversion to carbohydrate in photosynthesis:

$$CO_2 + \text{Ribulose 1,5-bisphosphate} \xrightarrow{\text{Rubisco}} 2 \times \text{glycerate 3-phosphate}$$

- Carbonic anhydrase catalyses the conversion of hydrogen carbonate to CO_2.

$$HCO_3^- + H^+ \xrightarrow{\text{Carbonic anhydrase}} CO_2 + H_2O$$

- Carboxysomes therefore increase the local concentrations of both rubisco and CO_2 and so enhance carbohydrate formation.

Aerobic photosynthesis is revised in more detail in Chapter 10.

Archaeal surface membranes

The general structure of the archaeal surface membrane is outlined in Chapter 3.

- In many species of archaea, the membrane is a single layer of glycerol–hydrocarbon–glycerol molecules, linked together by ether bonds, although parts of the membrane can be bilayers (Figure 3.13).
- Some archaeal membrane lipids contain both five- and six-carbon rings.
- The essential architecture of archaeal membranes is the same as in other organisms: an inner hydrophobic region flanked by outer hydrophilic surfaces.
- Proteins associate with the membrane in a similar way to the proteins of other types of membranes.
- The roles of archaeal cell surface membranes are akin to those of bacterial types.

➔ *The structure and activities of biological membranes are reviewed in Chapter 3.*

4.3 PROKARYOTIC CYTOPLASM

Bacterial and archaeal cytoplasm is protein-rich (approximately 20%) and therefore gel-like. In the electron microscope large areas are granular due to the presence of ribosomes.

- Ribosomes function in protein synthesis as described in Chapters 5 and 7.

➔ *Protein synthesis (translation) is revised in Chapters 5 and 7.*

- The nucleoid or bacterial chromosome is a prominent ribosome-free area (Figure 4.1).
- Many of the proteins present are enzymes that catalyse metabolic pathways or are involved in producing macromolecules.
- Metabolic pathways of bacteria include glycolysis (Chapter 7), the hexose monophosphate shunt, the TCA cycle (Chapter 11), and fermentations.

➔ *Details of metabolic pathways can be reviewed in the companion text,* Thrive in Biochemistry and Molecular Biology.

- Major processes involving macromolecules are the replication and the transcription of DNA and the translation of mRNA to form polypeptides (see Chapters 5–7).
- The cytoplasm also contains the nucleoid or bacterial chromosome and different types of microcompartments and **inclusions** and numerous ribosomes.
- An area free of ribosomes called the nucleoid or **bacterial chromosome** contains the genomic DNA.
- Small circular DNA molecules called **plasmids** (see later in chapter) are also found free in the cytoplasm.

Cytoplasmic microcompartments and inclusions

Cytoplasmic microcompartments are separated from the rest of the cytoplasm by some sort of membrane.

- Membranes include invaginations of the cytoplasmic surface membrane or of its inner leaflet or are formed of a protein shell.
- Bacterial photosynthetic chromatophores and **magnetosomes** are examples of microcompartments enclosed by invaginations of the surface membrane.
- The carboxysomes of cyanobacteria are an example of a protein-enclosed microcompartment, as are the **gas vacuoles** found in some aquatic bacteria.

Magnetosomes

- Magnetosomes are chains of 50-nm diameter beads of magnetite (Fe_3O_4) or greigite (Fe_3S_4) enclosed within an invagination of the bacterial surface membrane (Figure 4.11).

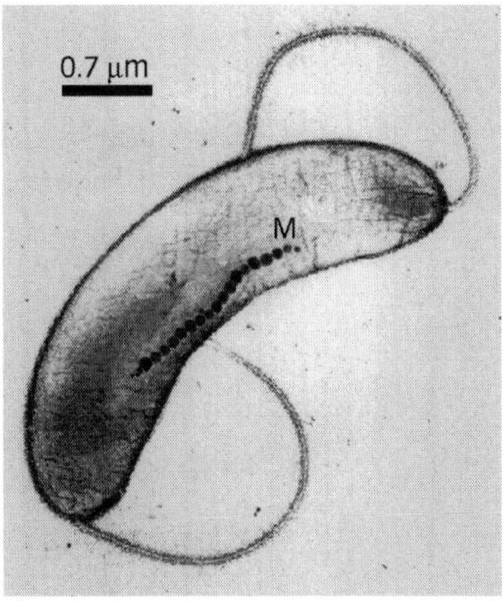

Figure 4.11 Electron micrograph of the bacterium *Aquaspirillum magnetotacticum*. Note the membrane-limited magnetosome (M) in the cytoplasm. Courtesy of Dr R.P. Blakemore, Department of Microbiology, University of New Hampshire, USA.

- Iron oxides/sulphides are magnetic and allow the bacteria to orientate themselves relative to the earth's magnetic field so that they can swim downwards to favourable anaerobic sediments in bog and marsh environments.

Gas vacuoles

- **Gas vacuoles** are found in some prokaryotes, both bacteria and archaea; a single cell can contain several hundred.
- Gas vacuoles are enclosures in the cytoplasm filled with gas of the same composition as found in the environment surrounding the cell, and are characteristic of planktonic prokaryotes, particularly cyanobacteria.
- Gas vacuoles provide buoyancy and counteract the natural tendency of the cells to sink.
- The envelope or membrane of gas vacuoles is made of rigid hydrophobic structural proteins.
- Vacuoles vary in size, with lengths ranging from 300 nm to over 1000 nm and diameters of about 45 nm to over 100 nm.

Looking for extra marks?

Some authorities call bacterial microcompartments 'bacterial organelles'. If you use this term emphasize how they differ from the eukaryotic organelles, for example, in size, differences in membranes.

Prokaryotic cytoplasm

Cytoplasmic inclusions

- Cytoplasmic inclusions are often energy stores.
- Their nature depends upon the species and includes:
  - particles of β hydroxybutyrate polyphosphates enclosed within a protein shell
  - lipid droplets
  - sulphur grains
  - various carbohydrate stores including starch or glycogen.

Bacterial cytoskeleton

Cytoskeletons are complex organizations of insoluble proteins fibres ramifying throughout the cytoplasm.

- The best described are those of eukaryotic cells, which are described in Chapter 7, but prokaryotic cells also have cytoskeletons that help maintain cell shape, position bacterial subcellular structures, and assist in chromosome segregation and cell division (Chapter 5).
- The insoluble fibres are formed by polymerizing soluble subunits present in the cytoplasm.
- Bacterial cytoskeletons are a complex organization of three types of proteins:
  - bacterial proteins resembling **actins**
  - bacterial proteins resembling **tubulins**
  - a family of **ATPases.**
- Bacterial actin-like proteins can form fibres that help maintain cell shape, catalyse the segregation of DNA, and participate in cell division (Chapter 5).
- Bacterial homologues of eukaryotic tubulins, such as the **FtsZ protein** discussed in Chapter 5, are essential for cell division and are known to participate in plasmid replication.
- Two subfamilies of a large group of ATPases called ParA and MinD are integral features of bacterial cytoskeletons.
- ParA ATPases are essential if chromosomes and plasmids are to be appropriately segregated during binary division.
- The MinD subfamily of enzymes specify the position at which FtsZ proteins assemble, which, in turn, determines the position of cell separation in binary division.

➔ *Binary division and the eukaryotic cytoskeleton are reviewed in Chapters 5 and 7 respectively.*

4.4 NUCLEOID AND PROKARYOTIC DNA PACKAGING

Prokaryotes generally have their genomic DNA packaged into a single chromosome or nucleoid.

- Bacterial chromosomes contain a single circular molecule of DNA but in some it is a linear molecule. Prokaryotic cells are therefore always **haploid**.
- The nucleoid lies free in the cytoplasm and is not membrane bound.
- The position of the DNA inside the nucleoid is maintained by a series of fibres associated with the bacterial chromosome.
- Since the DNA molecule is considerably longer than the bacterial cell, it must be condensed in size by careful packaging called **supercoiling**.

Supercoiling

An uncoiled or **relaxed** DNA double helix is capable of being negatively or positively supercoiled as shown in Figure 4.12.

- Bacterial chromosomes are made up of negatively supercoiled loops; each loop may be 50,000–100,000 base pairs (bp) in length.
- Negative supercoiling promotes unwinding of the DNA double helix, allowing proteins and enzymes easier access to each strand; positive supercoiling induces a tightening of the double helix and reduces access.
- Supercoiling is catalysed by two major types of **topoisomerases**.
- *Type I* topoisomerases relax DNA by hydrolysing one strand and allowing the double helix to rotate. The uncut strand is then passed through the break which is then resealed.
- *Type II* topoisomerases differ in using the free energy of ATP hydrolysis to cut *both* strands of the DNA. A section of DNA is then passed through the cut which is then resealed.

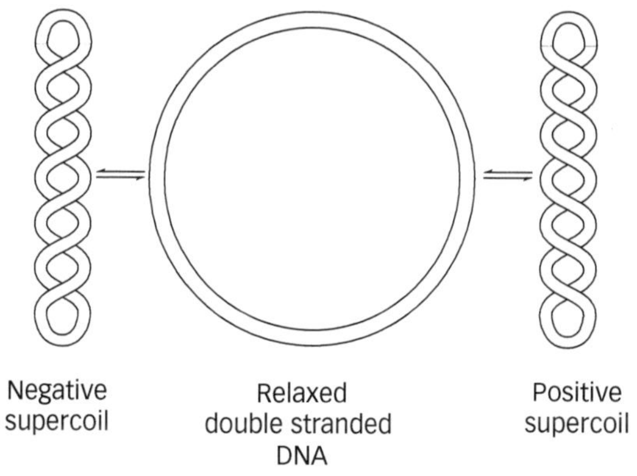

Figure 4.12 Schematic illustration of DNA supercoiling.

Prokaryotic genomes

- Prokaryotes have a type II topoisomerase called **DNA gyrase** that is able to induce negative supercoiling to assist the access of enzymes and proteins in DNA replication (Chapter 5).

Histones

- In eukaryotes, supercoiling is achieved by winding the linear DNA molecules around clusters of basic proteins called **histones** (see Chapter 6 and Figure 6.4).
- Archaea have both gyrase activity and express histones and so must condense their DNA by gyrase activity (as in bacteria) or by winding it around histones (as in eukaryotes) or by a combination of the two.
- The supercoiled loops are organized into bead-like structures that contain basic proteins, and may be anchored to the surface membrane by a protein core.

4.5 PROKARYOTIC GENOMES

In general, the genomes of prokaryotes contain 1000–6000 genes.

- The genome of *Escherichia coli* is typical of prokaryotic types in its size and number of genes (Figure 4.13), being 4,639,221 bp in circumference.
- Approximately 99.5% of the genome encodes genes or regulatory sequences.

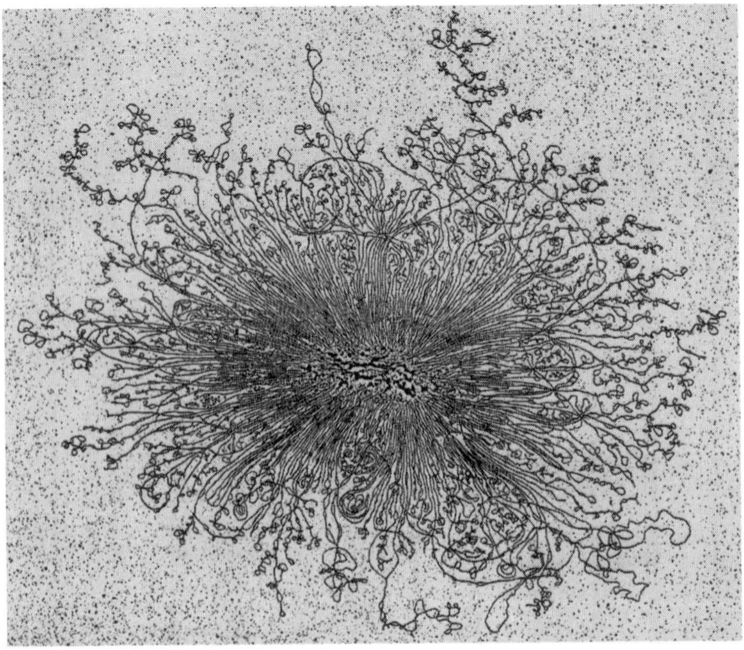

Figure 4.13 Electron micrograph of the chromosome of *Escherichia coli*. Courtesy of Bluegenes ©1983 Designer Genes Posters Ltd, USA.

  - 4288 genes code for proteins.
  - Promoters, operators, and the origin and terminus of replication (Chapter 5) form approximately 10% of the sequence.
  - Genes of tRNA and rRNA molecules (Chapter 5) comprise about 1% of the genome.
- This is a much smaller genome than those found in eukaryotic cells: the yeast genome is packaged into 16 chromosomes and its total nuclear DNA is approximately 13.4×10^6 bp long.

Check your understanding

4.2 What is the approximate mean number of amino acid residues in the proteins of *Escherichia coli*, if three bases in its genomic DNA is equivalent to one amino acid residue in a polypeptide chain?

Plasmids

The cytoplasm of prokaryotes and some yeasts, which are of course eukaryotes, contain extrachromosomal DNA structures called **plasmids**.

- Plasmid DNA resembles bacterial chromosomes in being circular and double stranded (Figure 4.14), but is generally much smaller with M_r of 2×10^6 to 20×10^6, which corresponds to 3000–30,000 bp.
- Bacterial plasmids normally contain genetic information for proteins that confer specialized and often protective roles on the organism; they often, for example, impart antibiotic resistance.

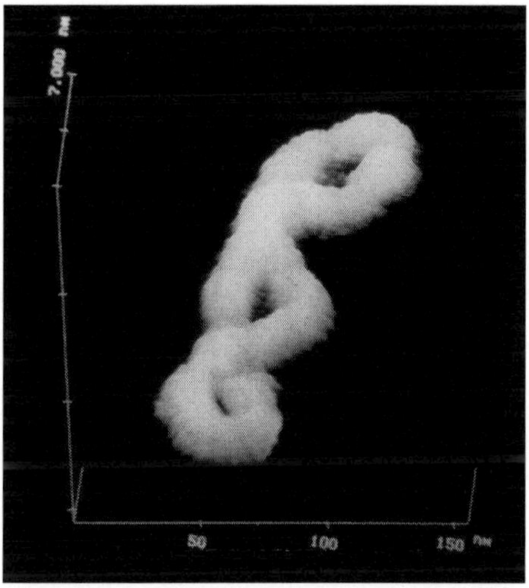

Figure 4.14 Bacterial plasmid visualized by atomic force microscopy. Courtesy of Digital Instruments, California, USA.

- Plasmids are self-replicating and their DNA is duplicated prior to cell division (Chapter 5) with copies of the plasmid being segregated to both daughter cells.

4.6 PROKARYOTIC SURFACE STRUCTURES

Prokaryotic surface structures include **capsules** or **slime layers, fimbriae** and **pili**, and **flagella.**

Capsules and slime layers

All prokaryotic cells secrete a capsule or slime layer.

- These vary in thickness and stiffness; traditionally, a thinner, stiffer layer is called a capsule; thicker, less organized material is a slime layer.
- Both consist of only 1–2% carbohydrate but this binds considerable quantities of water forming a gel coating on the cell surface.
- Capsules and slime layers protect the cell from host immune defences and dehydration, and aid the formation of biofilms, for example, plaque on teeth.

Fimbriae and pili

Fimbriae and pili are protein fibres that extend from bacterial cells.

- Fimbriae are approximately 2 μm long with diameters of 2–10 nm, and are composed of about 1000 subunits of the protein, **pilin** (M_r 15,000–22,000) arranged in a helical extended conformation.
- Fimbriae also contain glycoproteins called **lectins** that can bind to sugars and allow bacteria to adhere to surfaces.
- Fimbriae are more numerous than pili; up to 300 can be present on an *Escherichia coli* cell (Figure 4.15).
- Generally only one or two pili, which are longer than fimbriae, may be present on a bacterial cell.
- Pili have at least two essential roles:
  - They allow pathogenic bacterial cells to bind to specific sites on the plasma membranes of eukaryotic cells as a first step to invading the host.
  - Conjugation or F-pili allow tubes to form between different bacteria through which plasmids carrying genetic information (DNA) can unidirectionally pass from donor (F^+) bacterial cells to recipient ones (F^-).

Check your understanding

4.3 Which of the following statements is/are true?

a. Fimbriae have diameters of about 2–10 nm.

b. Fimbriae contain glycoproteins called lectins.

c. Pili are more numerous appendages than fimbriae on the bacterial cell surface.
d. Conjugation of F-pili is essential for the transfer of plasmids from donor (F^-) bacterial cell.
e. Pilin molecules can associate to form filaments with extended conformations.

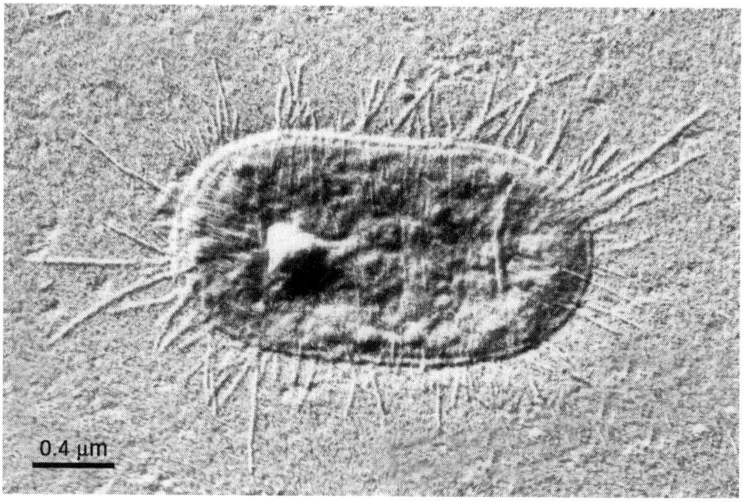

Figure 4.15 Electron micrograph of an *Escherichia coli* cell with numerous fimbriae.

4.7 BACTERIAL MOTILITY

Flagella are the commonest means of locomotion in prokaryotes, but not the only one. Some species also use **spirochete** and **gliding mechanisms**.

Flagella

Bacterial flagella are extracellular appendages about 2.5 µm long and 20 nm in diameter (Figure 4.16). They are very different in structure and action to the eukaryotic flagella reviewed in Chapter 7.

- The positioning on the cell and number of flagella varies between different species of prokaryotes.
  - Polar flagella are situated at the end(s) of a cell; if only one is present, it is described as **monotrichous**, if two then **amphitrichous**.
  - A group of flagella arising from a localized spot on the cell surface is called **lophotrichous**; if they are dispersed over the surface then **peritrichous**.

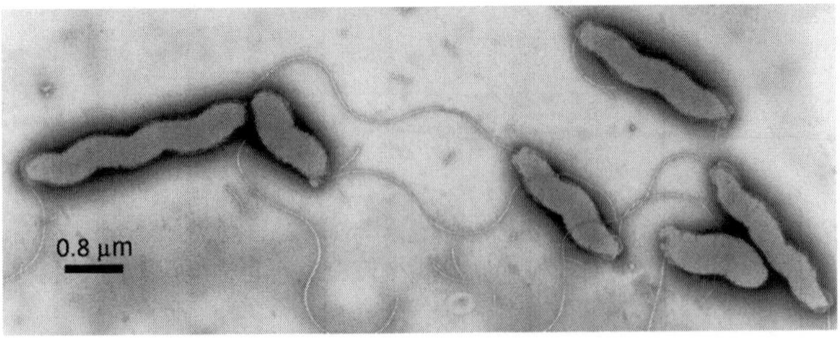

Figure 4.16 Electron micrograph of the flagellated bacterium *Campylobacter jejuni*. Courtesy of Dr A. Curry, Manchester Royal Infirmary, UK.

Flagellar structure

The structure of a bacterial flagellum is illustrated in Figure 4.17.

- Flagella are composed of three main parts:
  1. motor
  2. hook
  3. filament.
- The motor is situated in the surface membrane; it attaches the flagellum to the cell and drives the *rotation* of flagellum.
- The hook is a universal joint that allows the filament to rotate through 360°.
- The filament is composed of identical protein subunits called **flagellin** (M_r 40,000–60,000).
- These are aligned along the long axis of the filament to form a hollow helical tube.

Flagellar locomotion

- The motor is essentially an evolved H^+ uniport (Chapter 3) in the surface membrane.
- Electron transport chains in the membrane pump H^+ to the outside of the cell (see 'Chemiosmosis' section).
- Their return through the motor causes it to rotate: approximately 1000 H^+ are required to power one revolution.
- The hook imparts rotation to the filament, which drives the cell through the liquid medium.
- Flagella can rotate at approximately 300 revolutions per second, giving swimming speeds of 25 $\mu m\ s^{-1}$.
- **Swimming** or **run** movements of the cell occur when the flagella rotate *anticlockwise*. If several flagella are present, they come together to form a stable rotating bundle, which allows the cell to move in a straight line.

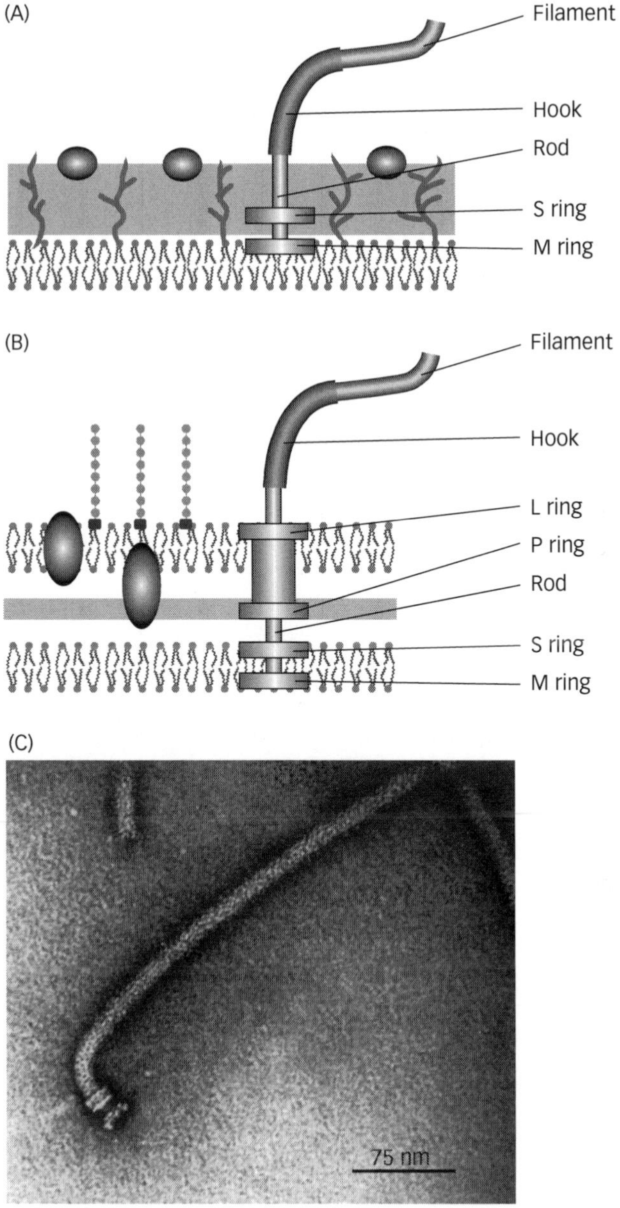

Figure 4.17 Simplified illustrations of (A) Gram-positive and (B) Gram-negative bacterial flagella. (C) Electron micrograph of a flagellum isolated from the Gram-negative bacterium, *Salmonella typhimurium*. Courtesy of Professor R. Macnab and Dr S.I. Aizawa, Department of Biophysics and Biochemistry, Yale University, USA.

- The direction of swimming can only be changed by reversing the rotation of the flagella to a *clockwise* direction, which causes the cell to **tumble** in a random manner.
- Simultaneously switching the flagella back to anticlockwise rotations allows swimming to start in a new direction.
- Alternating periods of swimming and tumbling allow for **chemotaxis**, the movement towards an attractant or away from a repellent chemical.

Check your understanding

4.4 Which of the following statements is/are false?

a. The arrangement of groups of bacterial flagella arising from a localized spot on the cell surface is called peritrichous.
b. The hook is a universal joint that imparts rotation to the filament driving the cell across solid media.
c. Filaments of bacterial flagella are composed of identical protein subunits arranged in an overall helical conformation.
d. The rotation of all prokaryotic flagella is thought to be driven by the hydrolysis of ATP.
e. Swimming occurs when prokaryotic flagella rotate clockwise.

Archaeal flagella

A number of archaea swim using flagella.

- Like bacterial flagella, archaeal types impart motility by rotating.
- However, they differ from bacterial types in a number of respects.
  - Their diameters are only 10–13 nm.
  - Several different types of filament proteins occur in the Archaea.
  - They are thought to be powered by the free energy of ATP hydrolysis.
- Archaeal cells are capable of swimming at speeds of only approximately 0.25 $\mu m\ s^{-1}$.

Spirochete movement

Spirochetes are bacteria with helical-shaped cells (Figure 4.18).

- They have a flagellum-like helical filament encased within a sheath running the length of the cell.
- Rotation of the filament between the cell and sheath causes the propagation of a helical wave through the spirochete which causes it to rotate and swim.

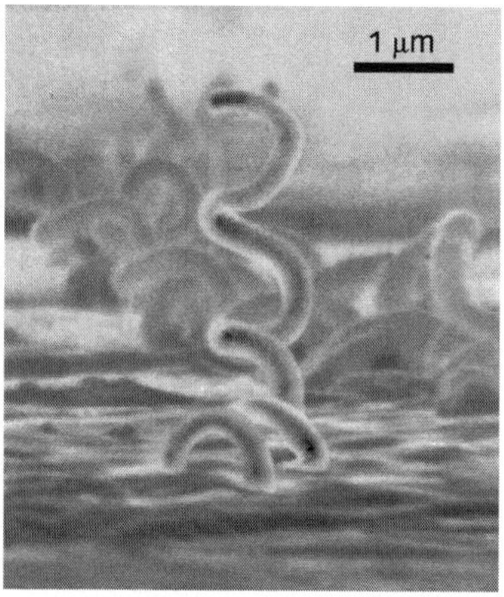

Figure 4.18 Electron micrograph of the spirochete, *Treponema pallidum*. Courtesy of Dr D. Cox, Public Health Image Library, Centers for Disease Control and Prevention, USA.

Gliding movements

Gliding movement is shown by some myxobacteria and cyanobacteria.

- Cyanobacteria secrete a glycoprotein-based slime that allows movement in a poorly understood manner.
- However, movement in some bacteria involves contact between the cell wall and substratum.
- Proteins of the cell wall temporarily attach to the surface and pull the bacterium forward a few nm at a time.
- Twitching motility is a peculiar type of gliding seen in some species of *Moraxella* and *Pseudomonas*, which possess special types of pili that cluster only at the poles of the cells.
- The pili are extended and adhere to the surface. Contraction of the pili then pulls the cell forward.

Revision tip

Do browse through the further reading references given for this chapter online. Go to http://www.oxfordtextbooks.co.uk/orc/thrive/

Check your understanding

4.5 Which of the following statements is/are true or is/are false?

a. All prokaryotic cells possess cell walls.

b. Cell walls directly influence the shape of prokaryotic cells.

c. All prokaryotic cells possess cell walls containing peptidoglycans which is a polymer of *N*-acetylglucosamine (β,1→4) *N*-acetyltalosaminuronic acid residues.

d. Only the cell walls of Gram-negative bacteria have a periplasm.

e. Chemiosmosis is the major mechanism in cells for forming ATP.

f. Purple sulphur bacteria use a proton motive force across the membrane to form ATP whose hydrolysis drives reverse electron transport to reduce $NADP^+$.

g. Thylakoids of cyanobacteria contain chlorophyll *a*.

h. Archaeal membranes can contain single and bilayers of glycerol–hydrocarbon–glycerol molecules.

i. Plasmids occur exclusively in the cytoplasm of prokaryotic cells.

j. Capsules are thinner than slime layers because they contain approximately 20% protein forming a relatively stiff gel.

4.6 Approximately how many H^+ will pass through a bacterial flagellum motor rotating at 240 rev s^{-1} in one minute?

5 Molecular biology of the prokaryotes

Key features of prokaryotic molecular biology

- The genomes of most prokaryotic cells are contained in only one chromosome that lies free in the cytoplasm and contains a single circular molecule of DNA.
- Prior to cell division, prokaryotic genomes must be replicated and each chromosome segregated to the daughter cells.
- Prokaryotic cells grow and divide by binary division.
- The genetic information of DNA is expressed by being transcribed to form messenger RNA (mRNA), transfer RNA (tRNA), and ribosomal RNA (rRNA) molecules, which are essential for the synthesis of proteins.

5.1 REPRODUCTION OF PROKARYOTIC CELLS

During binary division the growth of the cell proceeds to a point that stimulates division to give two daughter cells. This contrasts with eukaryotic cells that reproduce by mitosis or meiosis (Chapter 16).

- The time required for a prokaryotic cell to divide and form two cells is called the **generation time**.

- Generation times vary according to temperature and other environmental factors but growth can be rapid.
- In favourable conditions, bacteria like *Escherichia coli* can replicate their chromosomal DNA every 40 minutes and the following cell division takes approximately 20 more minutes.
- *E. coli* can have generation times as short as 20 minutes in laboratory conditions; in these circumstances, a new cycle of DNA replication must be initiated *before* the previous round is complete and cells can appear to have more than one chromosome.

5.2 BINARY DIVISION IN BACTERIA

Binary division in prokaryotes is best understood for bacterial cells (Figure 5.1).

- Binary division can be divided into four stages:
  1. DNA replication
  2. FtsZ ring formation
  3. FtsZ ring constriction and septum formation
  4. Cell separation.

DNA replication

DNA replication is best understood in *E. coli* on which this section will concentrate.

- Like most bacteria (and archaea) the genome of *E. coli* is a supercoiled circular DNA molecule made up of two complementary strands (Chapter 4).
- During replication, the DNA double helix must be unwound and separated.
- Supercoiling makes unwinding difficult. Therefore, topoisomerase (Chapter 4) activities are necessary to relax the supercoiling of the DNA prior to replication.

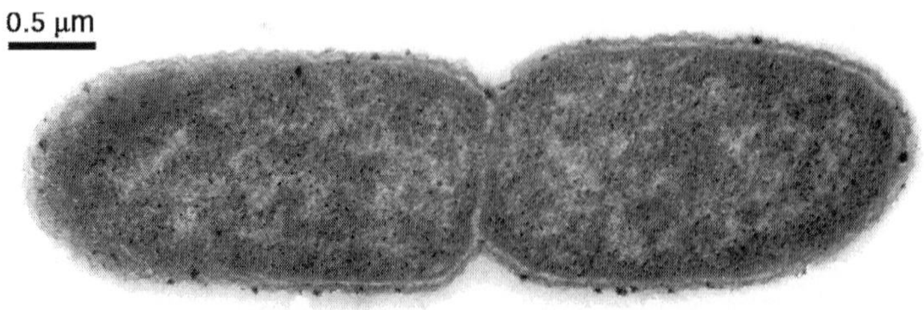

Figure 5.1 Electron micrograph of an *Escherichia coli* reproducing by binary division.

Key points of DNA replication in bacteria

DNA replication is **semiconservative** because each of the new molecules consists of one parental (original) strand and one daughter (new) strand of DNA.

- Replication of the bacterial chromosome begins and ends at unique sites on the chromosome.
- During replication, the parental DNA strands separate from one another to give two templates, each of which is copied.
- Copying a template gives a daughter strand that is complementary to the parental template.
- Enzymes capable of forming new DNA strands are called **DNA dependent DNA polymerases** (**DNA polymerases, DNA pols**).
- *Escherichia coli* has five DNA polymerases, designated I to V.
- All five polymerases form the new strands in the 5′→3′ direction.
- DNA replication is catalysed by DNA polymerase III.
- The site where parental DNA strands separate and replication occurs is called a **replication fork**.
- In addition to polymerase activity, many other proteins are essential in replicating DNA.
- All these enzymes and proteins form a large particle at the replication fork that is sometimes called a **replisome**.
- The length of DNA replicated in a single stretch by the polymerase is called a **replicon**.
- The 3′→5′ strand of parental DNA can be copied *continuously*, forming a new deoxyribonucleotide strand called the **leading strand**.
- The 5′→3′ parental strand can only be copied in relatively short discontinuous lengths as the single stranded template becomes exposed at a replication fork.
- These short lengths are called **Okazaki fragments** (after their discoverer) and are formed on the **lagging strand** (Figure 5.2). They are about 1000–2000 nucleotides long in *E. coli* but only about a tenth as long as this in eukaryotes.
- DNA replication can be divided into three phases:
  1. **initiation**
  2. **elongation**
  3. **termination**.

Initiation

- DNA replication begins at a single **point of origin** on the chromosome called ***oriC*** (Figure 5.3).
- From this point, two replication forks are formed and the DNA is replicated in both directions: that is, it is bidirectional.
- Replication is complete when both replication forks reach the single termination site (see 'Termination' section).

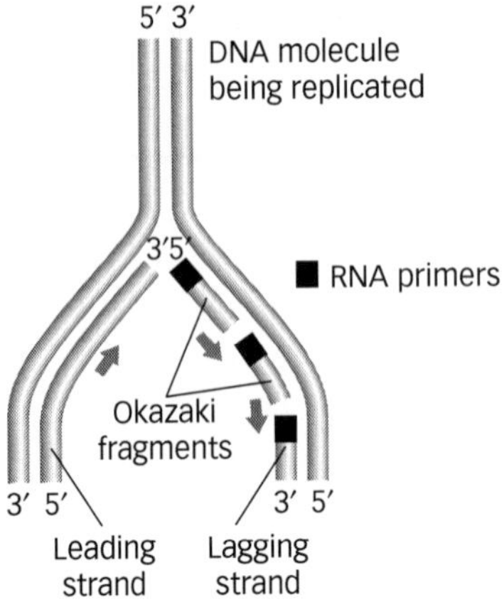

Figure 5.2 Leading and lagging strands and Okazaki fragments in the replication of DNA.

- The replication of circular DNA genomes is sometimes referred to as θ replication because the intermediate stages of replication produce structures resembling the Greek letter *theta*.
- The *oriC* site is a 245-bp sequence that is densely populated by A–T base pairs. These are stabilized by two hydrogen bonds, as opposed to G–C pairs with three.
- The weaker bonding of the A–T pair facilitates separation of the strands when replication begins.
- *OriC* contains three AT-rich, almost identical repeats and five AT-rich binding sites for DnaA proteins (M_r 52,000) known as the DnaA box (Figure 5.4).

Initiation from *oriC*

- Initiation of replication begins with the binding of multiple copies of **DnaA proteins** to the *oriC* site leading to the separation of an approximately 45-bp section of DNA.
- Separation requires ATPase activity and is also promoted by the profusion of A–T bp and the induction of negative supercoiling by DNA gyrase (Chapter 4).

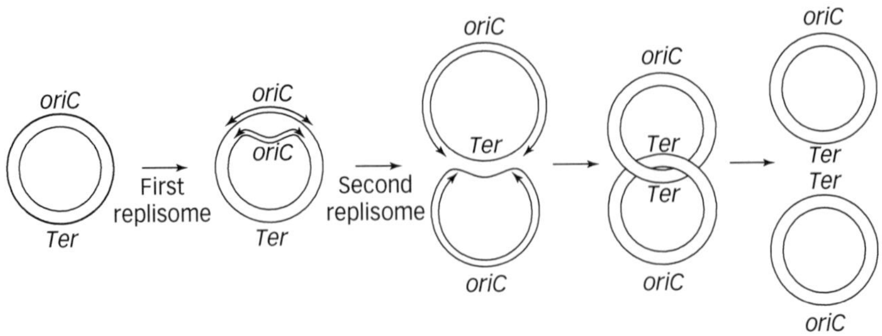

Figure 5.3 Bidirectional replication of a circular bacterial DNA molecule starting at the origin (*oriC*) and ending at the termination (*Ter*) site.

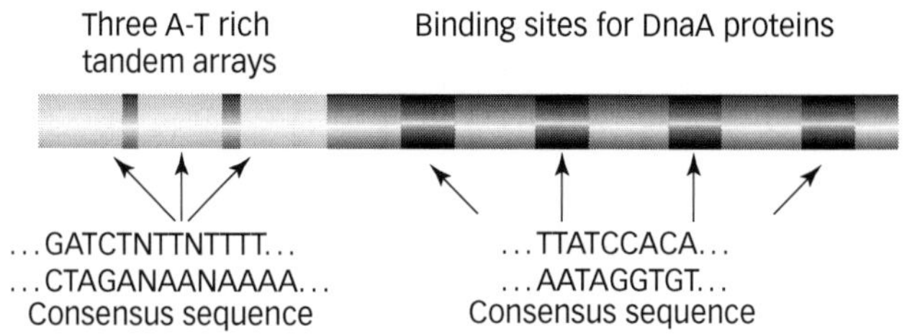

Figure 5.4 Schematic structure of the origin of replication (*oriC*) site of *Escherichia coli*.

- **Helicases**, also known as **DnaB proteins**, then bind to the DNA and unwind the double helix.
- DnaB is a ring-shaped hexamer that moves along the lagging strand (Figures 5.2 and 5.5) using free energy from ATP hydrolysis to break the hydrogen bonds between the strands.
- Spontaneous reannealing of the strands is prevented by the binding of **single-strand binding proteins** (Figure 5.5).
- Elongation of the DNA can begin in both directions at the 'Y'-shaped replication forks.

Elongation

- Figure 5.5 illustrates the elongation/replication of DNA in *E. coli*.
- DNA polymerase III cannot begin the synthesis of the daughter strand because it can only extend a pre-existing polynucleotide with a free 3′-OH.
- **Primase**, which is also called **DnaG**, and RNA polymerase are able to move along the opening replication fork and begin the *de novo* synthesis of short sequences of RNA called **primers**.
- The RNA formed is complementary to the parental strands.
- Only a single RNA primer is required for the leading strand but an RNA primer must be sequentially synthesized for each Okazaki fragment.
- Once an RNA primer has been formed for each parental DNA strand, **DNA polymerase III holoenzyme** can bind to the DNA and begin extending the RNA primers with DNA.

Activity of DNA polymerase III

- The holoenzyme of DNA polymerase III is made up of two DNA polymerase III molecules complexed with other proteins to form a replisome.
- Each DNA polymerase III binds to each side of the replication fork so that both parent DNA strands are complexed with the enzyme.
- The separated parental strands of DNA are read by DNA polymerase III in the 3′→5′ direction only, and so the newly synthesized daughter strands are produced in the 5′→3′ direction.

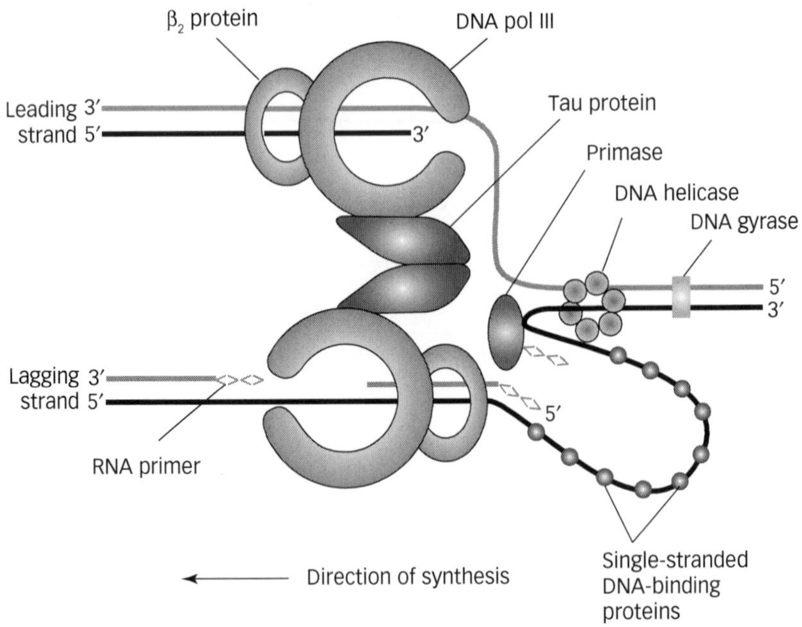

Figure 5.5 Schematic outline of the replication of the leading and lagging strands of DNA in *Escherichia coli* by DNA polymerase III holoenzyme. See text for details.

- DNA polymerase III has a high **processivity**—it is able to catalyse many consecutive reactions without releasing its substrate.
- This property depends on its β_2 subunit, which acts as a sliding clamp.
- This subunit has a ring-shaped conformation and is able to wrap around the double helix and move along it (Figure 5.5).
- This prevents the DNA polymerase III molecule diffusing away and keeps the enzyme firmly attached to its DNA substrate, greatly increasing its rate of activity.

Synthesis of the lagging strand

- Looping of the lagging strand allows the holoenzyme to synthesize the new strand in the 5′→3′ direction (Figure 5.5).
- When DNA polymerase III reaches the RNA primer of the previously synthesized Okazaki fragment, it detaches from the parent DNA strand, leaving a nick in the daughter strand followed by a short RNA primer.
- To produce a continuous daughter strand on the lagging parent template requires:
  - the RNA primers to be removed and replaced with DNA
  - the formation of phosphodiester bonds to produce a continuous strand.
- These steps are catalysed by DNA polymerase I and DNA ligase respectively.
- Synthesis of the leading strand is momentarily halted while these activities are performed.

Activities of DNA polymerase I and DNA ligase

- DNA polymerase I replaces III when it reaches an RNA primer.
- Unlike polymerase III, it possesses 5′→3′ exonuclease activity and is able to remove the RNA primer.
- The DNA polymerase I then fills in the gaps left by removing the RNA by synthesizing DNA in the 5′→3′ direction.
- Once DNA polymerase I has replaced the RNA primer, it detaches and DNA ligase seals the nick in the daughter strand by forming a phosphodiester bond between the 3′-OH of one deoxynucleotide and the 5′-phosphate of the next deoxynucleotide.

Revision tip

A simple animation showing the elongation stage of replication can be seen at: http://www.youtube.com/watch?v=-mtLXpgjHL0&feature=related

Termination

- Termination of replication begins when both replication forks meet at the *terminus of replication* or *Ter* site (Figure 5.6).
- The terminus site is opposite *oriC*, half way round the *E. coli* chromosome.
- The *E. coli* terminus is a 350-kbp region surrounded by seven termination sites; *TerE*, *TerD*, and *TerA* to one side of it and *TerG*, *TerF*, *TerB*, and *TerC* on the other.

Chromosome separation during termination

- Tus (terminator utilization substance) proteins bind to Ter sequences where they prevent helicase action and so arrest replication.

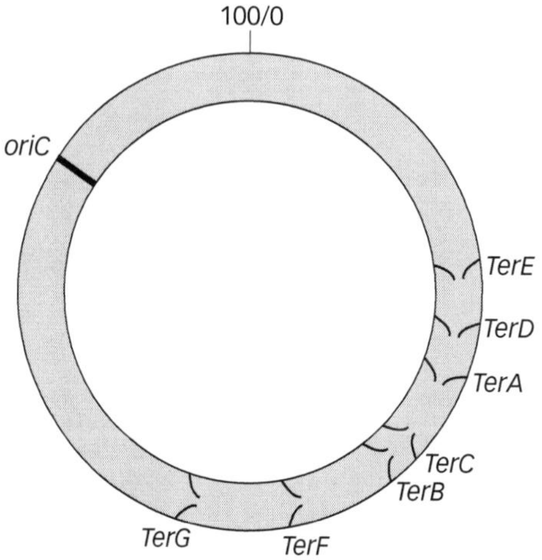

Figure 5.6 Relative positions of the origin of initiation (*oriC*) and terminus of replication (*Ter*) in the *Escherichia coli* chromosome. See text for details.

- Thus TerX–Tus complexes form one-way gates to a replication fork.
- A replication fork travelling clockwise can pass through the TerE, TerD, and TerA sites (Figure 5.6) but stops on encountering TerC–Tus (or TerB–Tus, TerF–Tus or TerG–Tus; TerB, TerF, and TerG act as back-up sites to TerC.)
- In a similar manner, an anticlockwise travelling replication fork will pass through the TerG, TerF, TerB, and TerC sites but stop on encountering TerA–Tus (or TerD or TerE, which are its back-up sites).
- The multiple Tus protein binding Ter sites within the bacterial chromosome ensure that the DNA is duplicated only once per replication cycle.
- The two copies of the bacterial chromosome are linked together in a concatenated structure, rather like two links in a chain.
- The two chromosomes are separated by the catalytic action of topoisomerase IV.
- When chromosomal replication is complete, it is essential that the cell divides and each daughter cell receives a copy of the chromosome.
- The process of cell division is assisted by the critical protein FtsZ, which orchestrates a number of the key events.

Check your understanding

5.1 Which of the following proteins is a helicase?

a. *OriC*

b. Gyrase

c. DnaA

d. DnaB

e. Topoisomerase IV.

Rate of replication, proofreading, and mismatch repair

The main polymerase in *E. coli* replicates DNA at a rate of about 1000 bp per second.

- The number of errors is fewer than the incorporation of one incorrect deoxynucleotide in a billion.
- Errors in DNA replication are limited by the:
  - specificity of the Watson and Crick base pairing rules
  - proofreading ability of DNA polymerase III.
- Proofreading is the removal of a deoxyribonucleotide that is not complementary to the parental strand, and its replacement with one that is.
- Many DNA polymerases have proofreading abilities that operate in the 3′→5′ direction, which is the opposite direction to that of replication.
- If the last incorporated deoxynucleotide is incorrect, the 3′→5′ exonuclease activity of DNA polymerase III is stimulated and the nucleotide is hydrolytically removed.

- The enzyme then replaces it with the correct deoxynucleotide in the 5′→3′ direction.

Check your understanding

5.2 How long would it take an *Escherichia coli* cell to replicate its genome?

Mismatch repair

- Parental strands are methylated (a $-CH_3$ group is added) at adenine bases found in ... GATC ... sequences shortly after replication ceases, allowing enzyme systems in *E. coli* to distinguish between the parental and daughter strands for short periods after replication.
- Any remaining incorrectly inserted bases can then be recognized (it is on the non-methylated strand) and repaired. This type of repair is called **methyl-directed mismatch repair**.
- The enzyme system involved in methyl-directed mismatch repair consists of:
  - Mut proteins (Mut S recognizes mutations in DNA; Mut H is an endonuclease linked to Mut S by Mut L)
  - the helicase, SSB
  - an exonuclease
  - DNA polymerase III
  - DNA ligase.
- Mut S recognizes mutations in the new DNA double helix, which allows Mut H to identify an unmethylated adenine in a GATC site on the new strand and hydrolyse the phosphodiester bond.
- This site may be 1000s of bp distant from the mutation.
- Cooperative action of SSB and the exonuclease then remove the entire length of DNA from the cleaved bond to beyond the mutation.
- This section is then replaced by DNA polymerase III, and integrated into the daughter strand by DNA ligase.
- Strict base pairing rules, proofreading, and mismatch repair reduce the overall error rate in DNA replication to less than one base in 10^{10}.

FtsZ ring formation

The formation of the FtsZ ring and septum, and division of the cell, is illustrated in Figure 5.7.

- FtsZ (*f*ilamentous-*t*emperature *s*ensitive) is one of a number of Fts proteins but has an essential role in binary division.
- FtsZ is an analogue of tubulin proteins that form microtubules of the cytoskeleton of eukaryotic cells (Chapters 4 and 7).

Binary division in bacteria

- It is found in all types of prokaryotic cells but other Fts proteins do not appear to occur in archaea.
- FtsZ and other Fts proteins form a cell division complex or **divisome** at approximately the middle of a cell (Figure 5.7).
- The site is determined by it having the minimum concentration of proteins called Min C, D, and E that inhibit the process. (Their highest concentrations are at the poles of the cell.)
- Monomers of FtsZ form a ring on the inner circumference of the cytoplasmic membrane that determines the future cell division plane.
- Other proteins are recruited to the divisome, including FtsA, a bacterial actin-like protein (Chapters 4 and 7) and those needed to synthesize peptidoglycans, for example FtsI, and new surface membrane materials.

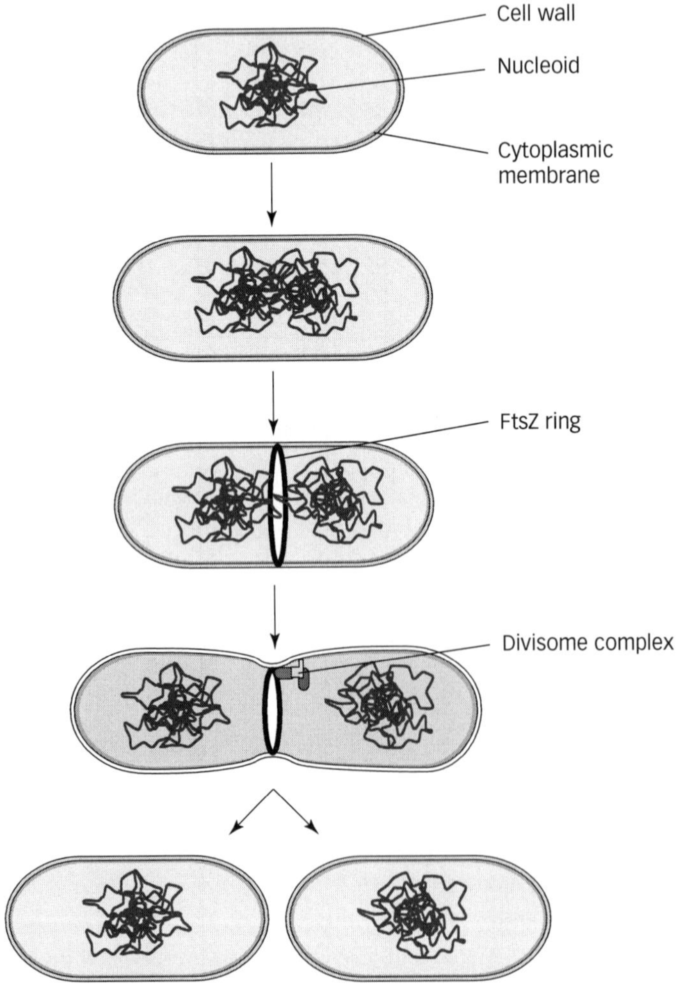

Figure 5.7 Schematic illustration of the formation of the FtsZ ring and septum, and division of a bacillus cell. See text for details.

Septum formation and cell separation

Constriction of the FtsZ ring and synthesis of the division septum begins as the cell elongates.

- These two processes are coordinated: the septum divides the cell in two as elongation doubles its length.
- During division, the chromosomes are segregated, one to each new cell.
- Division also more or less equally partitions the other contents of the cytoplasm to produce two cells that are immediately fully functional.

5.3 REPLICATION IN ARCHAEA

Archaea, like bacteria, have circular chromosomes and chromosome replication is bidirectional—but many other aspects of their molecular biology are more akin to that of eucarya than bacteria

- Unlike bacteria, archaea have one, two, or three origins of replication. (Eucarya can have thousands!)
- Archaeal proteins that recognize *ori* sites and which replicate DNA are more similar to their eucarya equivalents than they are to bacterial ones.
- Archaeal proteins are generally less complex than those of eukaryotic organisms. For example, eukaryotic helicases are hexamers of different subunits, while the archaeal enzyme consists of six identical polypeptides.

Check your understanding

5.3 Which of the following statements is/are false?

a. Fts is an abbreviation for filamentous-temperature sensitive.

b. FtsZ, like all Fts proteins, is found in all types of prokaryotic cells.

c. Chromosome segregation occurs during the coordinated formation of the septum and the elongation of the cell.

d. A cycle of DNA replication in *Escherichia coli* must be completed before a new one can be initiated.

e. Archaea and bacteria have circular chromosomes and therefore use similar molecular biologies to replicate their genomes.

Revision tip

Construct a simple table to summarize similarities and differences in the replications of bacteria and archaea.

5.4 KEY POINTS OF GENE EXPRESSION

Gene expression is the activation of genes leading to the production of the RNA or protein molecules they encode.

- The central dogma of gene expression (Figure 5.8) is that: *DNA is transcribed to give RNA, which is translated to give a polypeptide.*

Transcription

The 5′→3′ strand of DNA is the **coding** or **sense** strand; the 3′→5′ is the **non-coding** or **antisense** strand

- Only the *antisense* (3′→5′) DNA strand is used as a **template** for transcription to produce RNA molecules such as mRNAs, tRNAs, rRNAs.
- The sequence produced is an RNA copy of the sense strand of DNA (with U(racil) in place of T(hymine)) but is complementary to the *antisense* DNA strand.
- In some cases the RNA molecule first formed (the *initial* or *primary transcript*) may need to be modified to produce a functional RNA. These modifications are called **post-transcriptional modifications** or **RNA processing**.
- Transcription occurs in the cytoplasm in prokaryotes but in the nucleus of eukaryotes (Chapter 6).

Translation

Translation is the process by which mRNA is used as a template by ribosomes to produce polypeptide chains.

- In bacteria, translation occurs in the cytoplasm unlike in eukaryotes where it occurs on the cytosol (Chapter 7).

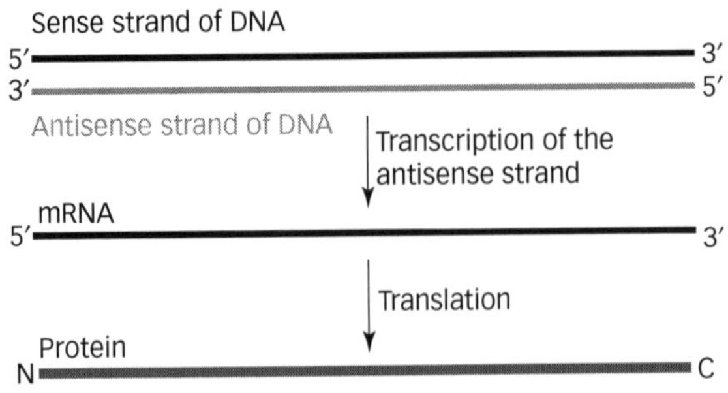

Figure 5.8 Gene expression illustrated by the flow of genetic information from double-stranded DNA to a protein. The antisense strand of DNA is transcribed to give an RNA molecule of complementary sequence. If the RNA is mRNA, it can be translated to produce a protein. See text for details of the processes involved.

- In prokaryotes, transcription and translation occur concurrently to produce polypeptides.
- Normally the initial polypeptide produced needs to be modified and folded to give a functional protein. These modifications are called **post-translational modifications** or **protein processing**.
- Post-transcriptional and post-translational modifications are more extensive in eukaryotes than prokaryotes, as outlined in Chapter 7.
- The folding of polypeptide chains to give proteins with native (active) conformations is assisted by other proteins called **chaperones.**
- Protein synthesis is perhaps the major task of cells in terms of energy devoted to it. Approximately one-third of all the proteins produced by bacterial cells are involved in producing other proteins!

➔ *Transcription and post-transcriptional modifications and translation and post-translational modifications in eukaryotes are reviewed in Chapters 6 and 7.*

5.5 TRANSCRIPTION IN BACTERIA

Transcription is carried out by a **DNA-dependent RNA polymerase** (**RNA polymerase, RNA pol** or **RNAP**), which produces a complementary copy of the DNA template.

- The requirements for activity are a template strand of DNA, the four ribonucleoside triphosphates (ATP, UTP, GTP, and CTP) and Mg^{2+}.
- RNA polymerases can only produce RNA in the 5′→3′ direction, so although genes can occur in either strand of the DNA they must always run in the 3′→5′ direction.
- Bacterial transcription is best understood in *E. coli* where it occurs in three phases:
  - **initiation**
  - **elongation**
  - **termination.**

Initiation

Initiation begins when RNA polymerase binds to **promoter regions** in the double stranded DNA.

- The structures of promoters are always given for the antisense (5′→3′) strand; a typical promoter is shown in Figure 5.9.

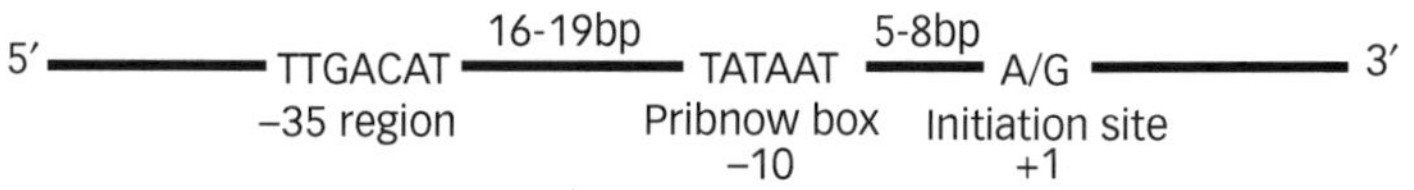

Figure 5.9 Schematic illustration of a typical bacterial promoter showing the −35 region and −10 Pribnow box upstream of the initiation site. See text for explanation of terms.

Transcription in bacteria

- The first base of the DNA to be transcribed is the initiation site or +1.
- Bases from +1 towards the 5′ end are called *upstream* and given negative numbers; bases towards the 3′ end are referred to as *downstream* and given positive numbers.
- Remember: it is the 3′→5′ strand that is transcribed by the polymerase.
- A typical bacterial promoter region comprises of a −35 region, and a −10 region called a **Pribnow box** (Figure 5.9).
- The Pribnow box has the *consensus* sequence of six ribonucleotides, … TATAAT …, located approximately 10 nucleotides upstream of the transcription initiation site.
- In some highly expressed genes there are also AT-rich sequences between –60 and –40 known as the upstream promoter regions or elements.

Check your understanding

5.4 Name two elements found in a typical bacterial promoter region.

RNA polymerase and promoter strengths

- Bacteria have a single RNA polymerase that catalyses transcription. It occurs in two forms:
  1. **apoenzyme** of composition $\alpha_2\beta\beta'\omega$, M_r approximately 460,000
  2. **holoenzyme** of composition $\alpha_2\beta\beta'\omega\sigma$, M_r approximately 530,000.
- The apoenzyme or **core** RNA polymerase binds non-specifically to DNA, but the **σ (sigma) subunit** or **factor** increases its specificity for promoter regions.
- The promoter consensus sequence given in Figure 5.9 is based on many promoters.
- The nearer the sequence of a promoter is to the consensus, the *stronger* it is and conversely, the greater the difference the *weaker* the promoter.
- The σ subunit binds tightly to strong promoters and so initiates transcription of their associated gene.
- The σ subunit can recognize only a few bases of a weak promoter so it is less likely to bind to it. Consequently, transcription of weak promoters occurs less often than with strong ones.

Looking for extra marks?

Bacteria, such as *E. coli* can express a number of different types of σ factors. The main type expressed is called σ^{70} (because its M_r is 70,000). Different σ subunits recognize different promoters and help regulate gene expression in differing circumstances. For example, if *E. coli* cells are starved of nitrogen, they express a specific σ subunit of M_r 34,000 (σ^{34}), which alters gene expression to produce proteins that help cope with the stress.

Elongation

Once the RNA polymerase holoenzyme has recognized and bound to a promoter, the σ factor dissociates from the RNA polymerase and elongation proceeds. Elongation of the mRNA is catalysed in the $5' \rightarrow 3'$ direction.

- Elongation by the core RNA polymerase begins at the initiation site, which is usually an A or a G (Figure 5.9).
- Although the core RNA polymerase has only a low affinity for DNA it is highly processive and well suited to elongating the RNA transcript (Figure 5.10).
- Elongation proceeds as the RNA polymerase moves along the DNA in a transcription 'bubble' of approximately 17 unpaired bases.
- The DNA within the bubble is unwound as the RNA polymerase moves along it and is rewound once that portion has been transcribed.
- The transcription bubble exposes unpaired deoxyribonucleotides to incoming ribonucleotide triphosphates that are polymerized to form the RNA transcript.
- As elongation continues there is a short sequence within the transcription bubble, approximately eight bp in length, that is an RNA-DNA heteroduplex (Figure 5.10)
- The sequence of the transcript is determined by that of the template according to the usual base pairing rules.
- Elongation then proceeds until directed to cease by a termination signal.

Termination

Termination of transcription in bacteria is caused by **termination signals** within the *DNA*, which are copied into the RNA transcript.

- Two types of termination signals are common:
  1. palindrome termination sequences
  2. ρ (rho) dependent termination.
- Palindromic sequences or inverted repeats rich in GC bases are one of the commonest termination signals.

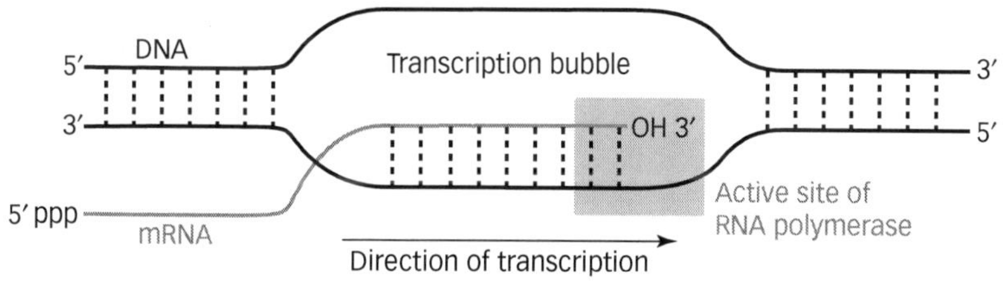

Figure 5.10 Schematic illustration of the transcription of a DNA template by RNA polymerase within a moving transcription bubble producing mRNA in the $5' \rightarrow 3'$ direction.

5′...C G T A C G C C C A G N N N N N N C T G G G C G T T T T T T ...3′
3′...G C A T G C G G G T C N N N N N N G A C C C G C A A A A A A ...5′

Transcription ↓

5′...C G U A C G C C C A G N N N N N N C U G G G C G U U U U U U – OH 3′

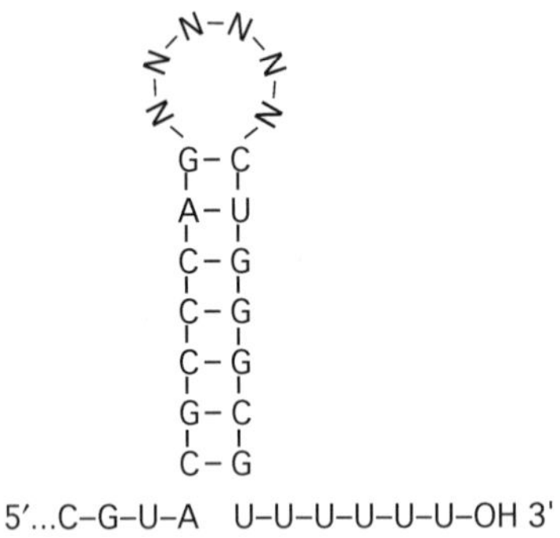

Figure 5.11 Schematic illustration of the formation of a hairpin bend by a palindromic inverted loop sequence involved in terminating transcription.

- Palindromes are stretches of nucleic acids that have the same sequences whether read in the 5′→3′ or 3′→5′ directions (Figure 5.11).
- Termination palindromes cause the RNA to form a stem-loop hairpin structure as illustrated in Figure 5.11. This structure causes the RNA polymerase to stop moving along the DNA template, upon which elongation ceases and the enzyme dissociates from the complex.
- This method of termination is known as ρ-independent termination to distinguish it from the second form

Check your understanding

5.5 Which of the following sequences are capable of forming a stem-loop structure?

a. UUCAAUGGCGGGAAAUCACGUCCGAUUACGG
b. AUGGGCCGCACUGGCAAAUCUCACGCCGGGUA
c. AAACCGCCCGGAUUUUAAUUCCCGGGAAAAU
d. UUACGGGCUUUACGGGGCCCCUATCCGGUUU
e. ACCGGAUCCCGGACUGACCACCGGGAUCCGGU

Rho-dependent termination

- Rho-dependent termination requires the presence of the ρ termination protein or factor and is used by about half the genes in *E. coli*.
- Rho factor is a hexamer of six identical subunits, each with an RNA binding site.

- Rho binds to 70–100 bp long regions of the RNA transcript that are rich in GC bases called **Rho utilization sites** (**ruts**) found approximately 100 bp upstream of the termination site.
- The RNA polymerase pauses at a region known as the **rho-sensitive pause site**.
- This allows the ρ factor to move along the RNA until it reaches the RNA–DNA heteroduplex and unwinds the RNA transcript from the DNA template, resulting in the termination of transcription.

Check your understanding

5.6 Which of the following statements is/are true?

a. During transcription RNA polymerase binds to double-stranded DNA.
b. The RNA polymerase holoenzyme is active during elongation.
c. RNA polymerase uses the DNA antisense strand as its template.
d. RNA polymerase has proofreading abilities.
e. RNA polymerase has a core structure made up of three subunits.

Rates of transcription and errors

Bacterial RNA polymerases transcribe the DNA template at a rate of approximately 50 bases per second.

- RNA polymerase does not have the proofreading ability of many DNA polymerases and so is unable to remove and replace incorrectly incorporated ribonucleotides.
- The rate of error is approximately one base in every 10^4, which is much higher than that in DNA replication.
- Errors in transcription are tolerated more than those in DNA replication because they are not passed down to progeny.

5.6 TRANSCRIPTION IN ARCHAEA

Transcription in archaea more closely resembles a pared-down version of the process in eukaryotic cells than that in bacteria.

➔ *Transcription in eucarya is revised in Chapter 6.*

- This section compares transcription in bacteria, archaea, and eukaryotes to emphasize the similarities between the latter two.

Initiation

- Archaeal promoters resemble those of eukaryotes.

- Two main sequences form archaeal promoters:
  1. **TATA box**
  2. **B recognition element**.
- The TATA box is a six to eight base sequence located 18–27 bases upstream of the initiation site.
- The B recognition element (**BRE**) is located upstream of the TATA box.
- The two sequences are recognized by proteins called **transcription factors**: a **TATA-binding protein** (**TBP**) and **transcription factor B** (**TFB**).
- TBP and TFB bind to the TATA box and BRE respectively and recruit the RNA polymerase to the promoter.

Archaeal RNA polymerase and elongation

Archaeal cells have only a single type of RNA polymerase composed of 11 or 12 subunits, which closely resembles eukaryotic RNA polymerase II in structure and antibiotic resistance.

- Once recruited to the initiation site by TBP and TFB, elongation of the RNA transcript begins.

Termination

Termination in archaea is less understood than in bacteria or eukaryotes.

- *Some*, but not all, archaeal termination signals are inverted repeats followed by AT-rich sequences similar to those described for bacteria.
- Others appear to consist of sequences of repeated thymines—but we do not yet understand how these function.
- Rho-like proteins have not been observed in archaea.

5.7 POST-TRANSCRIPTIONAL MODIFICATIONS

Post-transcriptional modifications or RNA processing is less common among prokaryotes than eukaryotes.

Bacterial RNA processing

Some evidence indicates that short stretches of adenine residues are added to the 3′ end of a few bacterial mRNA molecules forming a small polyadenylate tail (Chapter 6).

- The genes for bacterial rRNA molecules are linked together and transcribed in one stretch to produce an initial transcript about 5000 ribonucleotides long.
- This is hydrolysed (processed) to generate the three rRNA molecules found in bacterial ribosomes (see later and Table 5.2).

Archaeal RNA processing

Archaea have more extensive RNA processing capabilities than bacteria, although less than eukaryotes

- Modifications including **intron–exon splicing**.
- Some archaeal genes are **discontinuous** and contain coding and non-coding regions called **exons** and **introns** respectively. Discontinuous genes are the normal type in eukaryotes.
- Splicing is the removal of introns from pre-mRNA and the joining together of the newly formed ends to give a continuous mRNA molecule.

➔ *RNA processing or post-transcriptional modification are described more fully in Chapter 6.*

- Exons and introns are rare in archaeal genes that code for proteins, but are more common in genes that encode a number of tRNA and rRNA molecules. However, since they are processed by a different mechanism to the eukaryotic spliceosome described in Chapter 6, they should strictly be referred to as *archaeal introns*.
- Archaeal introns are hydrolytically removed by a specific endonuclease able to recognize the exon–intron borders.
- The exons are then ligated together to give a functional RNA molecule.

5.8 TRANSLATION IN PROKARYOTES

Translation brings together messenger RNA (mRNA), aminoacyl-tRNAs, and ribosomes to produce a polypeptide with a specific sequence of amino acid residues.

- Translation requires:
  - mRNA
  - amino acids and tRNAs
  - aminoacyl-tRNA synthetases
  - ribosome subunits
  - translation factors (TFs) and biological energy (ATP and GTP).
- The mRNA specifies the sequence of amino acids residues in the polypeptide.
- The sequence is determined by the order of **codons** in the mRNA.
- Codons are sequences of *three* nucleotides in the mRNA strand that code for one of the 20 amino acids specified by the **genetic code** (Table 5.1).
- Specific codons also specify where translation of the mRNA is to begin and end.
- Each amino acid is specified by more than one codon (other than the initiator codon (Table 5.1)). The genetic code is therefore described as **degenerate**.

Looking for extra marks?

The near universality of the genetic code is strong evidence that all present-day organisms evolved from a common ancestor.

First base of codon	Second base of codon U	C	A	G	Third base of codon
U	Phe Phe Leu leu	Ser Ser Ser Ser	Tyr Tyr Term[b] Term[b]	Cys Cys Term[b] Trp	U C A G
C	Leu Leu Leu Leu	Pro Pro Pro Pro	Hid His Gln Gln	Arg Arg Arg Arg	U C A G
A	Ile Ile Ile Met[a]	Thr Thr Thr Thr	Asn Asn Lys Lys	Ser Ser Arg Arg	U C A G
G	Val Val Val Val	Ala Ala Ala Ala	Asp Asp Glu Glu	Gly Gly Gly Gly	U C A G

Table 5.1 Codons of the genetic code. Each of the amino acids is specified by its standard three letter code. For example the two codons UUU and UUC specify the amino acid phenylalanine (Phe). Note how nearly all the amino acids have more than one codon.

[a] The start codon for translating an mRNA is usually AUG; this codes for methionine or its modified form, formylmethionine.
[b] The terminator or stop codons UAA, UAG and UGA indicate where translation of the mRNA must stop.

Check your understanding

5.7 If a protein is 600 amino acid residues long, how many nucleotides were needed to code for its amino acid sequence?

- During translation, amino acid residues are delivered to the ribosome by **tRNAs**.
- Different tRNAs have different **anticodons** and deliver different specific amino acids.
- An anticodon is a sequence of three bases on the tRNA that is complementary to a codon on the mRNA.
- Ribosomes are ribonucleoprotein particles composed of **ribosomal** (**r**) **proteins** and **ribosomal** or **rRNA** molecules.
- Polypeptides are formed on the ribosome by a step-wise addition of amino acids to the carboxyl terminus such that the direction of growth is from the amino to carboxyl terminus, N→C (Figure 5.12).

Transfer RNAs and aminoacyl synthetases

The role of tRNAs is to match an amino acid to its correct codon in the mRNA being translated.

- Transfer RNAs are approximately 76–96 ribonucleotides long.
- The structures of tRNAs are stabilized by internal base pairings that form stem-loops.

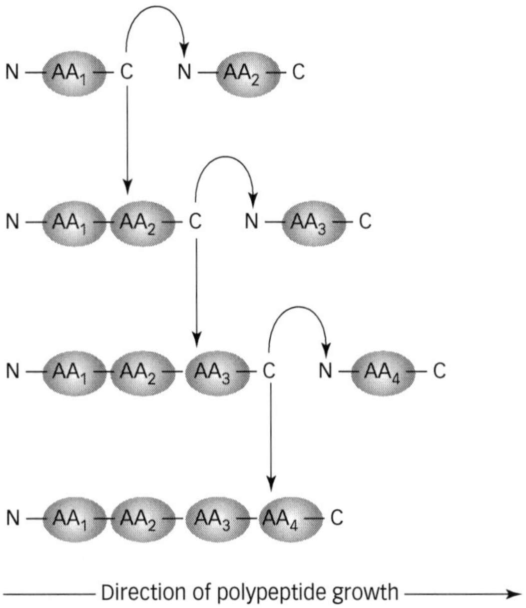

Figure 5.12 Step-wise synthesis of a polypeptide from the amino to the carboxyl terminus.

- Their secondary structures resemble a cloverleaf in appearance but their tertiary structures are roughly similar to a three-dimensional L-shape (Figure 5.13).
- The triplet of bases forming the anticodon is located at the apex on a hairpin loop so that they are available for pairing with the mRNA codon.
- A specific amino acid can only be bound to the 3′ CCA terminal sequence located at the opposite end of the tRNA (Figure 5.13).

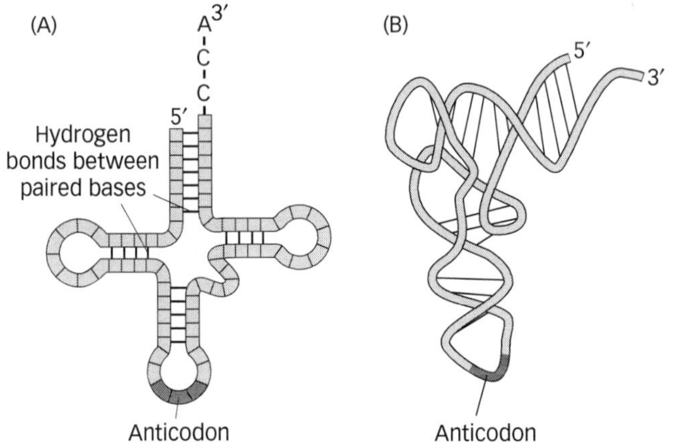

Figure 5.13 Structure of a tRNA. (A) Cloverleaf secondary structure showing the optimum intrachain hydrogen bonding, and the anticodon and amino acid attachment (CAA) sites. (B) The tertiary folded structure illustrating the anticodon and amino acid attachment site at opposite ends of the molecule.

- The corresponding tRNA for an amino acid is called its *cognate* tRNA and the two are covalently linked by the action of their cognate **aminoacyl synthetase**.

Aminoacyl synthetases

- Aminoacyl synthetases bind specific amino acids to their cognate tRNAs by catalysing the formation of an ester bond between the terminal 3′-OH of the tRNA and the carboxyl group of the amino acid.

  $$\text{Amino acid} + \text{ATP} + \text{tRNA} \rightleftharpoons \text{aminoacyl-tRNA} + \text{AMP} + \text{PP}_i$$

- The process is often referred to as *amino acid activation*.
- The equilibrium for this reaction is only about 50%. However, it is driven to completion by the subsequent hydrolysis of the pyrophosphate (PP_i) to phosphate.

  $$PP_i + H_2O \rightarrow 2P_i$$

- Each aminoacyl-tRNA synthetase can only recognize its cognate tRNA and amino acid so each cell must have at least 20 different enzymes.
- Synthetases are extremely specific for their amino acid and tRNA substrates and have proofreading properties: if an incorrect amino acid is bound, it is recognized by an additional site on the enzyme which cleaves the bond releasing the amino acid.
- It is essential that an incorrect amino acid is not added to a tRNA, otherwise an incorrect amino acid would be added to the growing polypeptide.

Ribosomes

The 'active' constituents of ribosomes that mediate interactions between the aminoacyl-tRNAs and mRNA and catalyse the formation of peptide bonds between amino acids are the *rRNAs*.

- Ribosomal proteins are thought to be structural and help maintain the conformations of the rRNA molecules in their active forms.
- Ribosomes from all three domains (Chapter 1) are made up of two subunits of different sizes called the *small* and *large subunits*.
- Prokaryotic subunits are generally smaller than eukaryotic types and have a simpler composition (Table 5.2).

	Ribosome
Intact dimensions (nm)	29×21
s_o (S)	70
M_r	2.8×10^6
rRNAs	Two of 23 S and 5 S in large subunit; one of 16 S in small subunit
Relative proportion of RNA in ribosome (%)	~65
Ribosomal proteins	34 in large subunit; 21 in small subunit
Relative proportion of protein in ribosome (%)	~35

Table 5.2 Sizes and composition of *Escherichia coli* ribosomes. See also Table 7.1

- Bacterial ribosomes are made up of 30 S (small) and 50 S (large) subunits, though they only combine to give the 70S ribosome when they are translating an mRNA (Figure 5.14).

Check your understanding

5.8 Which of the following statements is/are true?

a. The 70 S bacterial ribosome is made up of 30 S and 40 S subunits.
b. The 80 S bacterial ribosome is made up of 40 S and 60 S subunits.
c. The 70 S bacterial ribosome is made up of 60 S and 40 S subunits.
d. The 70 S bacterial ribosome is made up of 30 S and 60 S subunits.
e. The 70 S bacterial ribosome is made up of 50 S and 30 S subunits.

- Ribosomes have four key binding sites for translation:
  - mRNA binding site
  - aminoacyl (A) site
  - peptidyl (P) site
  - exit (E) site.
- The mRNA binding site is located in a channel formed between the small and large subunits; it holds the mRNA in place and allows the ribosome to move along the mRNA translating it in the 5′→3′ direction.
- The A site receives the incoming aminoacyl-tRNAs.
- The P site binds to the tRNA that is linked to the growing polypeptide.
- The E site transiently binds a free tRNA before it leaves the ribosome.
- **Polyribosomes** or **polysomes** are clusters of ribosomes that are simultaneously translating the same mRNA molecule.

Translation factors

Translation factors are soluble proteins of the cytoplasm that are associated with all stages of translation.

- Many translation factors are **G proteins.**
- G proteins are GTP-binding proteins with GTPase activities.
- In their *active* conformation they bind GTP and can associate with other molecules involved in translation, such as the ribosome (or its subunits), to form active complexes.
- Following the hydrolysis of the GTP to GDP, the factor changes conformation and becomes inactive and is able to dissociate from the complex.

➔ *G proteins are also involved in processes described in other chapters, for example, Chapters 7, 8, and 15.*

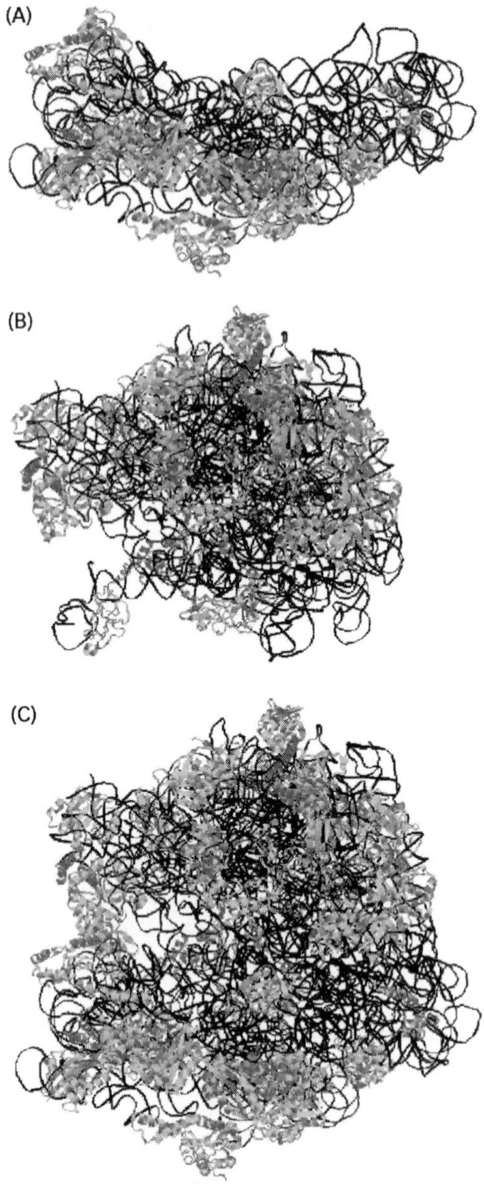

Figure 5.14 Structure of the 70 S *Escherichia coli* ribosome. Molecular models showing relative sizes of (A) the small and (B) large subunits and (C) the intact ribosome. Structures obtained using PDB file 3I1O. In all cases the rRNA is shown in black and the ribosomal proteins in the lighter tone. See also Figure 7.3.

5.9 TRANSLATION IN BACTERIA

In bacteria (and archaea) translation begins before transcription is complete. Thus, transcription and translation are closely coupled.

- Translation in bacteria (*E. coli*) is well understood. It consists of three phases:
  1. initiation
  2. elongation
  3. termination.

Initiation

Initiation of translation occurs when the ribosome becomes positioned over the initiator codon (AUG) of the mRNA, the first aminoacyl-tRNA binds to the mRNA, and the complete 70 S ribosome forms.

- The structure of a bacterial mRNA is given in Figure 5.15.
- About −10 bases upstream of the AUG start codon is a four to nine base sequence called the **Shine–Dalgarno sequence**, which is complementary to the 3′ end of the 16 S rRNA of the small ribosomal subunit.
- Each end of the mRNA is flanked by untranslated regions (UTRs).

Formation of initiation complexes

- The formation of a bacterial initiation complex is illustrated in Figure 5.16.
- The 16 S rRNA of the 30 S (small) ribosome subunit recognizes and base pairs with the Shine–Dalgarno sequence, which aligns the start AUG codon appropriately on the ribosome subunit.
- Initiation begins with the formation of a 30 S initiation complex (Figure 5.16).
- Initiation factors IF1 and IF3 bind to the 30 S ribosome subunit.
- In bacteria, the initiator amino acid is a form of methionine (Met) modified by the addition of a formyl group and called *N*-**formylmethionine** (**fMet**). Its cognate tRNA is referred to as **tRNA$_f$** and possesses the anticodon to AUG.
- The GTP-IF2 complex recognizes and binds to the initiating fMet-tRNA$_f$ and allows it to bind to the P site of the ribosomal subunit (Figure 5.16).
- (During elongation, internal Met residues are delivered by a different tRNA (with the same anticodon as tRNA$_f$) that like all other aminoacyl-tRNAs bind to the *A* site.)

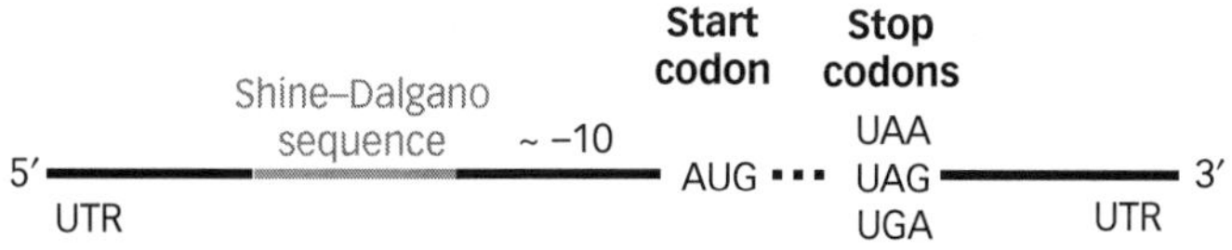

Figure 5.15 Simplified structure of a bacterial mRNA. UTR, untranslated regions. See text for details.

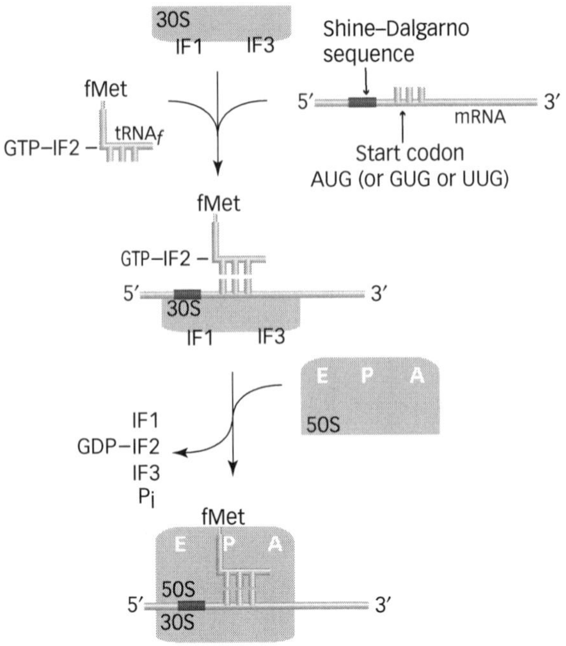

Figure 5.16 Overview of initiation of translation in *Escherichia coli*. See text for explanations.

- Binding of the fMet-tRNA$_f$ completes the formation of the **30 S initation complex**.
- Initiation factors IF1 and 3 are released and the 50 S (large) subunit is able to join the complex.
- The GTP bound to IF2 is hydrolysed allowing the release of IF2 and completing the formation of the **70 S initiation complex** (Figure 5.16).
- Elongation of the amino acid polypeptide can commence.

Check your understanding

5.9 Which initiation factors prevent the small and large ribosome subunits from attaching to each other before the initiation of translation?

Elongation

Elongation is the formation of the polypeptide using the mRNA as a template. The message is translated in the 5′→3′ and the polypeptide grows in the corresponding N→C direction.

- The first elongation step is outlined in Figure 5.17.
- At the start of elongation the fMet-tRNA$_f$ occupies the P-site of the ribosome and the A-site is vacant.

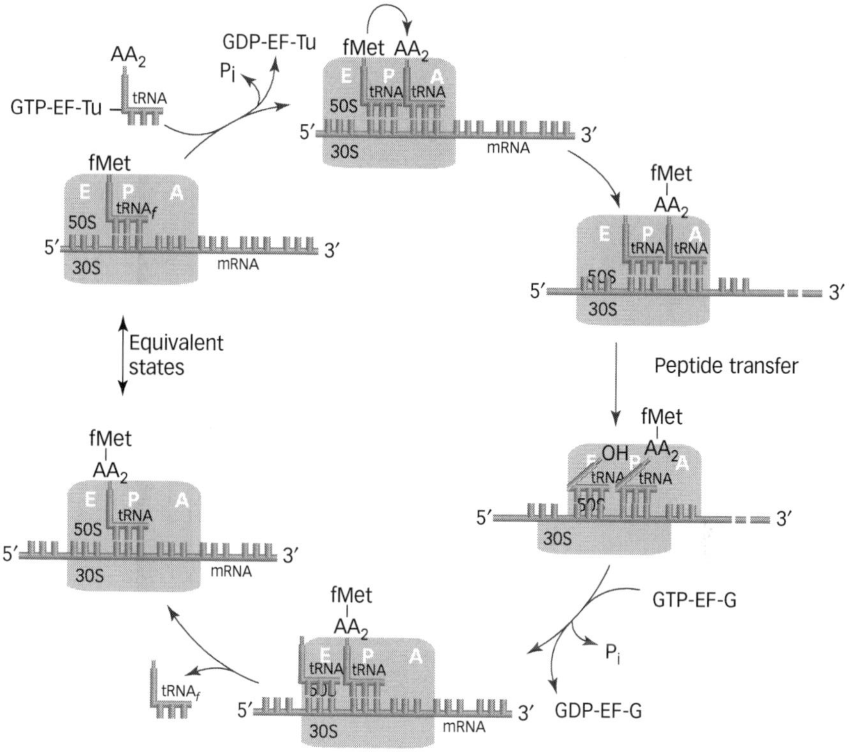

Figure 5.17 Overview of elongation in bacterial translation. See text for details.

- Elongation during translation requires a number of elongation factors (EFs).
- An aminoacyl-tRNA is delivered to the ribosome bound to the GTP activated elongation factor Tu (EF-Tu-GTP). This facilitates the correct positioning of the aminoacyl-tRNA in the A site.
- If the aminoacyl-tRNA is correct, its anticodon being complementary to the second codon of the mRNA, the EF-Tu-GTP hydrolyses its GTP and dissociates from the ribosome.
- The aminoacyl-tRNAs in the P and A sites are now correctly positioned next to the active site of a **peptidyl transferase**.

Looking for extra marks?

The amino acid selenocysteine, a component of a number of selenoproteins, is specified by the codon UGA that would normally function as a termination codon. However, in the presence of a selenocysteine insertion sequence element (SECIS) in the mRNA, UGA specifies selenocysteine. The characteristic nucleotide sequence and base-pairing patterns in the SECISs forms a secondary structure that changes the codon specificity.

Peptidyl transferase

- Peptidyl transferase activity of the ribosome is *not* associated with an enzyme but is a property of the **23 S rRNA** of the 50 S subunit.
- RNA molecules with catalytic properties are called **ribozymes**.
- Peptidyl transferase activity transfers the fMet in the P-site to the free amino group of the aminoacyl-tRNA forming a dipeptidyl-tRNA in the A site (Figure 5.17).
- The spent $tRNA_f$ then moves to the E site of the ribosome before being released from the ribosome-mRNA complex.

Translocation

- **Translocation** is the movement of the ribosome along the mRNA by one codon.
- During peptidyl transferase activity, each tRNA is thought to straddle two sites: the $tRNA_f$ the P and E sites, the second tRNA the P and A sites respectively (Figure 5.17).
- Translocation is completed by the simultaneous movements of the $tRNA_f$ and second tRNA to completely occupy the E and P sites.
- This movement is catalysed by elongation factor-G activated by a bound GTP (GTP-EF-G, Figure 5.17).
- Hydrolysis of the GTP is associated with the translocation, and the EF-G-GDP dissociates from the ribosome and the $tRNA_f$ leaves the E site.
- The first round of translocation is now complete and the dipeptidyl-tRNA occupies the P site and the A site is unoccupied (Figure 5.17).
- This is the equivalent state to the beginning of elongation.
- Subsequent rounds are repeats of the same steps, other than that a peptide is transferred from a peptidyl-tRNA in the P site to an aminoacyl-tRNA occupying the A site.
- The growing polypeptide exits the ribosome through an opening near the peptidyl transferase that leads into a tunnel passing through the 50 S subunit to the exterior.
- Amino acids are incorporated into the growing polypeptide at a rate of approximately 15 to 21 residues per second until a termination codon occupies the A site.

Check your understanding

5.10 If the A-site contains an aminoacyl-tRNA, which site of the ribosome contains the tRNA linked to the growing polypeptide chain?

Proofreading in elongation

- The high fidelity of translation is enhanced by a proofreading mechanism in the elongation process.

- The rate of error in translation is about one amino acid residue in 10^4, which is much lower than expected from the weak hydrogen bonding between the codons of mRNA and the anticodons of tRNAs.
- Kinetic proofreading involving GTP-EF-Tu increases the accuracy of translation.
- If the correct aminoacyl-tRNA is positioned over the codon in the A site, then the GTP is hydrolysed, EF-Tu-GDP is released, and the correct aminoacyl-tRNA then occupies the A site.
- If an incorrect aminoacyl-tRNA is positioned in the A site there is a weaker interaction (hydrogen bonding) between the codon and anticodon and GTP hydrolysis is delayed.
- This delay gives the incorrect aminoacyl-tRNA time to diffuse away from the A site before the amino acid can be incorporated into the polypeptide.

Termination

Termination of translation occurs when one of the three stop codons (UAA, UGA, UAG) is positioned in the A site of the ribosome (Figure 5.18).

- Stop codons are not recognized by aminoacyl-tRNAs but by protein **release factors**, **RF1** and **RF2**. Release factors have a similar conformation to tRNA molecules (see Figure 5.19).
- Release factor RF1 recognizes the termination codons UAA and UAG and binds to the ribosome when these are present at A site.
- Factor RF2 recognizes UAA and UGA.
- In both cases, a third release factor, RF3 binds to the ribosome as the complex, RF3–GTP along with RF1 or RF2 (Figure 5.18)
- In the presence of RF3–GTP, RF1 and RF2 bind to the stop codon at the A site, which allows a molecule of water to hydrolyse the ester bond linking the polypeptide to the 3′-OH of the tRNA
- Essentially, the polypeptide is transferred to the water molecule and its release from the ribosome terminates translation.

Check your understanding

5.11 Which three codons are recognized by release factors?

Ribosome dissociation

- The last tRNA used and the mRNA dissociate from the ribosome following termination.
- The binding of **ribosomal recycling factor** (RRF) and **EF-G** to the ribosome leads to the dissociation of the 70 S ribosome into subunits, at the expense of hydrolysing GTP.
- Ribosomal recycling factor has a remarkable similarity in overall conformation to tRNA molecules (Figure 5.19).

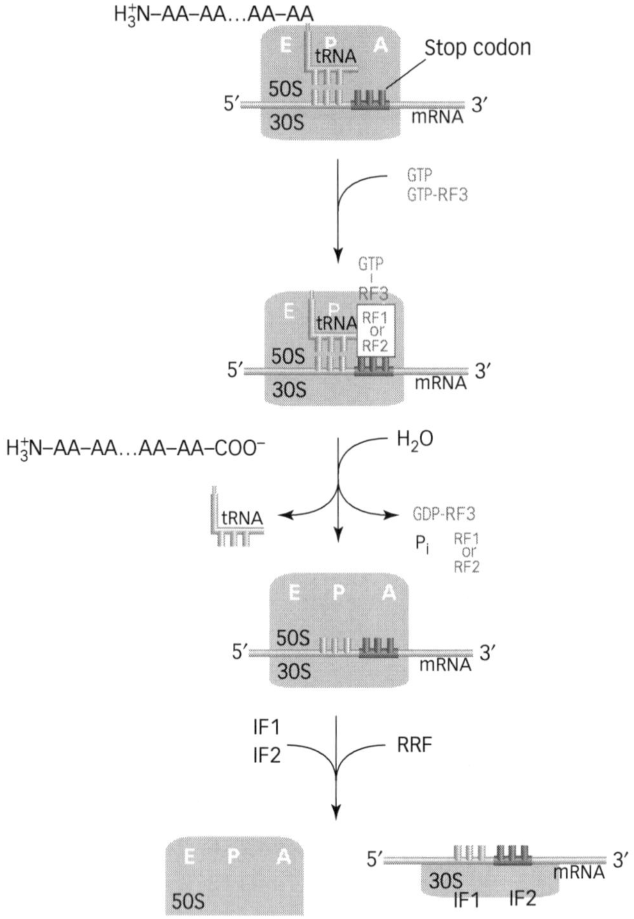

Figure 5.18 Overview of the termination of translation. See text for details.

- The rejoining of the two subunits is prevented by binding of initiation factors IF1 and IF3 to the 30 S subunit.

Check your understanding

5.12 Which three components bind to the ribosome during the termination of bacterial translation and release mRNA and tRNA?

Post-translational modifications

Following translation, the newly formed bacterial polypeptides undergo a number of post-translational modifications and folding before they become fully functional proteins.

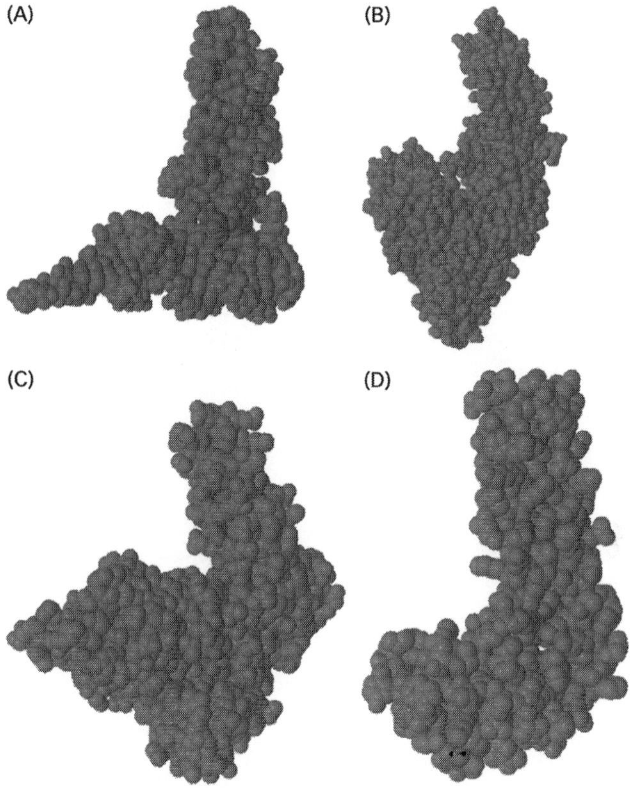

Figure 5.19 Molecular models to show the similar conformations of (A) a tRNA (B) EF-G, (C) ribosomal release factor 2 and (D) ribosomal recycling factor. PDB files 6TNA, 1EHZ, 1GQE and 1DD5 respectively.

- The first modification is the hydrolytic removal of the *N*-terminal formyl group by a formylase. In about 60% of cases the terminal methionine residue is also removed from the polypeptide.
- The transport of some polypeptides across the bacterial cell surface membrane involves the hydrolysis of signal peptide from their amino terminus.
- Phosphorylation is one of the commonest protein modifications and is essential in regulating the activity of some proteins.
- Glycosylation is the addition of carbohydrates to specific amino acid residues in the polypeptide chain to form glycoproteins.
- Enzyme activities associated with the glycosylation of proteins occur at the inner face of the surface membrane and in the periplasm.
- Many proteins that are secreted into the periplasmic space or from the cell are oxidized by the formation of disulphide bonds between specific pairs of cysteine residues.
- Post-translational modification are far more varied and numerous in eukaryotes than prokaryotes but in all cases they are necessary to produce soluble, active proteins.

➔ *Eukaryotic translation and post-translational modifications are discussed in Chapters 7 and 8 respectively, and in the companion volume,* Thrive in Biochemistry and Molecular Biology.

Polypeptide folding

- The unfolded polypeptide that is released from the ribosome must be folded into its specific active conformation.
- Folding is assisted by post-translational modifications but also requires the assistance of proteins called **chaperones**.
- The best known belong to two groups called Hsp 60 and Hsp 70 chaperones; some of their roles are discussed in Chapters 7, 8, 10, and 11.

Looking for extra marks?

Heat shock proteins (Hsp) are produced as a protective measure when the temperature at which the bacterium is growing is abruptly increased by a few degrees. In *E. coli* their formation is associated with a specific σ subunit of M_r 32,000 (σ^{32}).

5.10 TRANSLATION IN ARCHAEA

The molecular apparatus for translating mRNA molecules in archaea more closely resembles that of eukaryotes (described in Chapter 7), than that of bacteria.

- Like the eucarya, archaea use methionine as their initiator amino acid residue, not the formylmethionine of bacteria.
- Many of the archaeal ribosomal proteins are similar to those found in eukaryotic ribosomes, whereas those that resemble bacterial types are also found in eukaryotes.
- Translation factors are relatively similar in the archaea and eukaryotes compared to those of bacteria.
- Both archaea and eucarya use more translation factors than bacteria and those that are comparable with bacterial ones are composed of more subunits—but archaeal factors have fewer subunits than eukaryotic types.
- One way in which translation in archaea and bacterial do resemble one another is in the number of rRNA molecules in their ribosomes and the use of mRNA molecules that are complementary to the 16 S rRNA component.

Revision tip

Construct a simple table to summarize similarities and differences in the expression of genetic information in bacteria and archaea.

Looking for extra marks?

Overall, comparisons of the molecular biologies of replication, transcription, and translation support the case that archaea are more closely related to eukaryotes than to bacteria.

Revision tip

Do browse through the further reading references given for this chapter online. Go to http://www.oxfordtextbooks.co.uk/orc/thrive/

Check your understanding

5.13 Which of the following statements is/are true or is/are false?

a. DnaA binds to *oriC* and causes the DNA strands to separate.
b. DNA polymerase III does not have 3′→5′ exonuclease activity.
c. DnaA is responsible for forming RNA primers during DNA replication.
d. Okazaki fragments are synthesized on the lagging strand.
e. The DNA polymerase holoenzyme contains a single polymerase.
f. DNA polymerase I adds nucleotides to the 5′ end of the DNA.
g. DNA polymerase has 5′→3′ polymerase activity.
h. DNA replication is terminated when the replication forks meet half way round the bacterial chromosome.
i. Newly replicated linked bacterial chromosomes are separated by DNA ligase.
j. Archaea, like bacteria, have a single *oriC* site.

5.14 Which of the following statements is/are true or is/are false?

a. A replisome is the length of DNA replicated in a single stretch by the polymerase.
b. In DNA replication, the leading strand is synthesized in relatively short discontinuous lengths called Okazaki fragments.
c. In DNA transcription, the leading strand is synthesized *continuously* in a processive process.
d. In *Escherichia coli*, the *oriC* site is a 350-kbp region with a high proportion of A–T base pairs.
e. Helicases in *Escherichia coli are* also known as DnaA proteins and catalyse unwinding of the DNA double helix.
f. Generally, only a single RNA primer is required for the leading strand but each Okazaki fragment requires a new RNA primer.

g. DNA polymerase I possesses 5′→3′ exonuclease and 5′→3′ polymerizing activities.

h. A replication fork travelling clockwise passes through the *TerE*, *TerD*, and *TerA* sites but stops on encountering *TerC*–Tus.

i. A replication fork travelling anticlockwise will pass through the *TerG*, *TerF*, *TerB*, and *TerC* sites but stop on encountering *TerA*–Tus.

j. Is the following structure a consensus sequence of bacterial promoter on the template strand?

	−35 region		−10 region (Pribnow box)		+1 (Initiation site)		
3′ . . .	TTGACA	. . . 16–19 bp . . .	TATAAT	. . . 5–8 bp . . .	A/G	. . .	5′

5.15 How long would it take RNA polymerase to produce a transcript 2000 bases long?

5.16 How many errors would you typically expect to find in an RNA transcript 50,000 bases long?

5.17 If the amino acid residues are incorporated into a polypeptide at a rate of 15 residues per second, how long would the polypeptide be after 5 minutes?

6 The nucleus

Key features of the nucleus

- Nuclei vary in size and shape between different types of cells. However, they are often rounded or oval in shape and approximately 5 μm in diameter (Figure 6.1).
- The ground material enclosed within the nucleus is often referred to as **nucleoplasm** to distinguish it from the **cytoplasm** that surrounds the nucleus.
- Other than for a few chloroplast and mitochondrial genes (Chapters 10 and 11 respectively), all the genes of the eukaryotic cell that control its activities and development (its genome) are found in chromosomes in its nucleus.
- Chromosomes occur in pairs in the nucleus, with the numbers present varying between different species—for example, human diploid cells contain 23 pairs of chromosomes (Chapter 16).
- Each chromosome contains a single deoxyribonucleic acid (DNA) molecule in a nucleic acid–protein complex called **chromatin**.
- When examined by electron microscopy (Chapter 2), chromatin in the interphase nucleus appears as two separate phases:
  - genetically active **euchromatin**
  - inactive **heterochromatin**.

continued

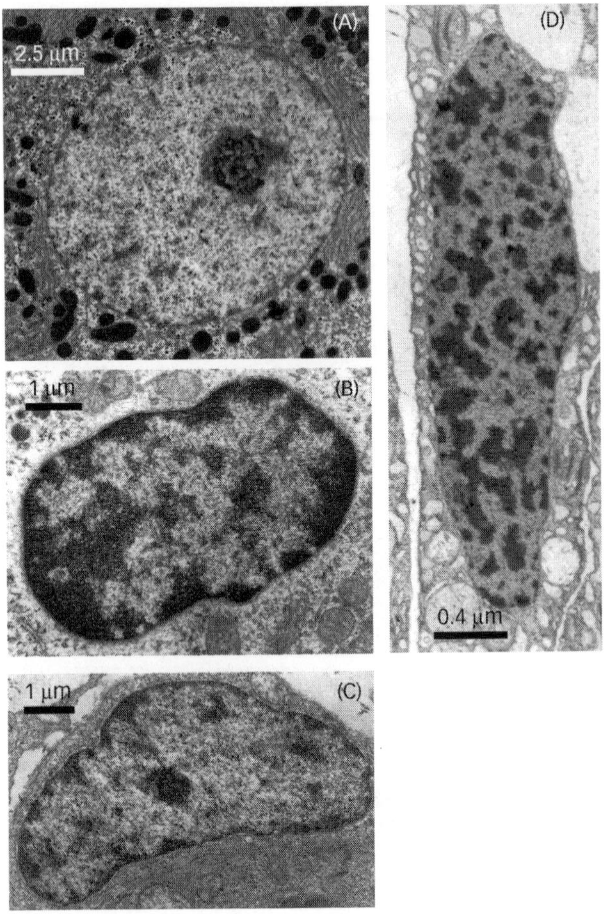

Figure 6.1 Electron micrographs of nuclei from a variety of cells. (A) Rat hepatocytes. (B) Endothelial cell. (C) Human monocyte. Courtesy of Dr A. Curry, Withington Hospital, Manchester, UK. (D) Pea cotyledon cell (*Pisum sativum*). Courtesy of Professor D.G. Robinson, Pflanzenphysiologisches Institut der Universität Göttingen, Germany.

- The nucleus is the site of DNA replication prior to mitotic or meiotic cell division (see Chapter 16) and of the transcription of DNA to form a variety of types of ribonucleic acids (RNAs).
- Newly formed RNA transcripts are subjected to RNA processing or post-transcriptional activities, which are necessary to produce mature RNA molecules.
- All nuclei contain at least one **nucleolus**, which are the sites where the genes for pre-ribosomal RNA molecules (rRNAs) are transcribed and where ribosomal subunits are assembled from rRNA and ribosomal proteins.
- In addition to the nucleolus, a number of other subnuclear non-membrane bound structures with a variety of functions can also be observed in nuclei using appropriate microscopy techniques. These include:

- Cajal bodies
- gemini of Cajal bodies
- speckles
- promyelocytic leukaemia bodies.
- Nuclei are limited by a **nuclear envelope**, which is punctured by **nuclear pores**, whose **nuclear pore complexes** mediate the movement of many materials through the envelope.
- The size and shape of the nucleus is maintained by a **nuclear skeleton** composed of fibrous proteins, which also provides an organizing frame to which the chromatin fibres are attached.

Check your understanding

6.1 Determine the percentage volume occupied by a nucleus of radius 2.5 μm in a cell of 20 μm diameter.

6.1 CHROMOSOMES AND CHROMATIN

The nucleus is approximately only 5–6 μm in diameter yet the total length of DNA molecules in a diploid human cell (that is, one containing 46 chromosomes) is approximately two metres, so careful packaging of the DNA within the nucleus is necessary.

- Nuclear or genomic DNA is divided between a number of chromosomes, with each one containing a single linear molecule of DNA (unlike the circular prokaryotic chromosomes described in Chapter 4).
- Chromosomes are not randomly distributed within the nucleus but confined to discrete locations called **chromosome territories**. Chromatin-free channels separate one territory from another (Figure 6.2).
- The existence of chromosome territories was confirmed relatively recently using *in situ* hybridization whereby multiple short lengths of radioactive or fluorescently-labelled DNA fragments (called **probes**) were hybridized with the chromosomes in a nucleus to indicate their positions.
- Mitotic and meiotic chromosomes are the most highly packaged (or folded) form of chromatin and are visible using light microscopy.

➔ *Mitosis and meiosis and the structure of chromosomes are described in Chapter 16.*

Check your understanding

6.2 Calculate the length of a mitotic chromosome, if its DNA molecule is 20×10^7 bp long and its packing ratio is 20,000.

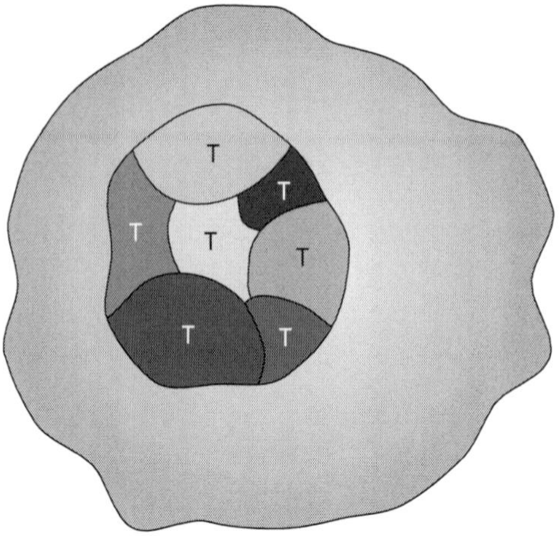

Figure 6.2 Schematic illustration of chromosome territories within a nucleus. Each separate chromosome is indicated with a letter T.

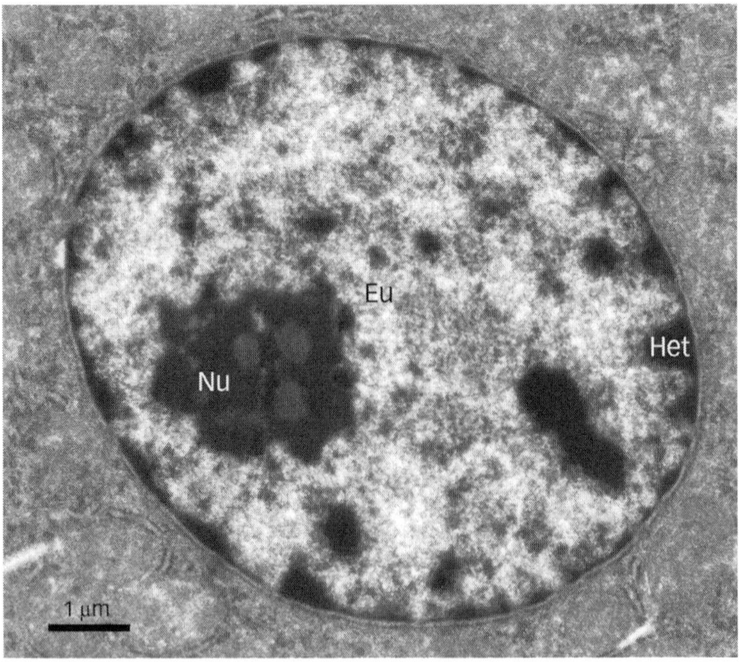

Figure 6.3 Electron micrograph of a nucleus of a chicken liver cell. Eu, euchromatin; Het, heterochromatin; Nu, nucleolus. Courtesy of Professor K.A. Brasch, School of Natural Sciences, California State University, USA.

- Within the nucleus, the chromosomes are present in a variety of differently folded forms (see 'Packaging of DNA' section) whose detailed structure depends heavily on the requirements of the cell and its stage in the cell cycle.
- Two main forms of chromatin visible within the nucleus using electron microscopy are **euchromatin** and **heterochromatin** (Figure 6.3).
- Euchromatin is less densely stained by electron microscopy techniques. It is a relatively loosely packed form of chromatin and is genetically active—that is, it is transcribed.
- Heterochromatin is visible as densely stained regions by electron microscopy. It occurs in two forms:
  - constitutive heterochromatin
  - facultative heterochromatin.
- Constitutive heterochromatin is found in all types of cells. It contains short sequences of tandemly repeated DNA and may be almost as tightly packed as that found in mitotic chromosomes (see 'Packaging of DNA' section), and so is not transcribed.
- Facultative heterochromatin can be interconverted with euchromatin. Its amount and distribution varies from cell type to cell type and also within a single cell depending upon its stage in the cell cycle (Chapter 16).

Packaging of DNA

The packaging or folding of chromatin within the nucleus is achieved by it being twisted into a series of ever increasing tighter coils, ultimately compacting the approximately 2 m of DNA into chromosomes 2–10 μm long.

- The degree of shortening during packaging is described by the **packing ratio**, which is the initial length of the DNA molecule divided by the length of the packaged form. Mitotic chromosomes have typical packing ratios of 15,000–20,000.
- There are several stages of packaging that lie between the two extremes of unpacked DNA molecules and mitotic chromosomes, as described in Table 6.1 and Figure 6.4.
- The major proteins involved in the packaging of DNA in eukaryotes are called **histones**, although a number of other proteins, such as DNA-binding transcription factors are also necessary.

Histones

- Histones are essential for the first stage in the folding of DNA molecules during their packaging in chromatin (Table 6.1 and Figure 6.4) in which DNA is bound to histones at a ratio of approximately 50:50 on a weight for weight basis.
- The structure of histones has been highly conserved during eukaryotic evolution (indeed, histones are found in some archaea).

Chromosomes and chromatin

Structural level	*Description*
'Beads on a string'	Approximately 10 nm in diameter Packing ratio of approximately 7–10 Loosely packed and unfolded and found in areas of active transcription
Chromatin fibre	Approximately 30 nm in diameter Packing ratio of approximately 42 Tightly packed fibre of nucleosomes
Looped chromatin domains	Approximately 300 nm in diameter Packing ratio of approximately 680 Loops of chromatin fibre held in place by a scaffold of proteins
Condensed-scaffold associated chromatin	Approximately 700 nm in diameter Packing ratio of approximately 1.2×10^4 Tightly coiled loops that are densely packed and transcriptionally inactive regions of DNA

Table 6.1 Outline of structural levels associated with the packaging of DNA and chromatin structures. See also Figure 6.4.

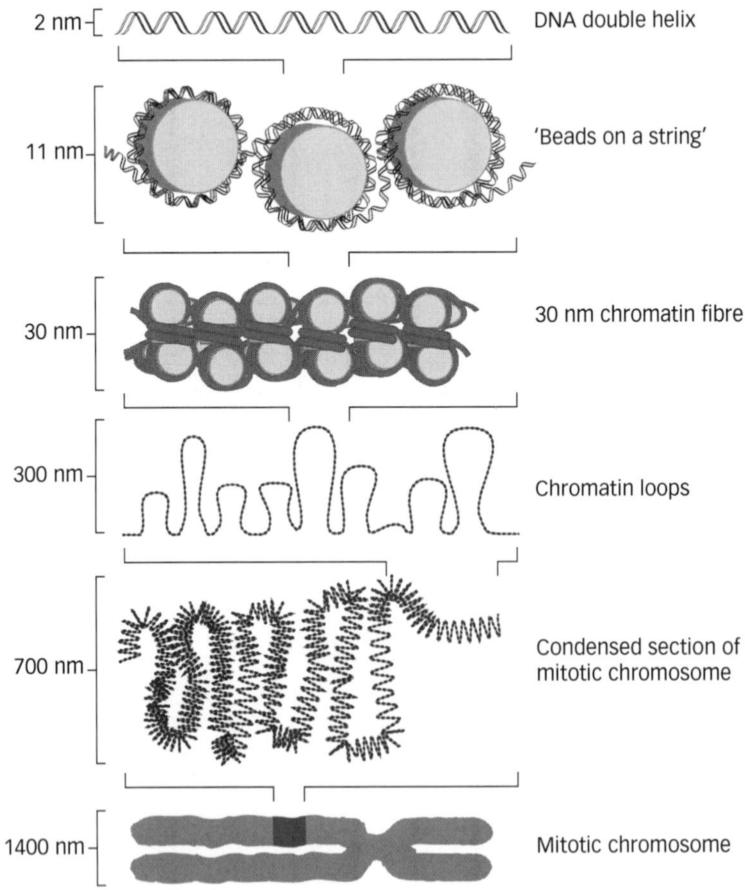

Figure 6.4 Illustration of the levels of the hierarchical folding of chromatin. See also Table 6.1.

- All histones contain a high proportion of lysine and arginine residues, which have positively-charged side chains at physiological pH.
- Histones are able to form ionic bonds with the negatively-charged phosphate groups on the DNA strands, allowing the otherwise rigid rod of the double helix to bend and fold.
- Five main types of histones are involved in DNA packaging: H1, H2A, H2B, H3, and H4. H1 is a linking histone and the remaining units are called **core** histones.
- Two units each of H2A, H2B, H3, and H4 bind together to form an octamer called a **nucleosome core**.
- Two loops of DNA approximately 146 bp in length wind around each nucleosome core to form a nucleosome, with neighbouring nucleosomes being linked by short lengths of DNA (Figure 6.4).
- Histone H1 is not part of the octamer complex, but is bound to the DNA between the nucleosomes, locking the DNA in place.
- The DNA–histone complex resembles a string of beads of approximately 10 nm in diameter.

Check your understanding

6.3 (a) Which histones make up the nucleosome core? (b) How many of each are present?

- Subsequent folding of the 'beads on a string' produces ever tighter coils of increasing diameters and therefore shorter lengths, allowing the chromosomes to be effectively packed within the nucleus while remaining accessible to the enzymes involved in transcription and replication.
- Euchromatin is composed of the loosely-folded forms ranging from 'beads on a string' to 30-nm chromatin fibre.
- Heterochromatin consists of the highly-folded helices of looped domains to regions almost as tightly packed as mitotic chromosomes.

Revision tip

The following website contains a remarkable animation showing the stages involved in folding DNA: http://www.dnai.org/c/index.html

Check your understanding

6.4 Describe the key characteristics of 'beads on a string' chromatin.

6.2 REPLICATION OF NUCLEAR DNA

Replication is the process by which cellular DNA is copied to form two complete daughter strands of DNA. The process occurs during interphase when the cell produces two copies of its chromosomal DNA before cell division begins.

Key features of DNA replication

- During the S phase (synthesis phase) of the cell cycle nuclear DNA is replicated to form an identical copy.
- The two complementary strands of a DNA molecule separate from each other during replication. **Helicase** activity breaks the hydrogen binds between the complementary bases and uncoils the parent DNA strands.
- Each DNA molecule therefore produces two templates on which complementary daughter strands of DNA can be formed.
- This process is called **semi-conservative replication** because each new double helix consists of one parental strand of DNA and one newly formed daughter strand.
- DNA replication is performed by a **DNA dependent polymerase** (**DNA polymerase** or **DNA pol**), which synthesizes the newly formed DNA in the 5′→3′direction.
- The length of DNA replicated in a single stretch is called a **replicon**.

Mechanism of nuclear DNA replication

The mechanism of DNA replication is similar in all eukaryotes and in many aspects resembles that of prokaryotic DNA.

- For example, it requires template DNA to copy, the four dNTPs, Mg^{2+}, and polymerases and other enzymes.
- As in prokaryotic cells, synthesis of the new DNA strands proceeds in the 5′→3′ direction.

➔ *You can read about prokaryotic DNA replication in Chapter 5 and the companion volume,* Thrive in Biochemistry and Molecular Biology.

Initiation of DNA replication in eukaryotes

- In eukaryotic linear DNA molecules, the replication of DNA occurs at *multiple* **origins of replication** (Figure 6.5) throughout the genome.
- These are rich in A–T base pairs and facilitate unwinding of the double helix because of their weaker hydrogen bonding compared to G–C pairings.

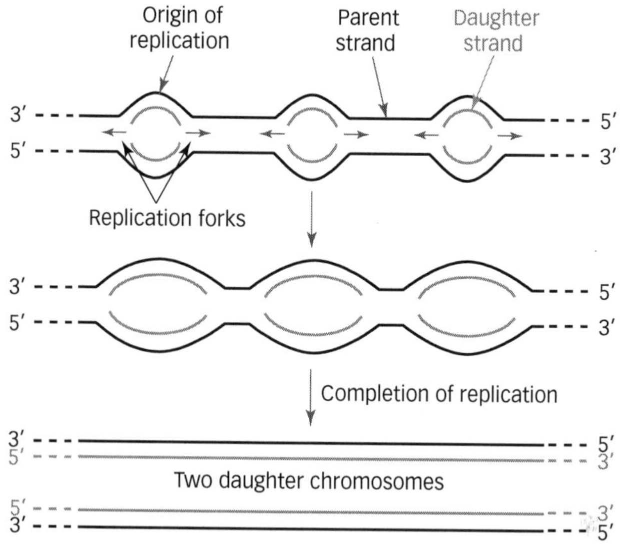

Figure 6.5 Multiple origins of replication (only three are shown) associated with the replication of eukaryotic chromosomal DNA. Note two replication forks are formed at each site. Compare with Figure 5.2.

- Replication of human DNA involves about 30,000 replication forks: multiple points of origin are essential if the huge eukaryotic DNA molecules are to be fully replicated in reasonable physiological times.
- By contrast, bacterial circular chromosomes have only a *single* point of origin (See Chapter 5, Figure 5.3) but archaea can have more than one.
- Each origin of replication allows two **replication forks** to be formed.

Eukaryotic prereplicative and preinitiation complexes

- Replication at each origin of replication (OoR) begins with the formation of a **prereplicative complex** (Figure 6.6).
- The OoR contains the binding site for a six subunit **origin recognition complex** (**ORC**) of proteins.
- Once bound, the ORC remains in place throughout the replication cycle.
- So-called **licensing factors** then bind to the ORC. These include **helicase loading proteins** that assist the binding of **helicases** (called MCM proteins in humans) to the site forming a **prereplicative complex** as shown in Figure 6.6.
- The origin is now said to be *licensed* for replication.
- A new prereplicative complex cannot form at this site until the current cell cycle is complete.
- Cyclin-dependent kinases (CDKs, see Chapter 16) then catalyse phosphorylations of components of the prereplicative complex leading to:

Replication of nuclear DNA

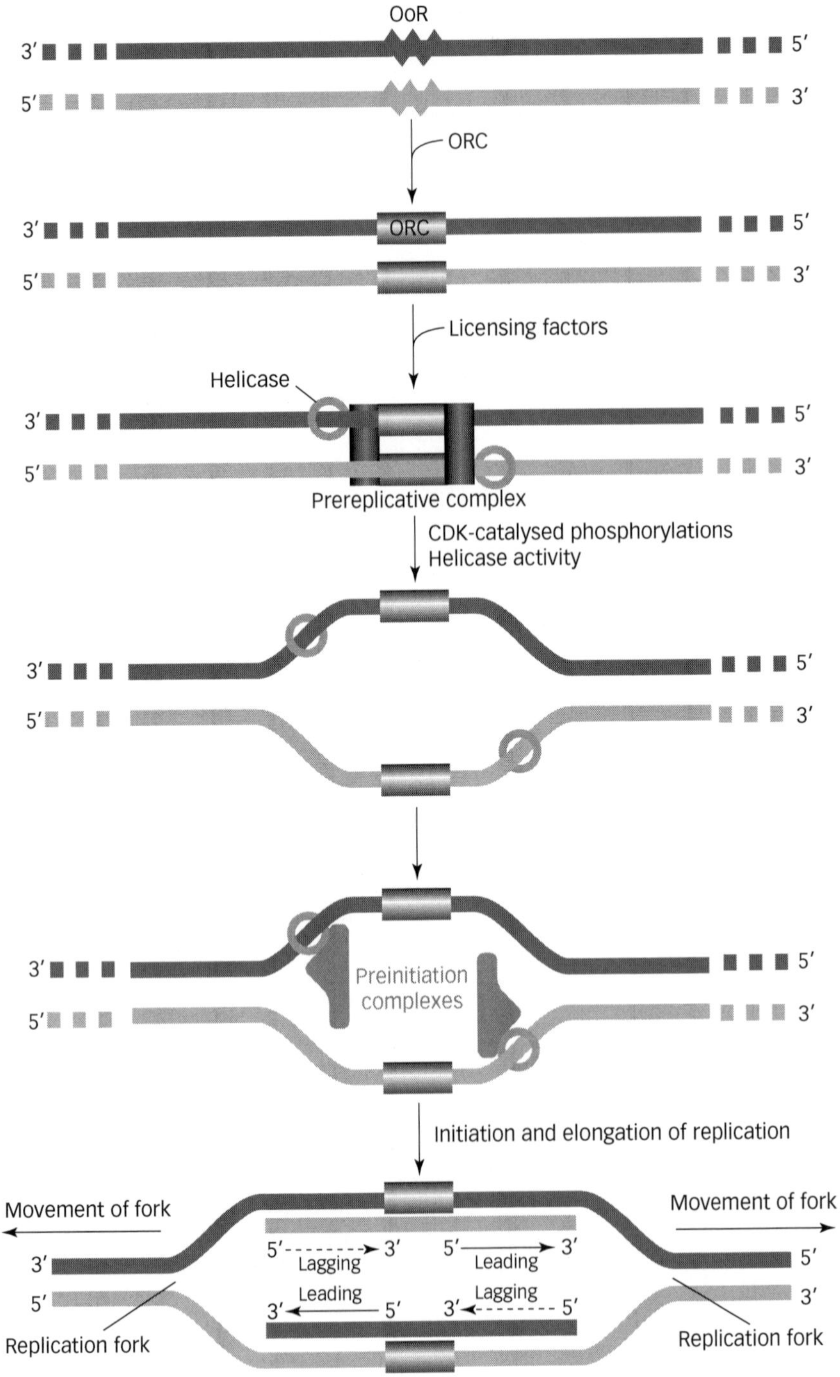

Figure 6.6 Schematic illustration showing the formation of eukaryotic prereplicative and preinitiation complexes at a eukaryotic origin of replication. OoR, origin of replication; ORC, origin recognition complex. See text for explanations.

  - the release of the helicase loading proteins
  - activation of the helicases
  - unwinding of the DNA at the origin.
- The separated DNA strands are stabilized by the binding of **replication protein A**, a **single stranded binding protein** (not shown in Figure 6.6 for reasons of simplicity).
- Unwinding allows other proteins involved in DNA replication to form **preinitiation complexes** at each putative replication fork (Figure 6.6).
- Each replication fork initiates the synthesis of complementary daughter DNA strands from the origin in both directions.

RNA primers and DNA replication

- Nine eukaryotic DNA polymerases are known and these are identified by Greek letters.
- Replication begins with the binding of **DNA polymerase** α to the replication fork. DNA polymerase α can initiate synthesis by producing a length of RNA about seven to ten ribonucleotides long called the **primer**, which is complementary to the parent strands.
- Remember in bacteria, the RNA primer is synthesized by the *RNA* polymerase activity of a **primase.**

➔ *Revise DNA replication in bacteria and archaea in Chapter 5.*

- The same polymerase (α) then extends the primer with a stretch of DNA about 20–30 deoxynucleotides long.
- **Polymerase switching**, which is the replacement of DNA polymerase α by DNA polymerase δ, then occurs.
- Switching occurs when:
  - DNA polymerase α is replaced by the protein, **proliferating cell nuclear antigen** (**PCNA**).
  - PCNA is an annular clamp that fits around the DNA double helix with a central hole sufficiently large to allow it to slide along the DNA strands (PCNA therefore has the same functions as the β protein in bacterial DNA replication (Chapter 5)).
  - DNA polymerase δ binds to the PCNA and is then free to copy the parental strand, extending the growing daughter chain for long stretches of nucleotides (a process usually referred to as **processive replication**).

Looking for extra marks?

Highly processive copying of DNA is only possible because sliding clamp molecules keep them firmly bound to the parental DNA.

Replication on the leading and lagging strands

- DNA polymerases can only synthesize DNA in the 5′→3′ direction.
- The **leading** strand is formed as a continuous length, but the other **lagging** strand is replicated as series of short lengths called Okazaki fragments (See Chapter 5, Figure 5.2 and Figure 6.6).
- Following initiation of the leading strand, DNA polymerase α moves to the replication origin on the lagging strand, where it acts with other proteins to prime the formation of Okazaki fragments.
- Replication continues in both directions until adjacent replicons meet.
- The RNA primers and the initial deoxynucleotides incorporated by DNA polymerase α are then enzymatically removed and the resulting gaps in the lagging strand are filled by the synthesis of DNA.
- The DNA polymerases other than α and δ, are thought to be involved in the processing of Okazaki fragments and the repair of DNA.
- The newly formed fragments of DNA are joined together by a **DNA ligase** to form continuous daughter strands.

DNA proofreading

Proofreading by DNA polymerase δ ensures an error rate of only one incorrectly incorporated nucleotide in 1 in 10^8–10^{10} during DNA replication. Proofreading is performed by the **exonuclease** ability of the DNA polymerase δ.

- Many DNA polymerases have such proofreading abilities that operate in the **3′→ 5′** direction—the *opposite* direction to that of DNA replication.
- If the last incorporated deoxynucleotide is incorrect (that is, is *not* complementary to the parental strand), the polymerase exonuclease activity is stimulated and the offending nucleotide removed.
- The DNA polymerase is then able to replace it with the correct deoxynucleotide in the 5′→3′ direction.

Looking for extra marks?

Despite the differences in detail in the ways DNA is copied in bacteria and eukaryotes note the many points of similarity.

Terminal ends of DNA replication

The terminal 5′ ends of the linear daughter DNA strands occur at the **telomeres** of each chromosome. Telomeres are highly repetitive sequences of DNA containing short G–C-rich repeated sequences.

- The actions of DNA polymerases and the use of RNA primers means the replicated daughter strands are always shorter than their parent templates.
- Telomeres are non-coding regions of DNA and their presence ensures that genetic coding regions are not lost during each DNA replication cycle. However, given

they progressively shorten at each successive DNA replication—and there is a limit to how short the telomeres can become before replication becomes impossible—there is therefore a limit to the number of mitotic cycles any cell can undergo.

Revision tip

Construct a simple table to summarize differences in DNA replication in prokaryotes and eukaryotes (hint: consult Chapter 5).

6.3 TRANSCRIPTION IN THE NUCLEUS

Transcription is the copying of a 3′→5′ strand of DNA, nucleotide by nucleotide, by a **DNA-dependent RNA polymerase** (usually abbreviated to **RNA polymerase**, **RNA pol**, or **RNAP**) to produce a 5′→3′ RNA molecule that is complementary to the template DNA sequence.

➔ *You can read about prokaryotic transcription in Chapter 5 and the companion volume,* Thrive in Biochemistry and Molecular Biology.

Key features of eukaryotic transcription

- Transcription in eukaryotes in some ways resembles that for prokaryotes (Chapter 5). Thus, transcription requires:
  - a template DNA strand running in 3′→5′ direction (antisense strand)
  - the four NTPs
  - Mg^{2+}
  - more polymerases and other enzymes.
- As in bacteria, the first base is designated +1. Bases towards the 5′ end or upstream are often identified using negative numbers; bases towards the 3′ or downstream end are given positive numbers.
- In eukaryotes, transcription is more complex than in prokaryotes with major differences including:
  - more types of RNA polymerases are involved
  - eukaryotic RNA polymerases are larger than those of prokaryotic cells (M_r greater than 600,000)
  - eukaryotic promoter sites differ from prokaryotic types
  - eukaryotic gene transcription involves enhancer sequences
  - initiation requires numerous transcription factors
  - the method of termination in eukaryotes differs.
- In addition, the initial RNA transcripts formed need modifying to become fully functional.

continued

- The best characterized eukaryotic RNA polymerases are RNA polymerase I/A, RNA polymerase II/B and RNA polymerase III/C, each of which is responsible for transcribing at least one class of RNA molecules (Table 6.2).
- In eukaryotes, RNA polymerase II is responsible for transcribing the genes of the precursors of mRNA molecules.
- Binding of RNA polymerase to transcription sites is controlled by proteins called **transcription factors.**
- Unlike the σ factor in bacteria (Chapter 5), transcription factors bind directly to the DNA *before* the RNA polymerase. It is the presence of these transcription factors that determines which gene is transcribed.
- Transcription factors may also bind to other proteins such as other transcription factors.
- Following the termination of transcription, processing at the 3′ end of the transcript determines its final length, rather than the point at which transcription is terminated.

Eukaryotic promoter regions

Eukaryotic promoters can be divided into three main groups, one group for each RNA polymerase respectively. This section will concentrate on the promoter region for RNA polymerase II.

- Eukaryotic promoters can be upstream or downstream of the transcription start; not only do they vary between the various RNA polymerase types, there is also variation within promoter types.
- Eukaryotic promoters are similar to those of bacterial cells in consisting of two regions:
    - a **TATA box** region at approximate position −25 bases from the start codon with the consensus sequence TATAAAA
    - many promoter regions also contain **CAAT** and **GC** boxes (consensus sequences GGNCAATCT and GGGCGG respectively) approximately 40–150 bases upstream of the initiator site.
- Unlike prokaryotes, eukaryotic transcription factors bind to the TATA box and surrounding area to facilitate the binding of the RNA polymerase and so ensure transcription begins at the start site (+1).

Name	*Number of subunits*	*Synthesizes*
RNA pol I/A, Pol I	14	Most pre-rRNAs (precursors of rRNAs)
RNA pol II/B, Pol II	12	Pre-mRNAs (precursors of mRNAs)
RNA pol III/C, Pol III	15	Precursors of 5S rRNA, tRNAs, numerous sn- and scRNAs

Table 6.2 Major eukaryotic RNA polymerases (examples from the yeast, *Saccharomyces cerevisiae*)

Transcription factors

- Transcription factors are proteins that help RNA polymerases to bind to promoter regions. Transcription factors that are *always* used to help *initiate* transcription are called **general** or **basal** transcription factors.
- Transcription factor IID (TFIID) is the principal component for initiating transcription because it contains a subunit that can recognize the TATA box. The subunit is known as the **TATA-binding protein** (**TBP**).
- TATA-binding protein can also bind to promoters without TATA box sequences through protein–protein interactions.
- Other specific transcription factors recognize and bind to DNA sequences called **enhancers**, which stimulate RNA polymerase II activity and increase the effectiveness of promoters. The positions of enhancers are not fixed relative to the gene promoter affected.
- Enhancers are cell specific: one type of enhancer will only work in one particular type of cell.
- Transcription factor IID is joined by other transcription factors and RNA polymerase II on the DNA to form a **preinitiation complex** (Figure 6.7).
- Initiation of transcription occurs when other transcription factors, such as TFIIH, bind to the preinitiation complex and cause the ATP-dependent phosphorylation of the RNA polymerase.
- Phosphorylation of RNA polymerase changes its conformation and causes it to separate from the complex of transcription factors. Transcription is initiated and **elongation** of the RNA transcript by the polymerase can occur.
- Elongation proceeds in a so-called **transcription 'bubble'** where the DNA is unwound as the RNA polymerase moves along it. This is similar to the situation in bacteria (see Figure 5.10).
- The DNA is rewound once that portion has been transcribed.
- **Termination** of transcription occurs approximately 10–35 nucleotides downstream of a polyadenylation sequence (AAUAAA) on the RNA transcript. Hydrolysis at this site by a specific endonuclease releases the RNA transcript and leaves it with a free 3′-OH.
- The site of cleavage is later subjected to the addition of a poly(A) tail (see 'Addition of poly(A) tail' section).

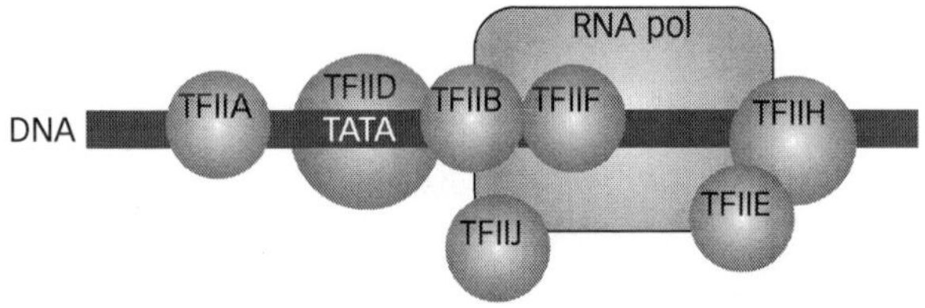

Figure 6.7 Schematic illustration of a eukaryotic transcription preinitiation complex.

- Dephosphorylation of the RNA polymerase causes its release from the DNA, and so it can participate in the formation of another preinitiation complex.

Check your understanding

6.5 Which of the following is/are true?

a. The first base in DNA to be transcribed is designated +1; bases towards the 5′ end or downstream stream are identified using negative numbers, bases towards the 3′ or upstream end are given positive numbers.

b. Eukaryotic RNA polymerases I/A, II/B, and III/C each transcribe at least one class of RNA molecules.

c. Eukaryotic promoters can be upstream or downstream of the transcription start.

d. The binding of transcription factor IID to the promoter on DNA forms a preinitiation complex.

e. Termination of transcription occurs approximately 10–35 nucleotides downstream of a poly(A) tail.

6.4 PRE-mRNA PROCESSING

Unlike prokaryotic mRNA, which is formed fully functional, the eukaryotic pre-mRNA formed by RNA polymerase II *does* require modification following its transcription. Other types of initial transcripts also need modifying but this section will concentrate on the changes to pre-mRNA transcripts that convert them to functional mRNAs.

- Pre-mRNA processing, also called post-transcriptional modifications includes:
    - capping
    - addition of a poly(A) tail
    - the removal of introns.
- Such processing of the pre-mRNA removes regions of RNA that are not required for translation, protects transcripts from degradation and allows them to enter the cytoplasm.

Information about translation can be found in Chapters 5 and 7.

Looking for extra marks?

In prokaryotes, both transcription and translation occur in the cytoplasm and are not separated temporally, thus extensive post-transcriptional activities are not possible.

Capping

- **Capping** is the addition of a methylated guanosine residue by its 5′ end to the 5′ end of the pre-mRNA forming a 'nonstandard' 5′-5′ phosphodiester bond (Figure 6.8).
- The 5′ cap:
  - protects the mRNA from 5′ exonuclease activity, which degrades foreign RNA
  - facilitates the transport of some mRNAs from the nucleus (see 'Active nucleocytoplasmic transport' section)
  - helps in ribosomal binding during translation (Chapter 7).

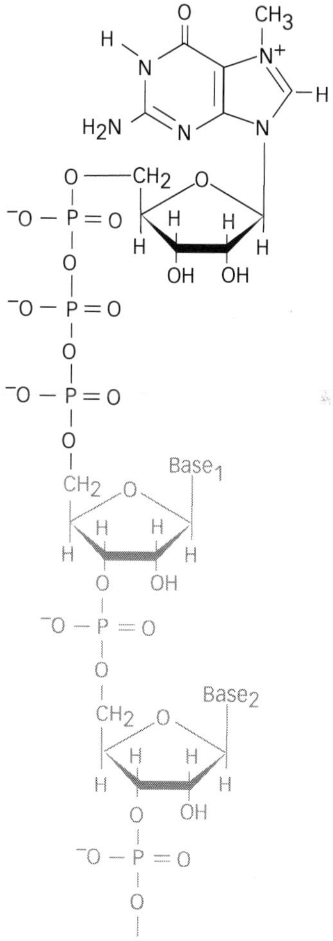

Figure 6.8 The 5′ cap of eukaryotic mRNA. See text.

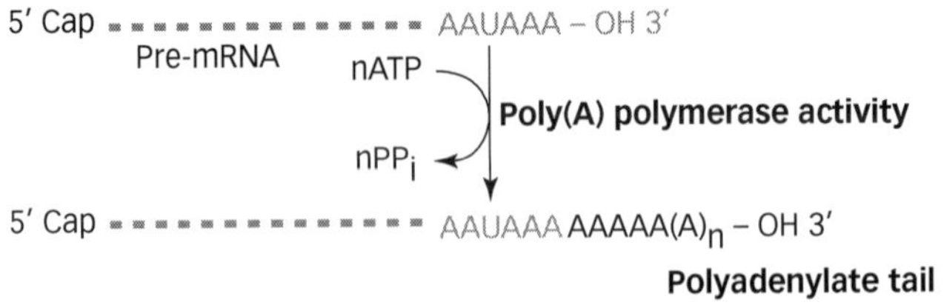

Figure 6.9 Polyadenylation. See text.

Addition of poly(A) tail

- The addition of a poly(A) tail, also known as **polyadenylation**, is the sequential adding of up to 200 adenine residues to the transcription termination site on the 3′ end of the RNA transcript (Figure 6.9).
- The poly(A) tail has a number of functions:
  - it protects the transcript from 3′ exonucleases
  - it assists in the export of some RNA molecules from the nucleus
  - its length can affect the life of the mRNA: the longer the tail, the longer the life span.

Splicing

- Eukaryotic cells have genes in which the coding regions or **exons** are separated from one another by non-coding regions called **introns**.
- Thus, pre-mRNA transcripts of eukaryotic **discontinuous genes** have coding and non-coding regions that require **splicing:** the removal of the introns from pre-mRNA and the joining together of the newly formed ends of the RNA molecules to give a continuous length of mRNA (Figure 6.10).
- Splicing follows the addition of the 5′ cap and the poly(A) tail.
- Splicing is performed by RNA-protein complexes called **spliceosomes**, which are composed of **small nuclear ribonucleoproteins** (**snRNPs**) formed from **small nuclear ribonucleic acids** (**snRNAs**) and proteins.

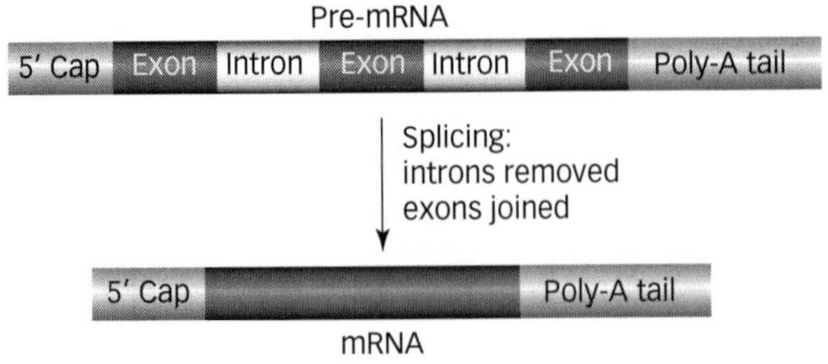

Figure 6.10 Schematic illustration of RNA splicing showing the removal of introns from a pre-mRNA transcript to produce a continuous mRNA transcript.

- There are five separate snRNPs called U1, U2, U4, U5, and U6.
- The positions on the pre-mRNA transcript at which the introns are spliced are called **splice sites**. Splice sites are located at either end of the intron.
- The 5′ splice site usually starts with a GU sequence and the 3′ splice site usually ends with an AG sequence.
- Introns also contain a sequence of nucleotides called a **branch site**, which is located at a point within the intron.
- The mechanism of splicing starts with binding of the U1 snRNP to the 5′ cleavage site of the intron, followed by the binding of a second snRNP, U2, to the branch site.
- The remaining snRNP units, U4, U5, and U6, then bind to U1 and U2, which causes the 5′ end of the intron to be pulled towards the branch site.
- The 5′ end of the intron is cleaved by a catalysed reaction and bound to the branch site to form a looped structure known as a **lariat** (Figure 6.11).
- The 3′ end of the intron is then catalytically cleaved and the free ends of the exons ligated together.
- The excised intron is then released for degradation.
- Thus, the reactions involved in splicing are catalysed by the RNA components of the spliceosome.
- Catalytic RNAs are called **ribozymes**.
- Although most RNA splicing involves a spliceosome to catalyse the cleavage reactions, some introns are self-splicing: removal of introns and ligation of exons is performed without the need of a spliceosome.

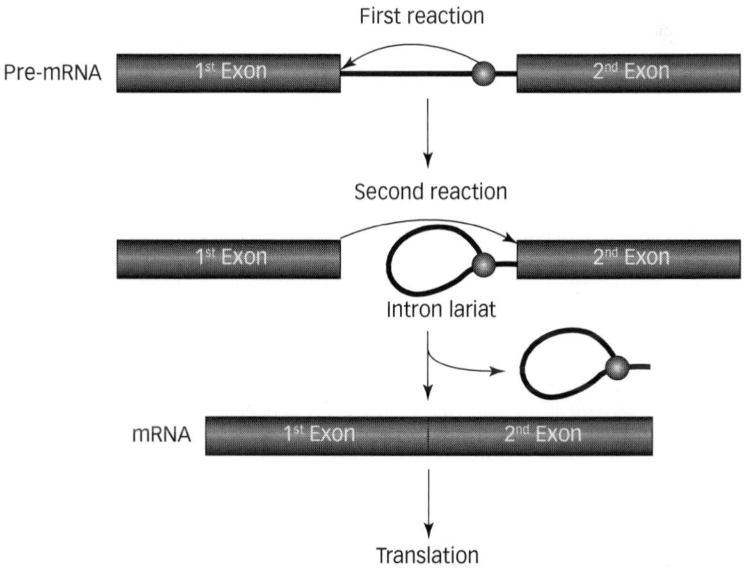

Figure 6.11 Overview of the mechanism of splicing. See text for explanation.

- Differential splicing, in which different combinations of introns are removed from a pre-mRNA transcript, means a variety of products can be produced from a single gene.

Revision tip

Construct a simple table to summarize differences in DNA transcription and post-transcriptional activities in prokaryotes and eukaryotes (hint: consult Chapter 5).

6.5 NUCLEOLI AND OTHER INTRANUCLEAR INCLUSIONS

Nucleoli (singular **nucleolus**) are prominent structures present in the nucleus (Figures 6.1, 6.3, and 6.12). At least one is present in all nuclei.

- Nucleoli are typically several μm in diameter but their size varies: they are larger in active cells; smaller in less active ones.
- The nucleolus is a dynamic structure that breaks down prior to mitosis but is reformed in the daughter cells.
- Nucleoli are formed in part from **the nucleolar-organizing** regions (**NORs**) of **nucleolar-organizing chromosomes** (**NOCs**), which are typically situated near the ends of NOCs.
- Humans have five NOCs: chromosomes 13, 14, 15, 21, and 22. However, the ten NORs in a diploid cell associate to form a single nucleolus.
- Nucleolar-organizing regions contain multiple copies of numerous genes for pre-rRNA.

Structure and functions of nucleoli

Nucleoli consist of several distinct regions of fibres and granules, that are distinguishable by electron microscopy (Figure 6.12), forming a central **fibrillar core** surrounded by a **granular region**.

- The fibrillar core contains the genes (NORs) that are transcribed by RNA polymerase I/A (Table 6.2) to give a single 45 S pre-ribosomal RNA transcript. This precursor is modified *in situ* to give functional 5.8 S, 18 S, and 28 S rRNA molecules.
- The 5 S rRNA molecule is synthesized by RNA polymerase III/C outside the nucleolus and imported into it during the formation of ribosomal subunits.

➔ *You can read about the structures of eukaryotic ribosome subunits in Chapter 7 and in the companion volume,* Thrive in Biochemistry and Molecular Biology.

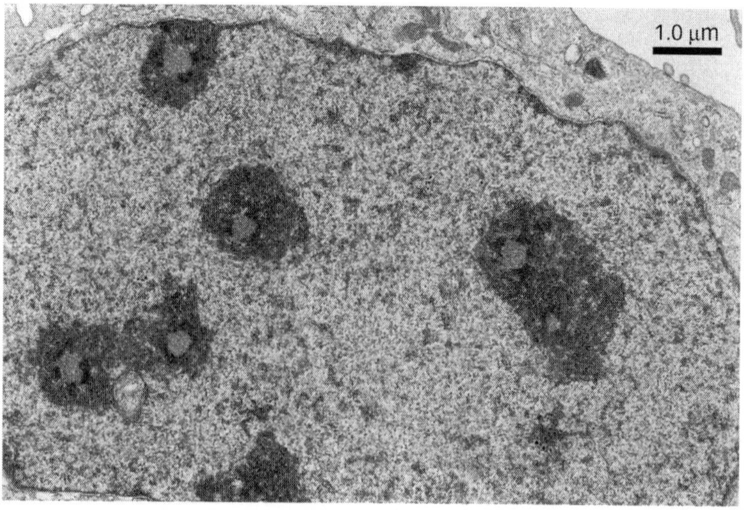

Figure 6.12 Electron micrograph of nucleoli in a bladder carcinoma cell. Note the different regions in each nucleolus. Courtesy of Professor U. Scheer and Dr D. Weisenberger, Department of Cell and Developmental Biology, University of Würzburg, Germany.

- The granules consist of rRNA molecules combined with ribosomal proteins to form ribosome subunits.
- Nucleoli also contain numerous non-ribosomal molecules. These include **nucleolin** and **small nucleolar RNA particles** (**snoRNPs**).
  - Nucleolin is the major protein present in the nucleolus. It may control transcription at the NORs.
  - Small nucleolar RNA particles are involved in transcription and RNA processing, and in regulating the assembly of ribosome subunits.
- Once formed, the ribosomal subunits are transported out of the nucleus to perform translation in the cytoplasm.

➔ *You can read about translation in bacterial cells and in eukaryotes in Chapters 5 and 7 respectively and in the companion volume,* Thrive in Biochemistry and Molecular Biology.

- Evidence is accumulating that the nucleolus also functions in regulating mitosis, cell cycle progression, and cell proliferation (Chapter 16).

Other intranuclear structures

In addition to the nucleolus, the nucleus encloses a number of non-membranous structures:

- Cajal bodies
- gemini of Cajal bodies
- speckles or interchromatin granule clusters
- promyelocytic leukaemia (PML) bodies.

- Cajal bodies were named after their discoverer, and gemini of Cajal bodies (GEMs) because of their similarity to them. Both are involved in the processing and maturation of small nuclear RNA (snRNA) and small nucleolar RNA (snoRNA) molecules.
- Speckles contain the proteins and RNA molecules necessary to catalyse the appropriate splicing activities essential to convert pre-messenger RNAs to functional mRNA molecules.
- Little is known of the activities of PML bodies.

Check your understanding

6.6 Which of the following statements is/are false?

a. Nuclei, nucleoli, and other intranuclear inclusions are membrane bound.
b. The number of nucleoli in a nucleus is numerically equal to the number of NOCs and NORs.
c. The number of NOCs and NORs is numerically equal.
d. Fully functional ribosomes are formed in the nucleolus.
e. Histones complex with rRNA to form ribonuclear proteins.

6.6 NUCLEAR ENVELOPE AND NUCLEAR PORES

The nuclear envelope protects the genome but presents a problem to the formation of chromatin and the replication and transcription of DNA since essential proteins and other molecules and ions must enter the nucleus, and its products, such as RNA molecules and ribosomal subunits, must leave.

- The nuclear envelope mediates the exchange of materials between the nucleoplasm and the cytoplasm.
- The nuclear envelope consists of **outer** and **inner nuclear membranes** separated by a **perinuclear space**, approximately 20–40 nm across (Figures 6.3 and 6.13).
- The outer membrane is continuous with that of the endoplasmic reticulum (ER) and the perinuclear space is continuous with the lumen of the ER (Chapter 8).
- Like the rough endoplasmic reticulum, the outer surface of the nucleus is often studded with ribosomes.
- The envelope contains numerous **nuclear pores**.

Nuclear pores and nuclear pore complexes

Nuclear pores are channels in the envelope where the inner and outer nuclear membranes fuse together.

- Nuclear pores are lined by elaborate structures called **nuclear pore complexes (NPCs)**.

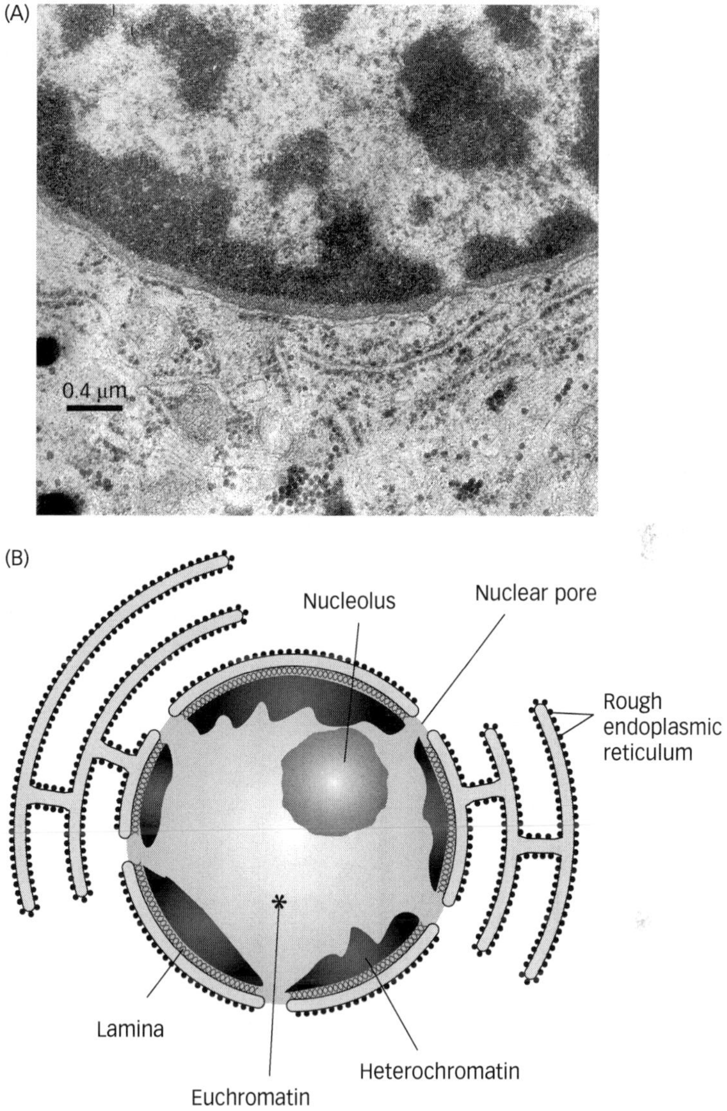

Figure 6.13 (A) Electron micrograph of portion of a nucleus. Courtesy of Professor D.W. Fawcett. (B) Schematic illustration of nucleus.

- The 5000–6000 pores of a typical mammalian nucleus are held in place by attachments to the nuclear envelope and to the nuclear lamina (see 'Nuclear skeleton' section).
- The best-studied NPCs are found in amphibian eggs, such as those of *Xenopus laevis*.
- Figure 6.14 shows the structure of amphibian NPCs as revealed by electron microscopy studies and the generalized structure of a single NPC.

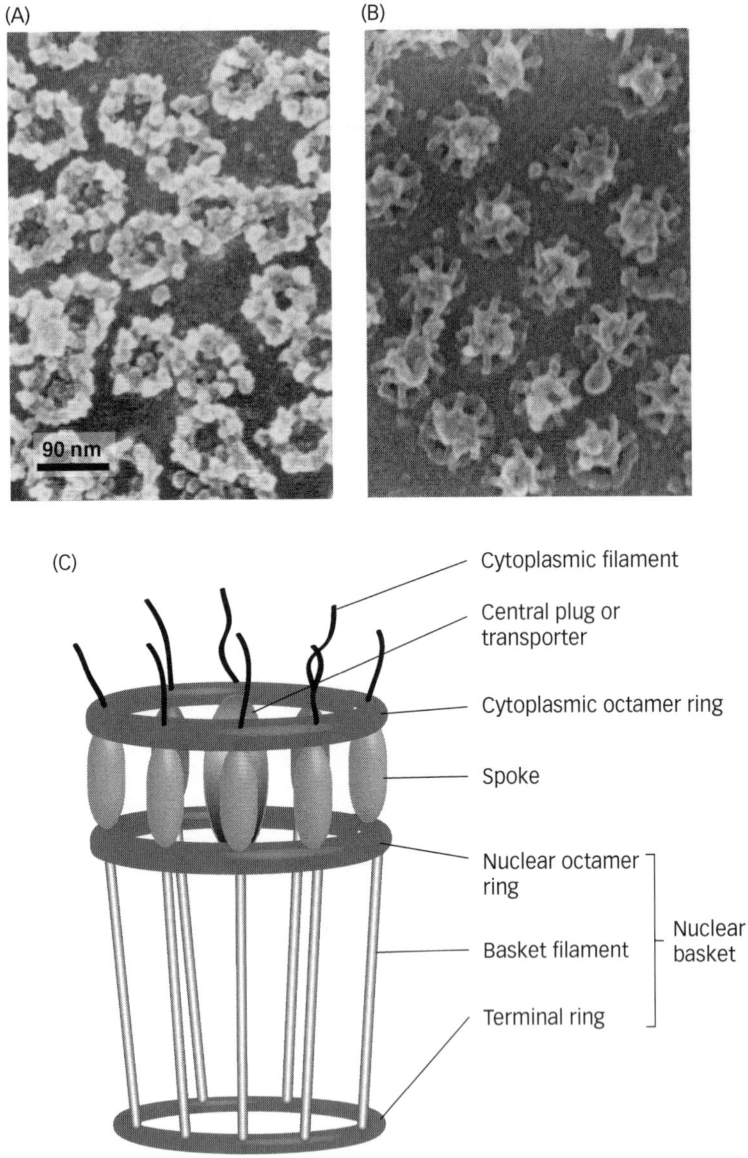

Figure 6.14 Electron micrographs showing the (A) outer (cytoplasmic) and (B) inner (nucleoplasmic) surfaces of nuclear pore complexes. Courtesy of Drs T.A. Allen and M.W. Goldberg, CRC Department of Structural Biology, Paterson Institute, Manchester. (C) Schematic representation of a nuclear pore complex. See text for an explanation of the structure.

- Nuclear pore complexes have ring-like structures about 80–120 nm in diameter and M_r of approximately 125×10^6; they comprise about 30 different types of proteins called **nucleoporins** (**nups**) depending upon the species.
- Nuclear pore complexes have an eightfold symmetry that can be readily seen in Figure 6.14.
- Thus, NPCs have eight, or a multiple of eight, copies of each nup, resulting in there being about 500–1000 nups in each pore.
- Protein fibrils extend from the outer surface of the NPC about 50–100 nm into the cytoplasm.
- In contrast, a basket-like structure is attached to the interior surface of the inner membrane and extends into the nucleoplasm.
- Nuclear pores and their complexes facilitate the movement of ions and small molecules and regulate the transport of macromolecules through the envelope.
- Ions and small molecules include Mg^{2+}, and dNTPs and NTPs used in replication and transcription of DNA respectively.
- Proteins imported into the nucleus include histones to form chromatin and the polymerases for replicating DNA and transcribing RNA, and transcription and splicing factors.
- Most RNA molecules, such as mRNAs and tRNAs, are exported from the nucleus in the form of RNA-protein complexes, usually called **ribonucleoprotein**. In addition, ribosomal subunits formed in the nucleolus are transferred to the cytoplasm through NPCs.
- Proteins may also be exported from the nucleus.

Nucleocytoplasmic transport

The movement between the nucleus and cytoplasm is generally called **nucleocytoplasmic transport**.

- Nuclear pore complexes are the sites of intensive activities. For example, histones are synthesized on ribosomes in the cytoplasm at a rate of approximately 300,000 per minute during the replication of chromosomes, and so must move through an individual NPC into the nucleus at rates of one to two per second.
- Ribosomal subunits are formed in the nucleolus (described earlier). Given a growing mammalian cell produces 10,000-or-so ribosomes per minute, then approximately 20,000 ribosomal subunits must be exported from the nucleus to the cytoplasm through the pores each minute.
- In addition to mediating the *active* transport of larger macromolecules and macromolecular complexes such as ribosomal subunits, NPCs also facilitate the transport movements of smaller molecules and ions by *diffusion*.

Nucleocytoplasmic transport by diffusion

Studies have suggested that NPCs have aqueous channels of approximately 9 nm diameter through which ions and small molecules can freely diffuse.

Nuclear envelope and nuclear pores

- The study of the movement of differently sized colloidal gold particles in cells using electron microscopy showed that particles of diameter less than 9 nm (equivalent diameter to a globular protein of M_r 40,000) are able to enter the nucleus and reach equilibrium concentrations between it and the cytoplasm. Smaller particles entered most rapidly, but particles with diameters greater than 10 nm did not enter the nucleus.
- Experiments using labelled proteins showed similar results. Smaller proteins (M_r up to about 15,000) rapidly reach equilibrium concentrations between the nucleoplasm and cytoplasm, but proteins of M_r up to about 50,000 equilibrate slowly. Those of about 60,000 barely enter the nucleus, while proteins of M_r 115,000 and above are excluded.
- It has been *suggested* that the aqueous diffusion pores may be the channels that occur near the periphery of the NPC between its eight spokes (Figure 6.14).
- In contrast to that of smaller molecules, the movement of larger particles through NPCs is by active transport probably through the central pore of the NPC. This channel seems able to accommodate the active passage of large complexes, up to 26 nm diameter, and includes ribosomal subunits.

Active nucleocytoplasmic transport

Active nucleocytoplasmic transport processes are rather complicated and only simplified views are given here. Both import (transport from cytoplasm to nucleus) and export (transport from nucleus to cytoplasm) are mediated by proteins called **importins** and **exportins** (also called **nuclear import/export receptors** or **karyopherins**) respectively.

Active cytoplasm to nucleus transport

- Proteins that enter the nucleus by active transport require a **nuclear locating signal** (**NLS**).
- The best characterized NLSs are rich in arginine and lysine residues; the large T antigen of the simian virus (SV) 40 has the sequence: …Pro–Lys–Lys–Lys–Arg–Lys–Val …
- When the SV 40 virus infects cells it induces the production of the viral T protein on ribosomes in the cytoplasm. The protein then enters the nucleus and promotes the synthesis of viral DNA.
- SV 40 mutants that produced a protein antigen incapable of entering the nucleus allowed the NLS sequence of the protein to be identified.

Looking for extra marks?

Numerous NLSs have subsequently been identified and many resemble that of the SV 40 antigen. However, others are chemically different.

- The active uptake of proteins into the nucleus is illustrated in Figure 6.15.
- Transport requires the participation of an importin that is able to recognize the NLS.
- A protein with an NLS is called a **cargo**. The best characterized importins belong to a class called β importins.
- The importin recognizes the NLS signal of the cargo and binds to it (Figure 6.15). The resulting complex can dock with an NPC by binding to receptors on its fibrils.
- The cargo–importin complex is then actively transferred through an NPC into the nucleus where it binds to a small protein called **Ran**.
- Ran is a **G-protein** with GTPase activity, which can bind and then hydrolyse the bound GTP to GDP. The conformation of Ran depends upon whether it has a

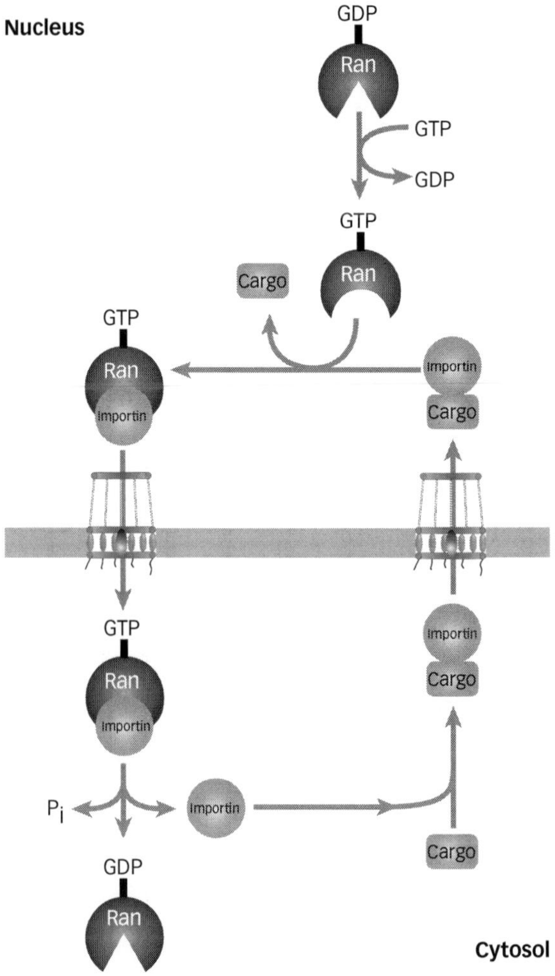

Figure 6.15 Active uptake of protein through an NPC. See text for explanation.

bound GTP or GDP. If GTP is bound, it is in its active conformation; if GDP then it is inactive.

➔ *More information about G proteins can be found in Chapters 7, 8, and 15.*

- The binding of Ran-GTP to the complex releases the cargo for use in the nucleus. The importin–Ran-GTP complex is then returned to the cytoplasm through a nuclear pore.
- Within the cytoplasm **Ran-GAP** (**GTPase-accelerating protein**) stimulates Ran to hydrolyse its bound GTP to GDP.
- The conformation of Ran-GDP has only a low affinity for the import receptor, which is then released and can participate in a new cycle of nuclear import.
- The Ran-GDP is returned to the nucleus following its binding to a specific **nuclear transport protein** (**NFT2**). In the nucleus it is reconverted to Ran-GTP under the influence of a specific **guanine nucleotide-exchange factor** (**Ran-GEF**).

Check your understanding

6.7 How many histone molecules are transported per second through an NPC, if 20,000 are formed in the cytoplasm per minute and the nucleus has 6000 pores?

Active nucleus to cytoplasm transport

- The active export of cargoes from the nucleus appears to involve similar mechanisms to that described earlier for nuclear import (Figure 6.16).
- The export of RNA molecules involves the participation of ribonucleoprotein complexes, whose protein portion contains a **nuclear export signal** (**NES**).
- The NES is recognized by an exportin that binds to it; a process promoted by GTP-Ran. The cargo–exportin-GTP–Ran complex is then transported through an NPC into the cytoplasm.
- Within the cytoplasm, Ran-GAP stimulates the GTPase activity of Ran, which converts it to the Ran-GDP conformation. The Ran-GDP conformation has limited affinities for the exportin and cargo, which are released into the cytoplasm. The exportin is then returned to the nucleus through an NPC.

The Ran cycle

- Figures 6.15 and 6.16 show how the hydrolysis of Ran-GTP to Ran-GDP provides the free energy needed to drive active transport through NPCs. It also shows that Ran directs the direction of transport, in a manner that has been called the *Ran cycle*.
- The direction of transport is governed by the concentration gradient of Ran-GTP across the nuclear envelope: highest in the nucleoplasm, lowest in the cytoplasm.

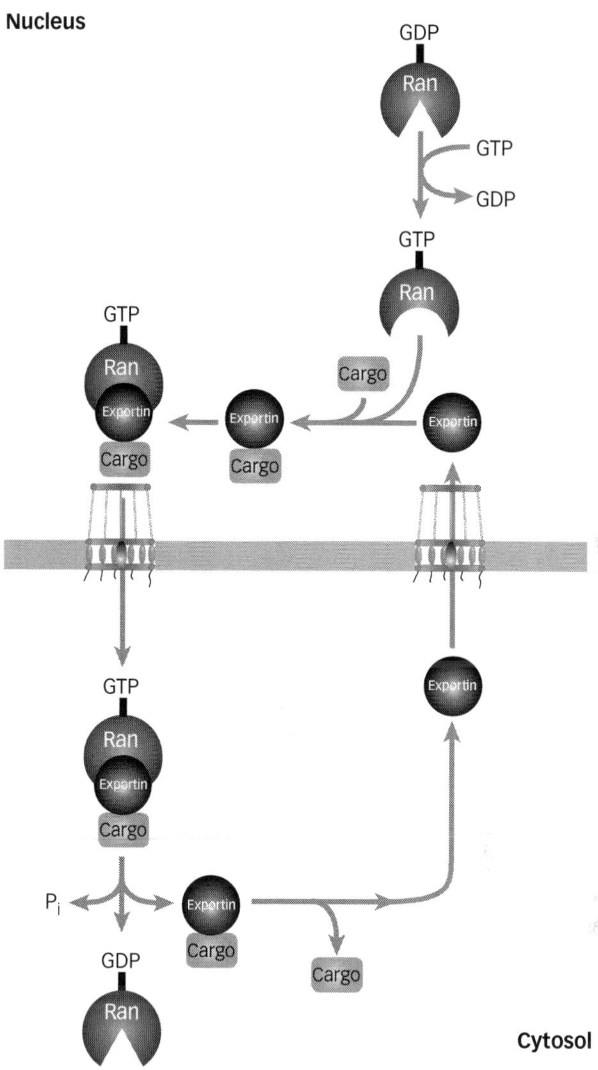

Figure 6.16 Active export of cargoes from the nucleus to cytoplasm. See text for explanation.

- The relatively high concentrations of Ran-GTP in the nucleoplasm are promoted by the action of Ran-GEF, while Ran-GAP ensures its concentration in the cytoplasm is much lower.
- The relatively high concentration of GTP in the nucleus encourages the release of NLS-containing cargoes but promotes the binding of NES-containing cargoes to exportins.
- Thus, the *direction* of transport is determined by the appropriate type of signal.

Check your understanding

6.8 If a cytoplasmic protein, such as hexokinase was modified by attaching an NLS to it, where in the cell would it accumulate at (a) 37°C and (b) 4°C?

6.7 NUCLEAR SKELETON

Nuclei are complex structures that are organized and held together by a **nuclear skeleton**.

- The nuclear skeleton consists of a proteinaceous nuclear **matrix** or **scaffold** that becomes visible when most of the nucleoplasm has been removed by detergent extraction to leave only the most insoluble residue.
- The matrix consists of protein fibres distributed throughout the nucleus, and may possibly:
    - constrain loops of DNA that contain five-or-so genes
    - constitute sites where the protein complexes involved in replication bind.
- Electron microscopy of nuclei shows a densely staining layer of fibrous proteins of the **intermediate type** (Chapter 7) called the **nuclear lamina** (Figure 6.13). These form a 10–40 nm thick layer that lines the inner face of the inner nuclear membrane, other than where nuclear pores occur.
- The lamina is composed of three proteins called **lamins A**, **B**, and **C**.
- The functions of the lamina include:
    - conferring mechanical strength to the nucleus
    - forming attachment sites for nuclear pores and probably also for chromatin fibres.

6.8 EVOLUTION OF THE NUCLEUS

The evolution of eukaryotic cells is the subject of much study and speculation.

- Evolutionary pressure leading to the nucleus could include the need to isolate splicing and other forms of RNA processing. However, organelles would appear to have arisen by a number of mechanisms.
- Chloroplasts and mitochondria almost certainly originated from endosymbionts in the evolving proeukaryote (Chapters 10 and 11).
- The nucleus, like chloroplasts and mitochondria, is surrounded by an envelope. However, its envelope is a single topological system that is continuous with the ER.
- This has led to the suggestion that the nucleus and the ER arose from invaginations of the cell surface membrane of the evolving eukaryotic cell.
- Other researchers believe that endosymbionts are involved, with different scenarios suggesting bacteria and archaea are implicated in the evolution of the nucleus (see Chapter 1).

Revision tip

Do browse through the further reading references given for this chapter online. Go to http://www.oxfordtextbooks.co.uk/orc/thrive/

Check your understanding

6.9 Which of the following statements is/are true or is/are false?

a. Chromosomes are randomly distributed in the nucleus in spaces called chromosome territories.

b. The packing ratio describes the degree of shortening of DNA during its packaging.

c. The beads on a string and the 30 nm fibre forms of chromatin have approximate packing ratios of 42 and 7–10 respectively.

d. The TATA-binding protein is found in Transcription factor II D.

e. Separated strands of DNA are stabilized by the binding of replication protein A, a single-stranded binding protein.

f. The leading strand in DNA replication is formed in the 5′→3′ direction.

g. During DNA replication, Okasaki fragments are formed in the 3′→5′ direction.

h. Exonuclease activity of DNA polymerase δ occurs in the 3′→5′ direction.

i. The 5′ cap on transcribed mRNA facilitates the transport of some mRNAs from the nucleus.

j. NFT2 is a specific nuclear transport protein that returns Ran-GTP to the nucleus.

7 The cytosol and cytoskeleton

Key features of the cytosol

- The composition of the cytosol is complex. It is a compartment where many metabolic reactions, protein synthesis and signal transduction pathways, and muscle contractile events occur.
- In some tissues, it is a dedicated site for storing materials, particularly energy reserves; it also contains numerous suspended ribosomes and organelles.
- The cytosol contains the cytoskeleton, a dynamic network of insoluble protein fibres that permeates the cytosol, supporting and imparting structural organization to it.
- The cytoskeleton is responsible for maintaining and changing the shape of the cell and for transporting organelles and vesicles throughout the cytoplasm.

7.1 COMPOSITION OF THE CYTOSOL

The cytosol is the gel-like, semifluid material that surrounds the membranous organelles and particles of the cytoplasm.

- It is composed of about 70% water. However, a significant proportion of the water is not unstructured or 'bulk' water but is associated with macromolecules and so is 'organized'.

- Approximately 5% of the water is bound to the large surface areas of protein and other macromolecules by ionic interactions. Other water molecules may form hydrogen-bonded clathrate structures surrounding macromolecules.
- Approximately 20–40% of the water may be bound to the insoluble protein fibres of the cytoskeleton (see 'The cytoskeleton' section).
- The cytosol contains a large number of small molecules (*Mr* less than 300) and ions. These maintain the ionic strength of the cytosol and provide a constant pH.
- The pH is buffered to approximately 7.2 (range 7.0–7.4), which is maintained by Na^+-linked antiporters in the plasma membrane (Chapter 3).
- The proteins of the cytosol form a 20–30% solution. This is extremely concentrated for proteins and gives it its gel-like consistency.
- Many of the proteins are enzyme molecules; others are concerned with regulating cellular processes such as metabolism, protein synthesis, signal transduction, or are soluble subunits of the cytoskeleton.

Looking for extra marks?

Cytosol is extremely crowded with molecules. Solutes therefore often do not behave as they would in dilute solutions.

Cytosolic pH

The high concentration of extracellular Na^+ relative to that in the cytosol allows two antiporters to increase the cytosolic pH.

- A Na^+/H^+ antiporter in the plasma membrane directly decreases the acidity (increases pH) of the cytosol by pumping out an H^+ in exchange for the influx of an Na^+.
- An Na^+-driven Cl^-/HCO_3^- exchanger pumps Cl^- and H^+ out of the cytosol coupled with an influx of Na^+ and HCO_3^-: effectively an HCl has been exchanged for $NaHCO_3$, increasing the pH of the cytosol.
- A third antiporter, the Na^+ independent Cl^-/HCO_3^- exchanger, decreases cytosolic pH by promoting the efflux of an HCO_3^- in exchange for a Cl^- entering the cell.
- The value of the cytosolic pH regulates the activity of all three antiporters leading to an increase or decrease in their activities as appropriate.

Check your understanding

7.1 Both the Na^+/H^+ antiporter and the Na^+-driven Cl^-/HCO_3^- exchanger reduce the acidity of the cytosol. Which of the two is the more efficient in this respect?

Storage in the cytosol

The cytosols of specific cell types store a variety of inclusions, including:

- Glycogen granules in hepatocytes (Figure 7.1) and skeletal muscle cells.
- Triacylglycerols in cells of white and brown adipose tissues (see Figure 14.15).

7.2 KEY FUNCTIONS OF THE CYTOSOL

The cytosol has a multiplicity of functions.

- It is the site of numerous metabolism pathways.
- It is involved in the transport of metabolites.
- It is the site of protein biosynthesis.
- It is involved in signal transduction.
- It forms the cytoskeleton.

Metabolism and the cytosol

Metabolism is the integrated network of reactions that support the life of the cell or organism.

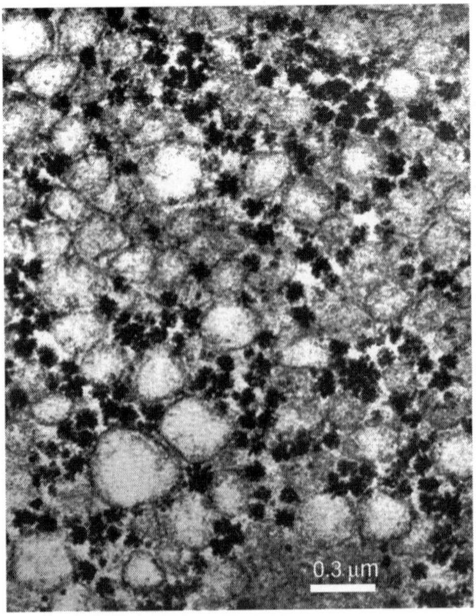

Figure 7.1 Electron micrograph of a portion of a liver cell showing numerous heavily stained glycogen granules. Reprinted from Cardell, R.R. and Cardell, E.L. (1990) Heterogeneity of glycogen distribution in hepatocytes. *Journal of Electron Microscopy Technique*, 14, 126–39.

- A large proportion of the metabolism of eukaryotes cells occurs in the cytosol (most metabolism in prokaryotes occurs in their cytoplasm).

➔ *You can read more about metabolism in the companion volume,* Thrive in Biochemistry and Molecular Biology.

- **Primary metabolism** is those metabolic pathways essential to the life of the organism. These are largely concerned with maintaining the chemical energy potential in the form of appropriate concentrations of ATP and the reduced coenzymes $FADH_2$ and NADH, and biosynthetic reducing power in the form of NADPH.
- Major metabolic pathways that occur in the cytosol are **glycolysis** (see Chapters 4 and 12), the pentose phosphate pathway, gluconeogenesis, and the synthesis of fatty acids (Chapter 8).
- The cytosol is the site of much **secondary metabolism**, the thousands of individual reactions that constitute the metabolic pathways that produce many of the small molecules needed by cells but which are not essential for life.
- Plant secondary metabolism is much more extensive than that seen in animals, and often produces minute amounts of a large array of small molecules such as antibiotics, insect deterrents, for example, various alkaloids, and pigments.
- In plants, sucrose is formed in the cytosol using the products of photosynthesis formed in chloroplasts (Figure 10.2).

Glycolysis

- Glucose is the nearest thing to a common metabolic fuel and most organisms seem able to carry out glycolysis.
- Glycolysis is the breakdown of glucose to pyruvate with the accompanying phosphorylation of ADP to ATP and the reduction of NAD^+ to NADH.
- It consists of 12 enzyme-catalysed reactions that convert one molecule of glucose to two of pyruvate with the production of two ATP and two NADP molecules, as illustrated in Figure 7.2.
- Glycolysis occurs in the cytosol of most organisms, although in a relatively few species of parasitic protozoa it is performed, at least in part, in microbodies called glycosomes (Chapter 12).

Revision tip

Explore aspects of human metabolism, including chemical, clinical, and molecular biology, and biochemistry data at the Human Metabolome Database (HMDB): http://www.hmdb.ca/

Transport of metabolites

- The transport of metabolites from their site of production to where they are used occurs in the cytosol.
- Such transport is relatively simple for water-soluble molecules, such as glucose and amino acids, which can diffuse rapidly through the cytosol, but is more of a

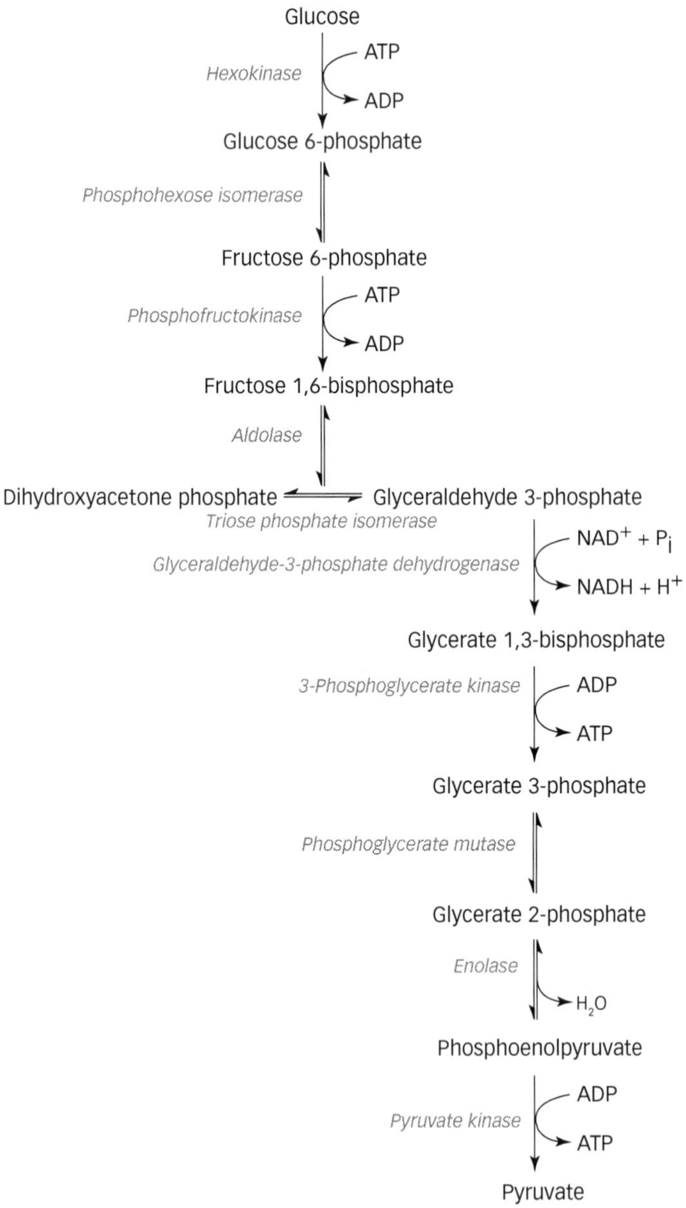

Figure 7.2 Overview of the reactions of glycolysis. See also Figure 12.8.

problem for hydrophobic molecules, such as fatty acids and steroids, because of their limited water solubility.

- Fatty acids are transported by specific binding proteins through the cytosol from the fatty acid synthase complex that take them to the smooth endoplasmic reticulum membrane.

- Steroid hormones enter the cytosol by diffusing through the plasma membrane and bind to cytoplasmic receptor proteins (also called nuclear receptors), which transfer them to the nucleus.
- Often materials to be transported through the cytosol are contained within small membrane-bound sacs called vesicles that are moved along microfilaments and microtubules of the cytoskeleton by motor proteins (see 'The cytoskeleton' section).
- Vesicles may contain substances taken into the cell by endocytosis, or material to be distributed between subcellular compartments or products to be secreted from the cell.

Protein synthesis

The term *protein synthesis* is often used loosely to describe the translation of an mRNA molecule by a ribosome to form the corresponding polypeptide. Translation in prokaryotes is described in Chapter 5. There are several key differences between the process in bacteria (and archaea) and eukaryotes, which are highlighted here.

➔ *Transcription in prokaryotes and eukaryotes is revised in Chapters 5 and 6 respectively. Chapter 5 also reviews translation in prokaryotes. See also the companion volume,* Thrive in Biochemistry and Molecular Biology.

Key differences between translation in prokaryotes and eukaryotes

- In prokaryotes, transcription and translation both occur in the cytoplasm but in eukaryotes transcription occurs in the nucleus, and translation on free ribosomes in the cytosol and on ribosomes bound to the endoplasmic reticulum.
- The structures of mRNA molecules differ between bacteria and eukaryotic cells: eukaryotic mRNA molecules do not have a Shine–Dalgarno sequence but are capped at their 5′ end and have a poly(A) tail.
- Translation in eukaryotes occurs after the mRNA is exported from the nucleus into the cytoplasm.
- Eukaryotic ribosomes are larger than prokaryotic ribosomes, but they share a similar structure (Figure 7.3 and Table 7.1).
- The 18 S rRNA of the eukaryotic 40 S (small) ribosomal subunit is homologous to the 16 S rRNA of the bacterial small subunit.
- The 28 S and 5 S rRNAs of the eukaryotic 60 S (large) ribosomal subunit are homologous to the bacterial 23 S and 5 S RNA molecules respectively. The 5.8 S rRNA of the 60 S subunit is homologous to the 5′ end of the 23 S molecule of bacteria.
- Eukaryotic ribosomal proteins, like prokaryotic types, function as supports for the rRNA molecules.
- Peptidyl transferase activity resides in the 28 S rRNA molecule, which like its prokaryotic counterpart is a ribozyme.
- The initiation codon, AUG, is recognized by a specific initiating tRNA, designated $tRNA_i$.

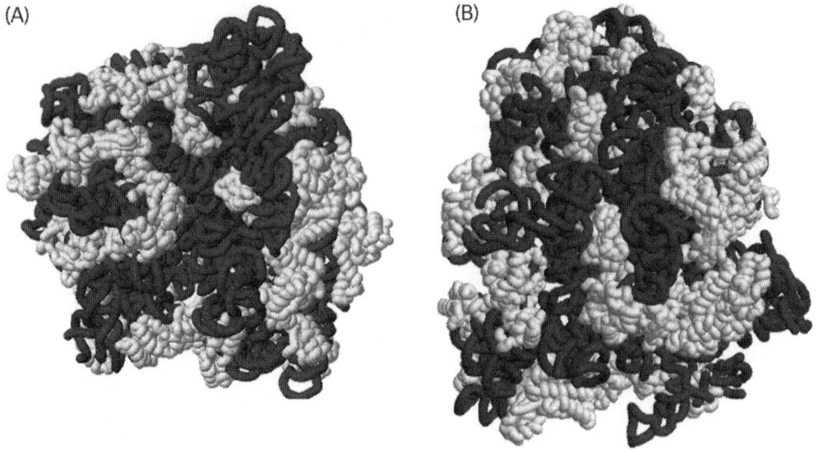

Figure 7.3 Relative sizes of the (A) 70 S *Escherichia coli* and (B) 80 S eukaryotic ribosomes. Structures obtained using the Protein database files 3I1O and 1S1I respectively. In both cases, the rRNA is shown in black and the ribosomal proteins in the lighter tone.

	Bacterial (Escherichia coli)	*Eukaryote (Saccharomyces cerevisiae)*
Intact dimensions/nm	29×21	32×32
s_o (S)	70	80
M_r	2.8×10^6	4.5×10^6
rRNAs	Two (23 S and 5 S) in large subunit One (16 S) in small subunit	Three (28 S, 5.8 S, and 5 S) in large subunit One (18 S) in small subunit
Ribosomal proteins	34 in large subunit 21 in small subunit	~49 in large subunit ~33 in small subunit
Relative proportion of RNA in ribosome (%)	~65	~50
Relative proportion of protein in ribosome (%)	~35	~50

Table 7.1 Sizes and composition of ribosomes

- The first amino acid residue incorporated into the polypeptide (the initiator amino acid residue) is methionine (Met) not *N*-formylmethionine (fMet) as in bacteria; it is delivered as the aminoacyl complex, Met-$tRNA_i$.
- Eukaryotic cells have more translation factors than prokaryotes and these are generally larger and more complex proteins.
- Abbreviations of eukaryotic initiation factors always begin with the letter 'e' to indicate their origin, making them easier to distinguish from prokaryotic factors.

Revision tip

To reinforce the differences in translation between bacteria and eukaryotes construct a simple table summarizing them.

Check your understanding

7.2 What are the sedimentation coefficients of the subunits of the eukaryotic 80 S ribosome?

Key points of eukaryotic translation

Translation in eukaryotes, as in prokaryotes, may be separated into three stages:

1. initiation
2. elongation
3. termination.

- Differences between bacterial and eukaryotic translation are most apparent in initiation and termination.
- Initiation in eukaryotes begins with the formation of a **pre-initiation complex (PIC)**, while its termination involves only a *single* termination factor.

Formation of the eukaryotic pre-initiation complex

- The formation of the PIC begins with the binding of eukaryotic initiator factor 3 (eIF3) to the 40S ribosomal subunit, which prevents it associating with the 60S subunit.
- Factor eIF6 binds to the 60S ribosomal subunit and prevents it from binding to the 40S subunit, unless the 40S subunit is appropriately positioned on the mRNA.
- Eukaryotic initiation factors 1 (eIF1) and eIF1A both bind to the free 40 S subunit (Figure 7.4).
- Initiating factor 2 (eIF2) is a G protein and in its active conformation has a bound GTP.

Recall the roles G proteins play in translation in prokaryotes (Chapter 5).

- In its activated state, eIF2 binds to Met-tRNA$_i$ and this complex combines with the eIF3-40 S subunit to form a 43 S PIC as shown in Figure 7.4.

Initiation of translation in eukaryotes

- Initiation of translation requires:
  - the involvement of several more eIFs
  - the positioning of the Met-tRNA$_i$ on the start AUG codon of the mRNA
  - the attachment of the 60 S ribosomal subunit.
- Factor eIF4E recognizes and binds to the 5′ cap and eIF4G binds to poly(A)-binding protein (PABPI), which itself binds to the poly(A) tail of mRNA.
- Mutual recognition of binding of eIF4E and eIF4G forces the mRNA into a circular conformation (Figure 7.5).

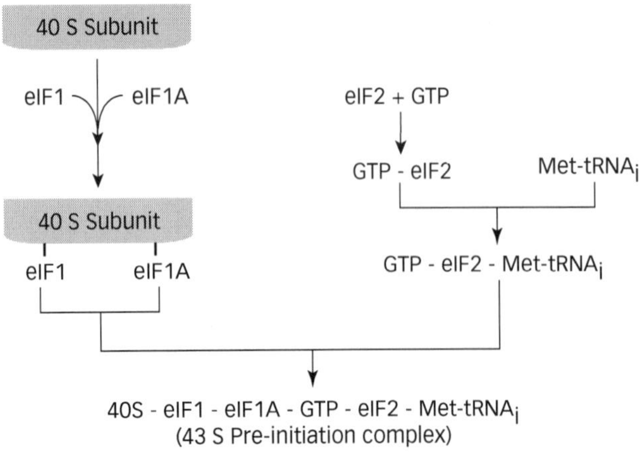

Figure 7.4 Formation of the eukaryotic pre-initiation complex. See text for explanations.

- Binding of eIF4G and circularization of the mRNA is essential for rapid initiation.
- Factors eIF4A (a helicase, see Chapter 5) and B unwind the mRNA and remove any unwanted hairpin bends.
- The next step is the positioning of the PIC so that the AUG start codon occupies the P site of the 40 S subunit.
- The PIC moves along the mRNA scanning it for the start codon, which is usually the first AUG encountered but is also normally found within the sequence … A/GNN*AUG*G …
- Factor eIF2 is a G protein and positions the Met-tRNA$_i$ in the P site of the ribosome at the expense of the hydrolysis of GTP.
- Factors eIF5A and eIF5B are necessary for the binding of the 60 S subunit to the complex. The former is a GTPase-activating protein, whereas eIF5B has GTPase activity.
- Binding of the 60S ribosomal subunit causes:
  - the hydrolysis of the GTP bound to eIF2

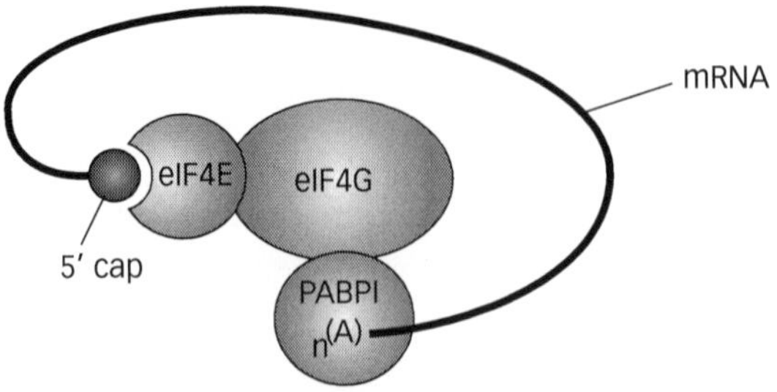

Figure 7.5 Formation of circularized mRNA by the actions of initiation factors 4E and 4G (eIF4E and eIF4G) and poly(A)-binding protein (PABPI).

- the release of all initiation factors
- completes initiation and allows elongation to begin.

Check your understanding

7.3 In eukaryotic translation, the initiator $tRNA_i$ is positioned into the ribosome by which eukaryotic initiation factor?

a. eIF3

b. eIF1

c. eIF2

d. eIF5A

e. eIF3J.

Elongation

- Elongation is the stage of translation where amino acid residues are added to the growing polypeptide using the mRNA as a template.
- Codons of three ribonucleotides specify each amino acid residue as described for bacterial elongation and as listed in Table 5.1 in Chapter 5.
- Elongation in eukaryotes is similar to that in bacteria (see Chapter 5, Figure 5.17).
- It proceeds from the 5′ end of the mRNA towards its 3′ terminus and the polypeptide chain grows from its amino terminus towards the carboxyl end.
- At the start of elongation the Met-$tRNA_i$ is positioned in the P site of the ribosome and the A and E sites are vacant.
- The second aminoacyl-tRNA enters the A site of the ribosome bound to elongation factor eEf1-GTP, which facilitates its correct positioning (Figure 7.6).
- If the correct aminoacyl-tRNA is positioned into the A site (that is, its tRNA anticodon matches the complementary codon on the mRNA) the eEF1 hydrolyses its GTP to GDP, and dissociates from the ribosome complex.
- The tRNAs in the P site and A site are now correctly positioned next to the active site of the 28 S rRNA molecule peptidyl transferase.
- Peptidyl transferase activity transfers the methionine residue from the Met-$tRNA_i$ by forming a peptide bond between it and the amino group of the second amino acid residue in the P site.
- **Translocation** then occurs, largely in the manner described for bacteria in Chapter 5 (Figure 5.17) but:
  - it is facilitated by eEF2
  - the ribosome moves along the mRNA by one codon
  - the spent second tRNA then moves to the E site the ribosome before its release from the ribosome-mRNA complex
  - the newly incorporated tRNA, attached to the growing polypeptide moves to the P site.

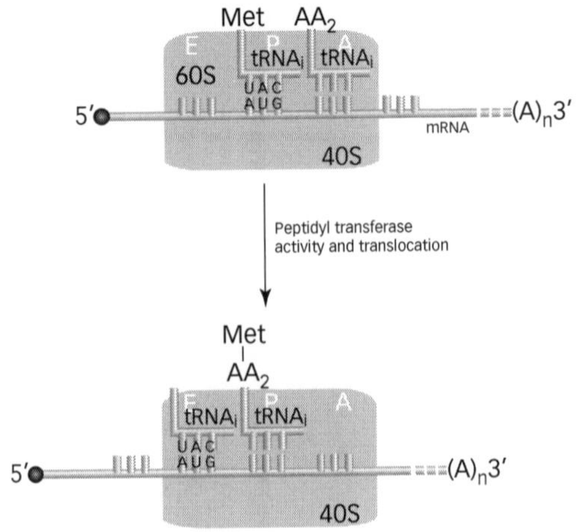

Figure 7.6 Schematic illustration of peptidyl transferase activity and translocation in eukaryotes. See also Figure 5.17.

- The next aminoacyl-tRNA is now able to enter the A site at the start of the next round with the growing polypeptide remaining in the P site bound to a tRNA throughout elongation.
- The rate of incorporation of amino acid residues into the growing polypeptide during elongation is two to five per second.

Check your understanding

7.4 Why do eEF-1 and eEF-2 correspond to the bacterial elongation factors EF-Tu and EF-G respectively?

Proofreading in elongation

- Selection and binding of the correct amino acid to the polypeptide chain is determined by the codon sequence on the mRNA but the high fidelity of translation is enhanced by a proofreading mechanism during elongation.
- Errors in translation occur at a rate of one amino acid residue per 10^4, which is lower than expected for codon-based control alone.
- Kinetic proofreading involving eEF2-GTP is responsible for this increased accuracy.
- Amino acid residues enter the A site of the ribosome as eEF2-GTP-aminoacyl-tRNAs complexes. If the correct tRNA is positioned over the codon, the GTP is hydrolysed and the elongation factor dissociates from the aminoacyl-tRNA.
- If an incorrect aminoacyl-tRNA is positioned at the A site, weaker hydrogen bonding occurs between the codon and anticodon and hydrolysis of the GTP is delayed.
- This delay gives the incorrect aminoacyl-tRNA time to diffuse away from the A site before its amino acid residue is incorporated into the growing polypeptide.

Termination

- Termination of translation occurs when a stop codon on the mRNA (UAA, UAG or UGA (see Chapter 5, Table 5.1)) is positioned at the A site of the ribosome.
- Eukaryotes have one release factor, eRF1, which recognizes all three stop codons, unlike bacteria that have three release factors (see Chapter 5).
- Like a number of molecules involved in translation, eRF1 mimics the structure of a tRNA (Figure 7.7).
- The factor eRF3 is a GTPase that stimulates the activity of eRF1.
- Termination also involves the DEAD-box protein, Dbp5, which supports eRF1 in recognizing stop codons.
- DEAD-box proteins are named from the … –Asp–Glu–Ala–Asp– … motif they contain. Using one letter amino acid abbreviations this is … DEAD …
- Once the stop codon is recognized, Dbp5 is released from eRF1, and eRF3 is then allowed to enter the termination complex.
- Once eRF1 has entered the A site of the ribosome it triggers hydrolysis and the release of the polypeptide chain from the tRNA occupying the P site.

Ribosome recycling

- Termination of translation is followed by a recycling of the ribosome.
- Once termination is complete, the ribosome, mRNA, and deacylated tRNA remain locked together in a post-translational termination complex.
- Initiation factor eIF3 binds to the 40S subunit of the ribosome and initiates the release of the 60S subunit.
- eIF3 remains bound to the 40S ribosome to prevent the 60S subunit from rebinding.
- eIF1 stimulates release of the tRNA from the P site whilst eIF3J stimulates the release of the mRNA from the 40S subunit.
- A member of a large family of ATPase called the ATP-binding cassette family, ABCE1 facilitates the recycling of the ribosomal subunits by hydrolysing NTPs.

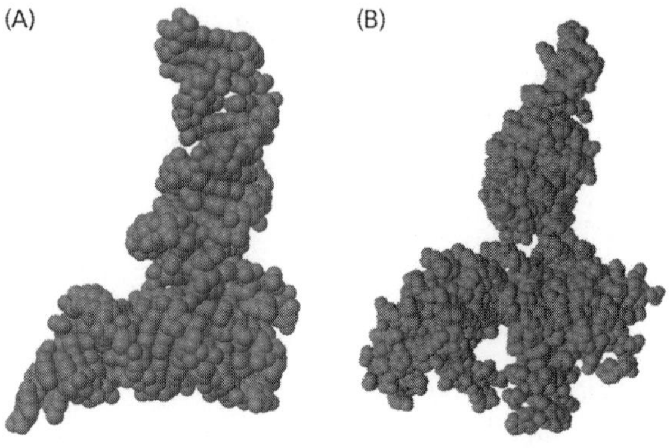

Figure 7.7 Molecular models to show the similar conformations of (A) a eukaryotic tRNA and (B) human release factor, eRF1. PDB files 2TNA and 1DT9 respectively. See also Figure 5.19.

Check your understanding

7.5 Which of the following is responsible for preventing the ribosomal subunits from rejoining before translation initiation?

a. eIF1

b. eEF2

c. eIF3

d. eIF3J

e. eIF5.

Looking for extra marks?

Note the broad similarities in translation in eukaryotes and prokaryotes (described in Chapter 5); most differences occur during initiation.

Post-translational modification in eukaryotes

After translation, the newly formed polypeptide undergoes a number of post-translational modifications and folding before a fully functional protein is formed.

- The range of protein modifications that occur in eukaryotes are far more numerous than those outlined for bacteria in Chapter 5.
- Some changes also modify identical proteins so they can play different roles in different cellular pathways.
- Although some post-translational modifications occur in the cytosol, many happen in the endomembrane system and other types in both regions.
- Different types of proteolytic cleavages occur in the cytosol, the endomembrane system, and even outside the cell.
- Phosphorylations and acetylations are generally cytosolic events, while methylation occurs in both the cytosol and nucleus.
- In contrast, glycosylations, sulphations, and oxidations normally occur in the endomembrane system.

Proteolytic cleavages

- The amino terminal methionine, with which all translation starts, is hydrolytically removed from nearly all eukaryotic proteins.
- Proteolytic cleavage also removes the amino terminal signal sequences found on membrane and secretory proteins (see Chapter 8).
- Some proteins are synthesized as inactive precursors that are activated by removal of a segment by proteolytic cleavage. Examples include zymogens, clotting proteins, and some hormones (see Chapter 9, Figure 9.4 and associated text).

Phosphorylation

- Post-translational phosphorylation is one of the commonest modifications to occur in the cytosol.
- Residues typically phosphorylated include serine, tyrosine, histidine, and threonine.
- Phosphorylation is commonly used to regulate enzyme activities and in intracellular signalling (Chapter 15). In general, addition of a phosphate group activates the signalling protein or enzyme and its removal decreases or eliminates activity as illustrated for the enzyme *phosphorylase* in Figure 7.8.
- Phosphorylase *b* is converted to the active *a* form by phosphorylation of specific serine residues in each of its subunits using ATP as a phosphate donor in a reaction catalysed by phosphorylase kinase.
- Phosphorylase *a* is converted to the inactive *b* form by phosphatase activity that hydrolytically removes the phosphates from the serine residues.
- The *processes* of phosphorylation and dephosphorylation are usually reversible but the *reactions* concerned are not.

Methylation

- Methylation is the addition of methyl groups ($-CH_3$), which are added to the terminal carboxyl group or to specific amino acid residue side chains.
- Enzymes concerned with catalysing methylation reactions are found in both the cytosol and nucleus.
- Methylation of carboxyl groups is reversible whereas methylation of the nitrogen atoms of the amino terminus or the side chains of arginine and lysine residues is irreversible.

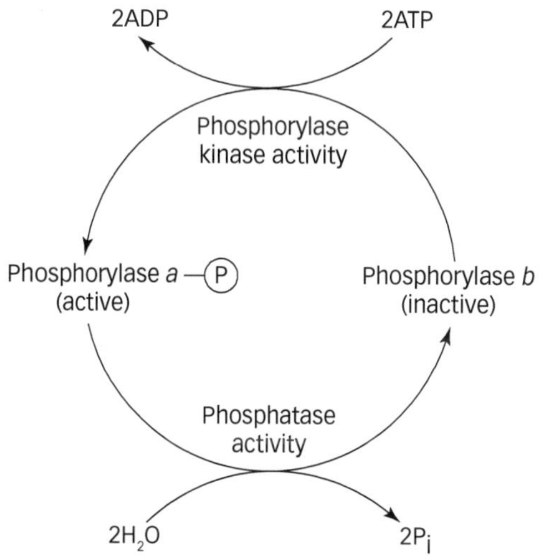

Figure 7.8 Interconversions of the phosphorylase *a* and *b* forms. The enzyme is a dimer and therefore requires two ATP molecules to convert it to its active form.

- Methylation makes the protein more hydrophobic and is used to regulate protein activity and gene expression.

Amino terminal acetylation

- Acetylation is the addition of an acetyl group, usually to the amino terminus of the protein; it involves removal of the N-terminal methionine and addition of an acetyl group donated from acetyl CoA.
- Commonly acylated residues include serine, threonine, alanine, and glycine.
- Acetylation is involved in protein–protein interactions and membrane targeting and occurs in the cytosol.

Signal transduction

- Signal transduction is the conversion of one type of signal into another (Chapter 15).
- The initial signal may be the entry of a steroid hormone into the cytosol or the activation of a receptor on the outer surface of the plasma membrane.
- This initial signal is converted to intracellular biochemical signals within the cytosol to give rise to specific intracellular responses in the cytosol, nucleus, or other organelles.

➔ *Signal transduction is introduced in Chapter 3 and discussed in more detail in Chapter 15.*

7.3 THE CYTOSKELETON

The cytoskeleton is a three-dimensional intricate array of protein fibres that extends throughout the cytosol, and is involved in supportive, locomotory, and transport roles.

- Organization is imparted to the cytosol (and indeed cytoplasm) by the dynamic actions of the cytoskeleton.
- Although the cytoskeleton does provide a supportive framework to the cell, the name is slightly misleading given it has a number of other essential roles, which include:
    - actively moving macromolecules, vesicles, and organelles around the cytosol
    - roles in cellular contraction and motility
    - the separation of chromosomes at cell division (Chapter 16).

Check your understanding

7.6 The overall shape of an animal cell is determined by which of the following?

a. Cell wall

b. Plasma membrane

c. Cytoskeleton
d. Mitochondria
e. Nucleus

- Three types of fibres occur in the cytoskeleton:
  1. microfilaments (MFs)
  2. microtubules (MTs)
  3. intermediate fibres (IFs).
- Each type of fibre is made of different proteins and is adapted for different roles in the cell.
- All three types of fibres are interconnected by linker proteins called **plakins**, and so they operate as an integrated network.
- Motile and transport activities of the cytoskeleton require the participation of **motor proteins.**
- The motor proteins associated with MFs are **myosins**; those of MTs are **kinesins** and **dyneins**, which can catalyse the hydrolysis of ATP.
- The cytoskeleton is *not* a static entity but is in a state of flux, being constantly reorganized as its protein fibres are assembled and reassembled.

Microfilaments

Microfilaments are contractile elements of the cytoskeleton. They often occur as conspicuous bundles near to the plasma membrane.

- Many MFs form attachments with the plasma membrane and their contractions can adjust the shape of the cell.

Structure of microfilaments

- Microfilaments are made of the protein, **actin**, which occurs in the cytosol as soluble monomers called **G actin** (M_r 42,000) that can polymerize to form long strands (polymers) called **F actin.**
- Both G and F actins associate with a large number of **actin-binding proteins.**
- Microfilaments are formed of two actin polymers twisted together to give a double helical structure with a diameter of approximately 7 nm (Figure 7.9).
- The assembly of microfilaments has a polarity:
  - one end, usually referred to as the **plus** (+) end, favours the addition of new monomers allowing it to grow in that direction.

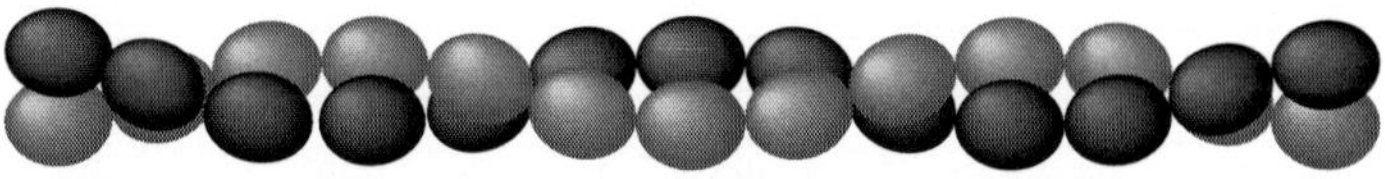

Figure 7.9 Schematic structure of a microfilament. See text for details.

  - the other, the **negative** (–) end, is usually where disassembly of the microfilament occurs.
- Growth of an MF requires:
  - an MF with a free (+) end
  - the presence of several other proteins, such as thymosin and profilin
  - the presence of free G actin molecule with a bound ATP (G actin-ATP).
- Polymerization begins with the G actin-ATP attaching to the (+) end of an MF.
- Binding triggers a latent ATPase activity in the subunit and the ATP is hydrolysed to ADP, upon which further G actin-ATP molecules cannot bind and the MF will tend to disassemble (depolymerize).
- Terminal F actin are the most likely to have bound GDP, and therefore this is the most likely end for disassembly.
- Continued growth of the MF requires a new G actin-ATP *be added before the previously added subunit hydrolyses its ATP*. Large proteins called **formins** move with the growing (+) end of the MF and prevent the cessation of growth until the target destination is reached.
- Complete extension of an MF occurs when it reaches its target destination: here specific capping protein stabilizes the MF structure.

Key roles of microfilaments

- Actin filaments are, perhaps, best known for their combination with myosin in muscle tissues (see 'Muscle contraction' section).
- Microfilaments also:
  - influence the shape of the cell
  - contribute to cytoplasmic streaming
  - form highly cross-linked bundles called **stress fibres**, that appear and disappear during the movements of some types of 'crawling' cells such as macrophages (Chapter 14)
  - help form the cleavage furrow in cell division (Chapter 16)
  - assist in amoeboid locomotion of leucocytes and macrophages and cells during embryogenenesis (Chapter 17).
- In addition to these roles in motility, permanent bundles of actin filaments also support the long microvilli of intestinal cells (Figure 7.10).

Skeletal muscle

Muscle tissue consists of bundles of fibres called **myofibrils** (Figure 7.11).

- Each myofibril is a **syncytium** that developed from the fusion of many embryonic cells and contains numerous nuclei.
- Myofibrils contain organelles that are often given muscle-specific names: *sarcolemma* for plasma membrane, *sarcoplasmic reticulum* for endoplasmic reticulum, and *sarcosol* not cytosol.

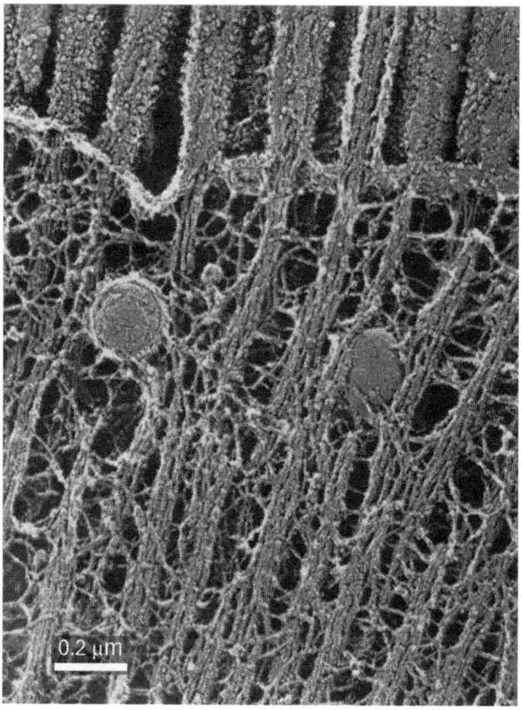

Figure 7.10 Electron micrograph showing the highly organized cytoskeletal network in the microvilli of chicken intestine. Reprinted with permission from Mooseker, MS., Bonder, E.M., Conzelman, K.A., Fishkind, D.J., Howe C.L., and Keller, T.C., 3rd (1984) Brush border cytoskeleton and integration of cellular functions. *Journal of Cell Biology*, 99, 104s–112s.

- Myofibrils are divided into segments called **sarcomeres**, which are limited by Z discs (from the German, *zwischen*, between) (Figure 7.11).
- Sarcomeres contain two types of interdigitating protein filaments that are cross linked to each other:
  - the thinner filaments are largely permanent actin filaments with typical diameters of 7–8 nm
  - one end of an actin filament is attached to a Z disc
  - the thicker mainly myosin filaments have diameters of 12–16 nm.
- This arrangement of fibres explains the striped appearance of muscle tissue when examined by microscopy (Figure 7.12).

Muscle contraction

- The two filaments are able to move towards each other in a sliding-type mechanism in which existing cross links are broken and new ones formed.
- This movement causes the sarcomere to contract, which, in turn, shortens the myofibril and therefore the muscle.

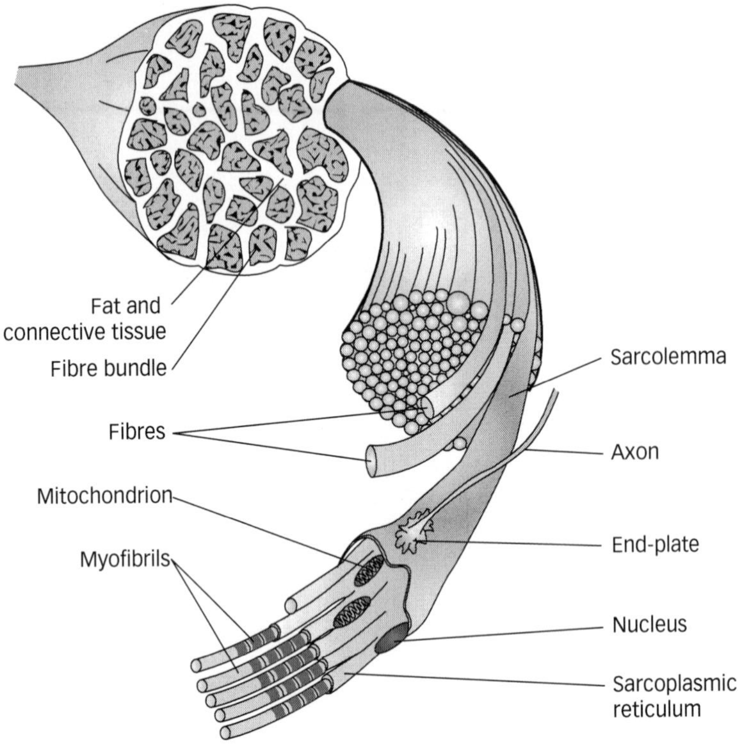

Figure 7.11 Schematic illustration of the relationship between muscle tissues and myofibrils.

- Thick filaments are composed of several hundred myosin dimers. These have long extended α helices coiled around each other to form an extended straight coiled coil or rod domain, each of which terminates in a globular head.
- Contraction of skeletal muscle fibres begins when a motor nerve impulse leads to the sarcoplasmic reticulum releasing Ca^{2+} into the sarcosol (Chapter 8).
- The myosin head is an adenosine triphosphatase (ATPase), so an ATP molecule bound to its active site can be enzymatically hydrolysed to ADP and P_i.
- When the ADP and P_i are *released* from the head, the free energy of the hydrolytic reaction is used to change the conformation of the myosin head.
- This conformational change results in the formation of a new cross bridge forming between the myosin and actin molecules and a shortening of their overall lengths by about 1 nm.
- Since the actin is attached to the Z disc, this movement is a contractile event.
- The release of ADP and P_i allows a new ATP to bind and a new contractile cycle to begin.

Control of contraction

- Calcium ions and regulatory proteins called the troponin–tropomyosin complex play key roles in muscle contraction and relaxation (Figure 7.13).

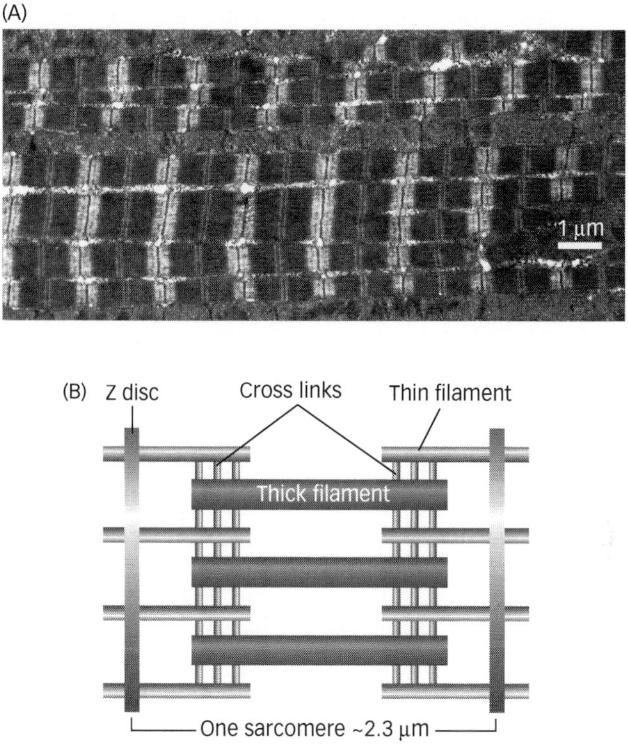

Figure 7.12 (A) Electron micrograph of muscle tissue from anchovy, *Engraulis encrasicolus*. (B) Schematic showing the arrangement of thin and thick filaments of a sarcomere.

- Tropomyosin consists of two long α helices that coil around each other in an extended coiled coil conformation. Each tropomyosin molecule is able to bind around seven actin molecules of the thin filament.
- Troponin is a trimer of three different subunits called T, I, and C respectively:
  - troponin T binds to tropomyosin
  - troponin I inhibits the interactions between actin and myosin essential for contraction
  - troponin C is able to reversibly bind Ca^{2+}.

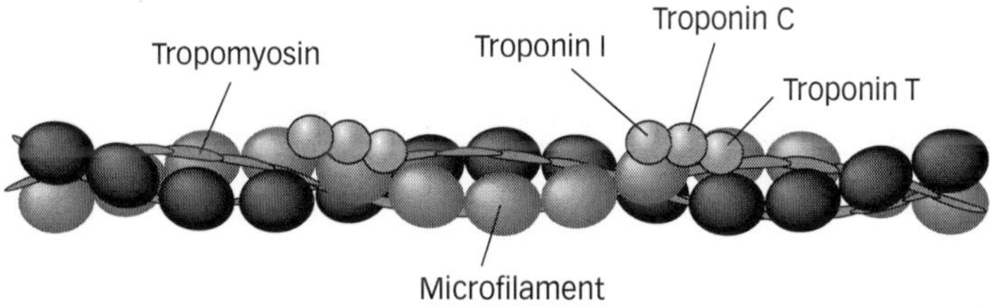

Figure 7.13 Schematic illustration of the troponin–tropomyosin complex and actin filament.

- If Ca^{2+} is not present, the binding of the troponin–tropomyosin complex to actin inhibits contraction.
- The arrival of a nerve impulse stimulates the release of Ca^{2+} from the sarcoplasmic reticulum, which are able to bind to troponin C, and change its conformation.
- The troponin–tropomyosin complex now has a reduced affinity for actin, the inhibition by troponin I is relieved, and muscle contraction occurs.
- The active return of Ca^{2+} to the sarcoplasmic reticulum reverses the contractile events and the myofibril relaxes.

Check your understanding

7.7 Which of the following is/are true?

a. The troponin-tropomyosin complex is associated with myosin filaments.
b. Myosin filaments consist of two α helices stabilized by hydrophobic interactions.
c. Muscle tissue converts chemical energy into kinetic energy.
d. Microtubules in sarcomeres are filaments with diameters of 7–8 nm.
e. The contraction and relaxation of skeletal muscle fibres is associated with the uptake and release of Ca^{2+} by the sarcoplasmic reticulum.

Microfilaments and intracellular transport

Microfilaments also function in transport activities within the cytosol.

- Contraction of MF happens when myosin II (the usual myosin of skeletal muscle) motor proteins aggregate to form filaments of approximately 16 molecules.
- The free energy from the hydrolysis of ATP by the myosin filament exerts a pulling force on adjacent actin filaments and these in turn pull on their target sites such as the plasma membrane.
- A family of *single* myosin motor proteins called **minimyosins** lack the extended tail of myosin II. However, the short tail domains of different members of the group are able to specifically bind to one of a number of different cargoes, for example different types of vesicles.
- Minimyosins do not stimulate MFs to contract but are able to move along an MF generally towards the (+) end at the expense of ATP hydrolysis.
- Movement of the motor pulls the attached cargo to its appropriate destination in the cell (Figure 7.14).

Microtubules

Microtubules are also transport elements of the cytoskeleton.

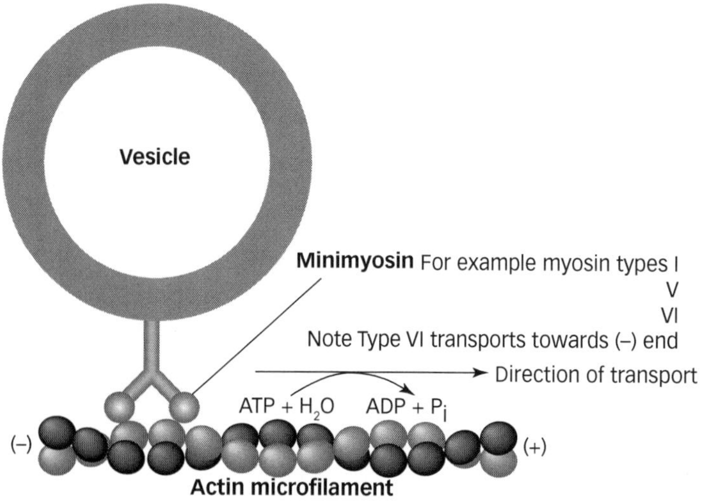

Figure 7.14 Schematic showing the transport of a cargo towards the positive end of a microfilament by a minimyosin.

Structure of microtubules

- Microtubules are hollow cylinders bounded by 13 longitudinal rows of protofilaments with a diameter of approximately **25 nm**.
- Microtubules are made of a dimeric protein called **tubulin**.
- Each tubulin dimer consists of an α and a **β tubulin** monomer, which are aligned in the same direction in **protofilaments** (Figure 7.15); this arrangement gives the protofilament a polarity, with opposite α and β ends.

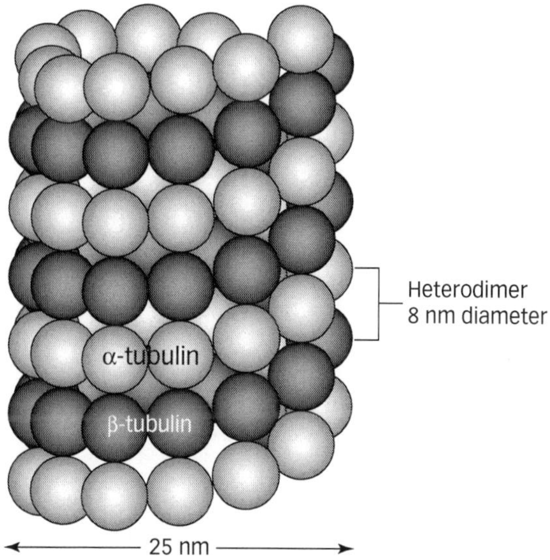

Figure 7.15 Schematic structure of a microtubule. See text for details.

- Although of identical M_r (50,000), α and β tubulins are different monomers.
- Like MFs, MTs are also polarized.
- Polymerization or growth of an MT resembles that of a growing MF described previously but uses GTP not ATP.
- Tubulin subunits are latent **GTPases**, and those attaching to the growing (+) end of an MT have bound GTP molecule.
- Once the monomer is added to the chain, the GTP is hydrolysed but with a sufficiently long delay to temporarily stabilize the growing (+) end of the MT until a further tubulin monomer (with its bound GTP) can attach. This allows growth of the MT in the (+) direction.
- Growth at the (+) end proceeds until a target destination is reached when **plus-end tubulin interacting proteins** (**+TIPs**) attach and stabilize the MT.
- The (–) end of an MT is protected from premature disassembly by being embedded in a **microtubule-organizing centre** (MTOC).

Microtubule-organizing centres

- Microtubule-organizing centres are **centrosomes** and the **basal bodies** of eukaryotic **cilia** and **flagella**.
- The centrosome is an MTOC found near the nucleus of plant and animal cells, which organizes the MT cytoskeleton.
- The centrosome of animal cells, protozoa, and some yeast cells contain two **centrioles**, which are hollow cylinders approximately 0.4 mm long and about 0.15 μm in diameter.
- The walls of centrioles are composed of nine sets of MTs, each comprising three fused MTs.
- The two centrioles in a centrosome lie at right angles to one another and are surrounded by a diffuse granular **pericentriolar material**.
- The pericentriolar material contains complexes that can initiate the formation of MTs. These complexes consist of a third type of tubulin (**γ tubulin**) and a number of proteins called **γ ring proteins** (**GRiPs**) that associate to form **γ tubulin ring structure complexes** (**γ TuRCs**) from which MTs grow.
- Hence, the centrosome contains γ TuRCs, initiates MT growth outwards from the pericentriolar material and protects and stabilizes their (–) ends.
- During cell division, the centrosome is replicated and each daughter centrosome migrates to an opposite pole of the cell, allowing the MTs to reorganize to form the **spindle apparatus** (Chapter 16).

Key roles of microtubules

- Long flexible MTs are essential in many locomotory and transport processes in the cell.
- Microtubules form tracks along which motor proteins can move macromolecules and subcellular organelles in the cytoplasm.
- The two best known types of motor proteins of MTs are the kinesins and dyneins whose actions are similar to that of minimyosins in transport along MFs (see

'Microfilaments and intracellular transport' section). Both kinesins and dyneins are capable of moving along MTs using the free energy obtained by their hydrolysis of ATP.

- Kinesins tow cargoes from the (–) to the (+) pole; dyneins in the opposite direction.
- Microtubules help the cell maintain its overall shape by influencing the distribution of both MFs and IFs (see later).
- In plant cells, MTs influence the plane of cell division and the direction in which cellulose fibrils are laid down during cell wall formation (Chapter 13).
- They are also responsible for the flexing actions of eukaryotic cilia and flagella, and the formation of the mitotic spindle (Chapter 16).

Eukaryotic cilia and flagella

- Eukaryotic cilia and flagella are extensions of the plasma membrane supported by an MT framework called the **axoneme**, which differs from cytoplasmic MTs in being a permanent structure (Figure 7.16).
- Both cilia and flagella have diameters of approximately 0.5 μm; cilia are 2–10 μm long and flagella 100–200 μm in length.
- Nine cross-linked pairs of MTs extend from the basal body along the full length of the cilium or flagellum, while two single MTs run partially down their centre giving a '9 + 2' arrangement.
- This contrasts with the hollow tubular structures of basal bodies and centrioles ('9 + 0').
- Cilia and flagella are able to beat: in the power stroke, the appendage is relatively stiff; it then relaxes in the recovery stroke prior to another power stroke (Figure 7.17).
- Dynein molecules are associated with MTs of cilia and flagella and move along one of the MTs in a pair, with the tail of one of the pair attached to its partner at the expense of hydrolysing ATP.
- The cross-linking of the MTs within a pair means they cannot slide past one another, which causes the cilium or flagellum to bend in its power stroke.
- The beating of cilia and flagella has various uses, which include:
  - locomotion in a variety of unicellular organisms
  - spermatozoa move using a single flagellum

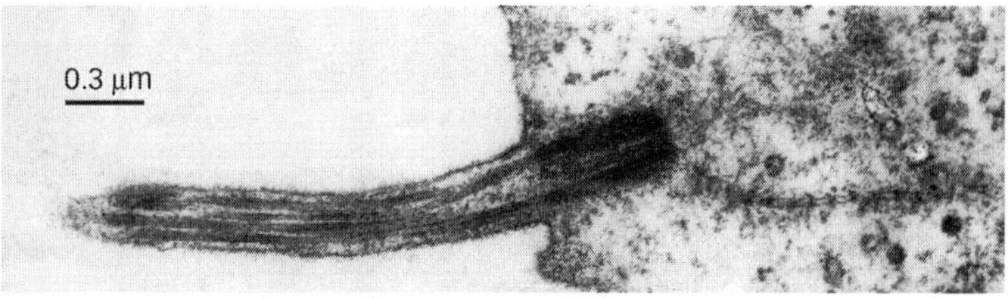

Figure 7.16 Electron micrograph through the longitudinal section of a cilium. Courtesy of Professor J.B. Tucker, School of Biological and Medical Sciences, University of St Andrews, UK.

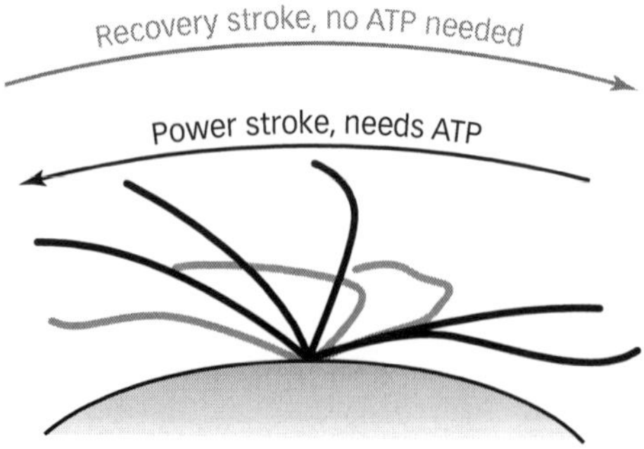

Figure 7.17 The beating action of a cilium. Note the relative stiffness of the cilium in the power stroke compared to the recovery.

- the coordinated beating of cilia in the air passages of vertebrate respiratory systems allows mucus and trapped particles to be swept out of the air passages and lungs.

Looking for extra marks?

Flagellum comes from the Latin for whip; so eukaryotic flagella are aptly named, not so prokaryotic types which rotate.

Intermediate fibres

Intermediate fibres are the supportive elements of the cytoskeleton. These fibres have diameters of 8–12 nm, values between that of MFs and MTs, hence their name.

Structure of intermediate fibres

- The protein components of intermediate fibres are tissue specific, unlike those of MFs and MTs. They can be divided into six classes (Table 7.2).

Intermediate filament	***Protein (M_r)***	***Associated tissues and cells (general role)***
I	Acidic keratins (40,000–57,000)	Tonofilaments in epithelial cells (mechanical strength; see Chapter 3)
II	Basic keratins (40,000–70,000)	Tonofilaments in epithelial cells (mechanical strength; see Chapter 3)
III	Desmin (53,000)	Muscle (support)
III	Glial fibrillary acidic (GFA) protein (52,000)	Glial cells, astrocytes (cell shape)
III	Vimentin (57,000)	Mesenchymal (cell shape)
IV	Neurofilament (NF) proteins L (62,000) M (102,000) H (110, 000)	Neurons of peripheral and central nervous system (support axons)
V	Lamins (60,000–70,000)	All cells (support nuclear envelope, see Chapter 6)
VI	Nestin (240,000)	Neuroepithelial stem cells (role unknown)

Table 7.2 Intermediate filaments and associated tissues

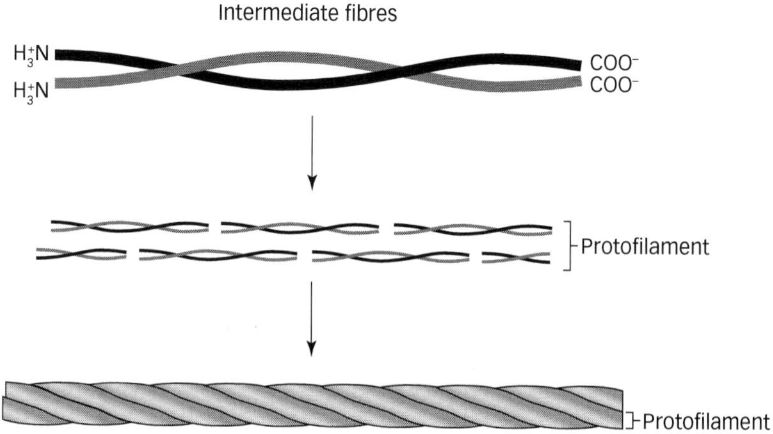

Figure 7.18 Schematic assembly and structure of an intermediate fibre. See text for details.

- In spite of this heterogeneity, all IFs appear to possess a common structure of a dimer of two intertwined IF polypeptides (Figure 7.18).
- Two dimers form a protofilament, which interact to form an IF composed of eight protofilaments.
- This organization produces an IF with:
  - a central rod-like domain
  - amino and carboxyl domains that flank the central rod and differ in sizes and composition.
- It is presumed that the diverse extremities of IFs account for the functional differences among the IFs.

Key functions of intermediate filaments

- Intermediate filaments form the most obvious skeletal role, that of a scaffold and to provide mechanical strength to support the structure of the cell (Table 7.2).
- They are often prominent in areas subject to extremes of tension where they are thought to function in resisting mechanical stress.
- For example, neurofilaments give mechanical support to the long axons of nerve cells, and keratins form **tonofilaments** in adhesive junctions (Chapter 3).

Revision tip

Do browse through the further reading references given for this chapter online. Go to http://www.oxfordtextbooks.co.uk/orc/thrive/

Check your understanding

7.8 Which of the following statements is/are true or is/are false?

a. The cytosol is the largest compartment in all eukaryotic cells and in animal cells the cytosol occupies up to about 55% of the cell volume.

b. The cytoskeleton is a three-dimensional network of protein fibres distributed throughout the cytoplasm.

c. The cytoskeleton is largely responsible for maintaining the shape of the cell and once formed is maintained largely unchanged.

d. The cytoskeleton is a dynamic network of insoluble protein fibres constantly being modified to suit the needs of the organism.

e. The pH of the cytosol is buffered to approximately 6.8 largely by the actions of Na^+-linked antiporters in the plasma membrane.

f. The growth of a microfilament requires soluble G actin-ATP.

g. Intracellular transport using actin filaments is associated with the hydrolysis of ATP by minimyosin motor proteins.

h. Tubulin dimer consists of α and β tubulin subunits that are aligned in opposite directions in protofilaments.

i. The beating of cilia and flagella is always used for locomotion.

j. The role of intermediate filaments conforms to that of a supportive skeleton in contrast to those of microfilaments and microtubules.

7.9 Which of the following statements is/are true or is/are false?

a. In eukaryotes, translation occurs only on free ribosomes in the cytosol.

b. Amino acids are lined up with their corresponding codons on the mRNA strands by an ribosome.

c. Eukaryotic ribosomes are smaller than those of prokaryotic cells.

d. The translation factor, eIF3 binds to the large (60 S) subunit and prevents it from binding to the 40 S ribosomal subunit, unless the 40 S subunit is bound to an mRNA.

e. The initiation complex scans the length of the mRNA molecules until the start AUG codon is positioned at the A site of the ribosomal subunit.

f. The translation factor, eRF1 has a similar structure as a tRNA.

g. Bacterial protein modifications are far more numerous than those of eukaryotes.

h. The reactions involved in the phosphorylation of proteins cannot be reversed.

i. Residues typically phosphorylated include: serine, tyrosine, histidine, and threonine.

j. Oxidation of cysteine residues can result in formation of disulphide bonds.

8 The endoplasmic reticulum

Key features of the endoplasmic reticulum

- The endoplasmic reticulum (ER) forms part of the endomembrane system of eukaryotic cells (Chapter 1).
- The ER membrane is continuous with the outer nuclear membrane and encloses a space called the lumen that is continuous with the perinuclear space (Chapter 6).
- The ER consists of two morphologically distinct forms or domains (Figure 8.1):
  - smooth ER (SER), which lacks ribosomes
  - rough ER (RER), which is studded with ribosomes giving it a rough appearance.
- Other areas of ER include transitional ER; the nuclear envelope is sometimes also included as part of the ER system.
- The ER is a dynamic system whose structure is constantly being rearranged to meet the needs of the cell; it provides a large surface area within the eukaryotic cell for the attachment of enzymes and, in the case of RER, ribosomes.

continued

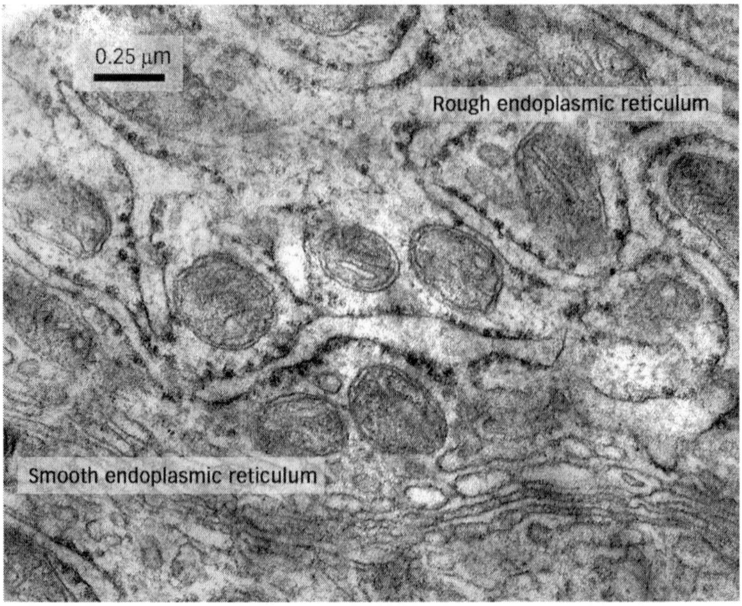

Figure 8.1 Electron micrograph showing smooth and rough endoplasmic reticulum.

- Cellular activities associated with the ER include:
  - forming an extensive transport system within the cell
  - the biosynthesis and distribution of lipids
  - the detoxification of certain drugs and poisons
  - the storage of Ca^{2+}
  - the synthesis, sorting, and transport of integral membrane and secretory proteins.

8.1 STRUCTURE OF THE ENDOPLASMIC RETICULUM

The endoplasmic reticulum (ER) is composed of a single membrane approximately 5 nm thick, which is organized into an interconnected series of flattened **sacs** or **cisternae** and **tubules**.

- In cells specialized for secretory activities, RER is a prominent feature and is generally formed of flattened cisternae that are organized into stacks, which are densely coated with ribosomes (Figure 8.2).
- In other cell types it can consist of a loose array of tubular cisternae carrying comparatively few ribosomes.
- Most cells have much less SER compared to RER, and usually it is largely composed of convoluted tubules.

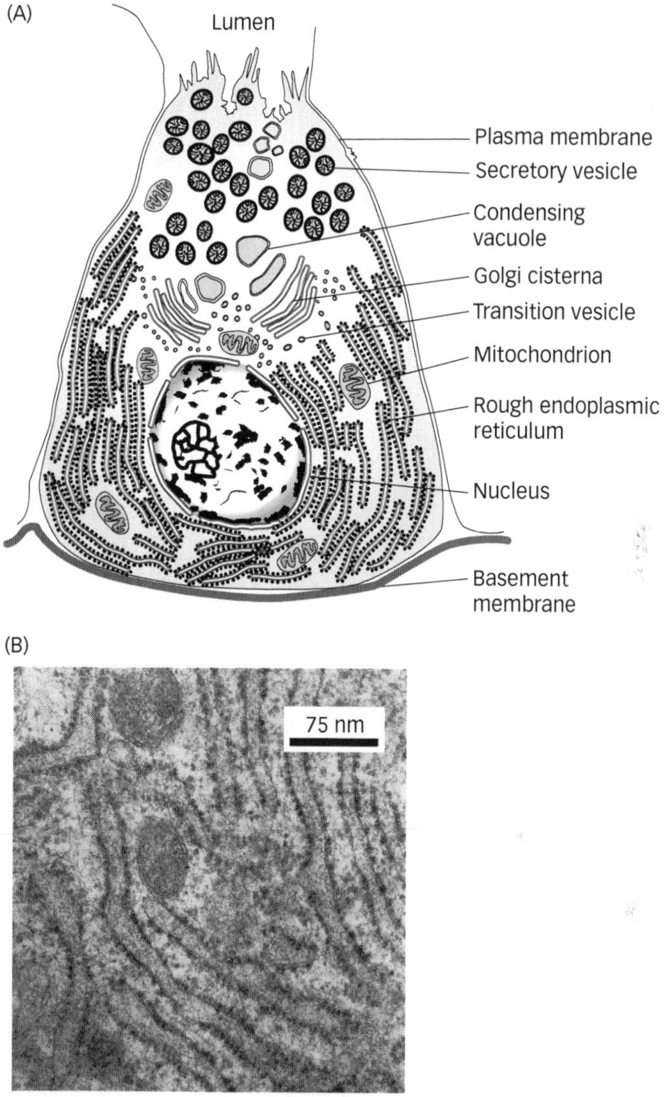

Figure 8.2 (A) Schematic illustration of a pancreatic acinar cell. (B) Electron micrograph showing the detailed structure of rough endoplasmic reticulum.

- Irrespective of their detailed morphologies, the ER membrane system is a continuous system that encloses a single lumen (Figure 8.3)
- In cisternal regions, the ER membranes may be separated by a space 20–30 nm wide. Tubules are approximately 30–60 nm in diameter and of indeterminate length.
- The lumen enclosed by the ER membrane has a high *oxidizing* potential, which contrasts with the *reducing* environment of the cytosol.

Structure of the endoplasmic reticulum

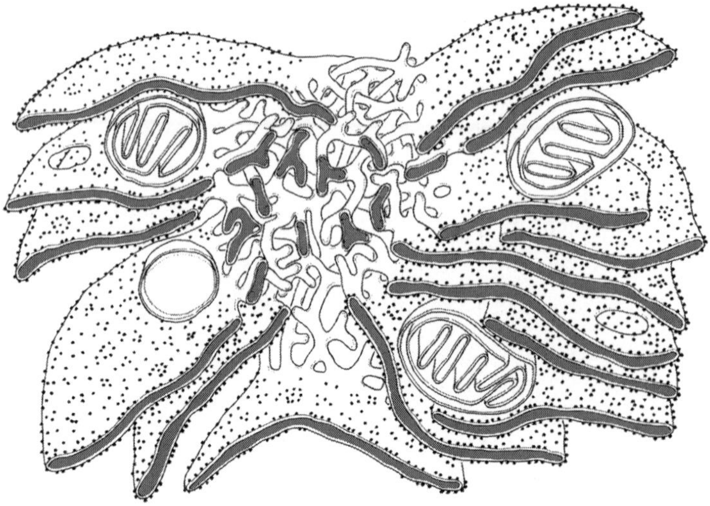

Figure 8.3 Three-dimensional representation of a portion of the endoplasmic reticulum of a rat hepatocyte. The exposed lumen has been lightly shaded. Redrawn from Krstic, R.V. (1979) *Ultrastructure of the Mammalian Cell*, Heidelberg: Springer-Verlag.

- The membranes of the rough and smooth endoplasmic reticulum are distinguishable morphologically by the presence or absence of ribosomes on their cytosolic surface respectively.
- Subdomains of the RER, called **transitional elements** (TEs) or **transitional ER** lack ribosomes and have a smooth appearance, and are sites where vesicles bud off from the ER and carry lipids and proteins to the Golgi apparatus (Chapter 9).

Revision tip

Remember: smooth ER is *not* simply RER without the ribosomes.

➔ *The interplay of vesicles between the endoplasmic reticulum and Golgi apparatus is revised more fully in Chapter 9.*

- In addition to their obvious morphological difference, the RER and SER differ in composition and have separate and distinctive functions.

Check your understanding

8.1 Hepatocytes have membranes with a total surface area of approximately 110,000 μm^2. If the SER and RER form about 16% and 35% of the total area respectively, what are their surface areas in μm^2?

8.2 KEY ACTIVITIES OF THE SMOOTH ENDOPLASMIC RETICULUM

Most cells have relatively little SER but it is extensive in cells that synthesize lipids (Figure 8.4), such as testes, ovaries, adrenal glands, and liver. Regions of the SER also accumulate Ca^{2+}.

Biosynthesis of lipids

The SER is the major site for the production of membrane lipids, particularly **phospholipids** and **steroids** in eukaryotic cells. It is also involved in the first stages of the formation of **sphingolipids**.

Phospholipid biosynthesis

- Fatty acids (FAs) used in the production of phosphoacylglycerols (and sphingolipids (see 'Sphingolipids biosynthesis' section)) are produced in the cytosol and, being insoluble, are transferred to the SER membrane by **fatty acid binding proteins** for incorporation into the cytosolic leaflet of the membrane.
- The enzymes that form phospholipids are all found in the cytosolic face of the SER membrane. Their active sites face the cytosol, where their substrates are found.

➔ *Information about lipids can be obtained in Chapter 3. Biochemical details of the biosynthesis of lipids can be found in the companion volume,* Thrive in Biochemistry and Molecular Biology.

- Once formed, phospholipids must be transferred between the bilayers of the SER membrane. Given their hydrophilic heads, this is thermodynamically unfavourable and does not occur spontaneously at a significant rate.

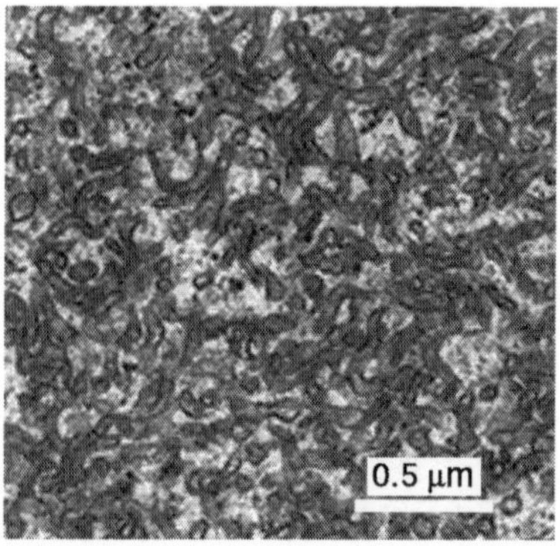

Figure 8.4 Electron micrograph showing the extensive smooth endoplasmic reticulum in a cell of the mouse retinal pigment epithelium. Reprinted with permission from Imanishi, Y., Gerke, V., and Palczewski, K. (2004) Retinosomes: new insights into intracellular managing of hydrophobic substances in lipid bodies. *Journal of Cell Biology*, 166, 447–53.

- However, **phospholipid translocators** or **flippases** catalyse their translocation from one side of the bilayer to the other within minutes; a rate that is approximately 10^5 times faster than would occur spontaneously.
- Phospholipid translocators are substrate specific, and thus generate the membrane lipid asymmetry between the two leaflets discussed in Chapter 3.
- Once formed in the SER, lipids are transferred to other members of the endomembrane system (for example, the Golgi apparatus, lysosomes) and plasma membrane in vesicles.
- Membranes of mitochondria, chloroplasts, and peroxisomes (and also the plasma membrane) also receive lipids from the SER but via **phospholipids exchange/ transfer proteins**.
- Specific exchange proteins acquire their particular phospholipids from the SER and carry it through the cytosol and deliver it to their target membrane.

Steroid biosynthesis

- Hepatocytes are the major producers of lipoprotein particles, which are released into the bloodstream and deliver lipids to other cells of the body.
- Hydroxymethylglutaryl-CoA reductase (HMG-CoA reductase), the key enzyme involved in the biosynthesis of steroids, is present in large amounts in the SER of hepatocytes.
- Large amounts of SER membrane are also found in adrenal gland cells that synthesize the steroid, cortisol, and the gonads, which produce sex hormones.
- Smooth ER is also associated with plastids in plant cells and is possibly involved in the biosynthesis of plant hormones.

Sphingolipids biosynthesis

- The SER also produces **ceramide**, which is the first stage in the production of sphingolipids (**sphingomyelins** and **glycolipids**).
- The ceramide is transferred to the Golgi apparatus in vesicles (Chapter 3).
- Within the Golgi apparatus, sugars are added to the ceramide to form **glycolipids** or phosphocholine is added to generate **sphingomyelin** molecules; both reactions are catalysed by enzymes attached to the *inner* face of the Golgi apparatus.
- Consequently, membranes containing glycolipids and sphingomyelins, such as the plasma membrane, never have them in their cytosolic surfaces.

Detoxification

Many poisons are relatively hydrophobic and so cannot easily be excreted by the kidneys.

- Detoxification by the SER, particularly in liver hepatocytes, converts them to compounds that are more water soluble and therefore more easily removed by the kidneys.
- Detoxification is achieved in two phases, I and II.

Phase I detoxification

- In phase I of the detoxification process, the drug is typically oxidized and/or hydroxylated by a member of the **cytochrome P-450 monooxygenase** family.
- These enzymes are localized in the smooth endoplasmic reticulum, and catalyse the hydroxylation of a wide variety of substances (R-H in the equations) by incorporating one of the oxygen atoms into the drug to form the hydroxyl group. The other oxygen atom is reduced to water (hence 'monooxygenases').
- NADPH supplies an electron to complete the phase I stage.

$$R\text{-}H + O_2 + NADPH + H^+ \rightarrow R\text{-}OH + H_2O + NADP^+$$

- The same enzyme system catalyses the conversion of unsaturated compounds to epoxides, which are substrates for epoxide hydrolase that catalyses their conversion to a glycol.

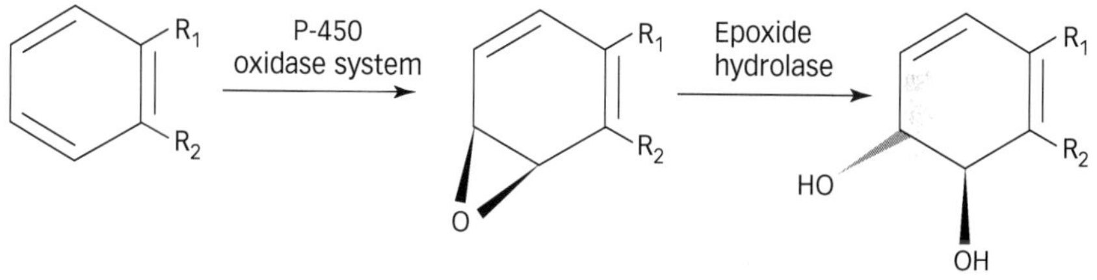

- The hydroxyl groups on products of these reactions makes them:
  - more water soluble compared to the original drug
  - form attachment points for the actions of phase II enzymes.

Phase II detoxification

- Phase II of the detoxification links polar compounds, such as a glucuronate or sulphate groups, to the oxidized product to form compounds that are even more water soluble. Enzymes that catalyse these addition reactions include glucuronosyl transferases of the SER membrane and sulphotransferase in the cytosol.
- These enzymes use UDP-glucuronate and 3′-phosphoadenosyl 5′-phosphosulphate (PAPS) as the respective donor substrates:

$$R\text{-}OH + UDP\text{-}glucuronate \xrightarrow{\text{Glucuronosyltransferase}} R\text{-}O\text{-}glucuronate + UDP$$

$$R\text{-}OH + PAPS \xrightarrow{\text{Sulphotransferase}} R\text{-}O\text{-}SO_3^- + PAP$$

- The water solubility of the products of these reactions facilitates their excretion mainly by the kidneys, although small amounts are also lost in bile to the faeces.
- Enzymes that catalyse both phases have broad specificities. At least 50 different members of the cytochrome P-450 enzyme family are found in the SER of hepatocytes enabling them to detoxify a wide range of organic toxins and drugs to which the body may be exposed.

Storage of Ca^{2+}

The concentration of free Ca^{2+} in the cytosol is generally kept low; about 0.1 μmolar or less. The variation of this concentration is essential to regulate many reactions, stimulate muscular contraction (Chapter 7) and secretion (Chapter 9), and mediate **signal transduction** (Chapters 3 and 15).

- Most cells have specific parts of the ER that can accumulate Ca^{2+}.
- The ER membrane possesses an **ATP-dependent Ca^{2+}-pump** that is able to sequester Ca^{2+} from the cytosol into the ER lumen.
- The lumen has a high concentration of Ca^{2+}-binding proteins, which allows its concentration to reach relatively high values (100 μm to 5 mmol dm^{-3}) compared to the cytosol.
- The subsequent release of Ca^{2+} can be used to regulate a number of cellular activities such as those involved in cell signalling.
- The ER is the most significant storage site of cellular Ca^{2+}, although other organelles, for example, mitochondria (Chapter 11), also have significant stores.
- In muscle cells, the SER has evolved and occurs in a modified form specifically for Ca^{2+} uptake and release called the **sarcoplasmic reticulum** (see Chapter 7).
- The release and reuptake of Ca^{2+} by the sarcoplasmic reticulum stimulates the alternate contraction and subsequent relaxation of muscle fibres respectively.

Revision tip

Remember, active transport can only be demonstrated if an ion moves against its electrochemical gradient.

Check your understanding

8.2 Which of the following statements is/are true?

a. Fatty acids are transferred from the cytosol to the SER membrane by fatty acid binding proteins.

b. Phospholipid exchange/transfer proteins catalyse the translocation of membrane lipids from one side of the bilayer to the other approximately 10^5 times faster than would occur spontaneously.

c. The cytochrome P-450 monooxygenase family of enzymes is localized in the smooth endoplasmic reticulum.

d. The phase II stage of SER-mediated detoxification involves glucuronosyltransferase and sulphotransferase activities.

e. The ER membrane possesses a G protein that is able to sequester Ca^{2+} from the cytosol into the ER lumen.

8.3 KEY ACTIVITIES OF THE ROUGH ENDOPLASMIC RETICULUM

Ribosomes of the RER produce polypeptides that will become part of the endosomal system (ER, Golgi and lysosomal system membranes), or integral membrane proteins (Chapter 3) or be secreted from the cell (Chapter 9).

- Hence the RER is particularly abundant in cells that produce and secrete large amounts of proteins such as pancreatic β cells (which release insulin) and plasma cells (which secrete antibodies).

Translation of mRNA by ribosomes of the RER

The production of *all* polypeptides begins on free ribosomes in the cytosol.

- Polypeptides destined for the endosomal or plasma membranes or for secretion possess an **amino terminal signal peptide** sequence (Figure 8.5) that directs the ribosome to the ER membrane. Ribosomes producing polypeptides that lack such signals remain free in the cytosol.
- Thus, both free and ER bound ribosomes are interchangeable and constitute a common pool; whether a ribosome binds to the ER membrane is determined by the *type* of polypeptide it is producing.

➔ *Details of translation are given in Chapter 7 (and for bacteria in Chapter 5) and in the companion volume,* Thrive in Biochemistry and Molecular Biology.

- The translocation of polypeptides across the ER membrane begins before translation is completed and is therefore a **co-translational event.**
- Translation begins in the usual manner and proceeds until the amino terminal signal peptide protrudes from the free ribosome (Figure 8.6).
- The signal peptide is recognized by a **signal recognition particle** (**SRP**) that binds to both it and the ribosome (Figure 8.6). The resulting change in the conformation of the ribosome arrests translation.
- The ribosome complex is then captured by a receptor on the cytosolic side of the RER membrane.

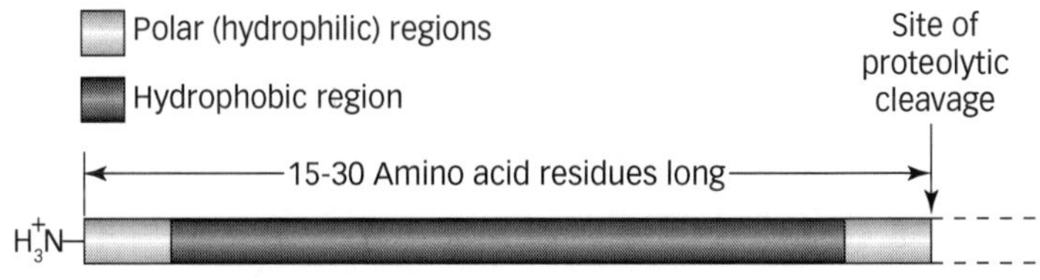

Figure 8.5 General structure of an amino terminal signal peptide sequence.

Key activities of the rough endoplasmic reticulum

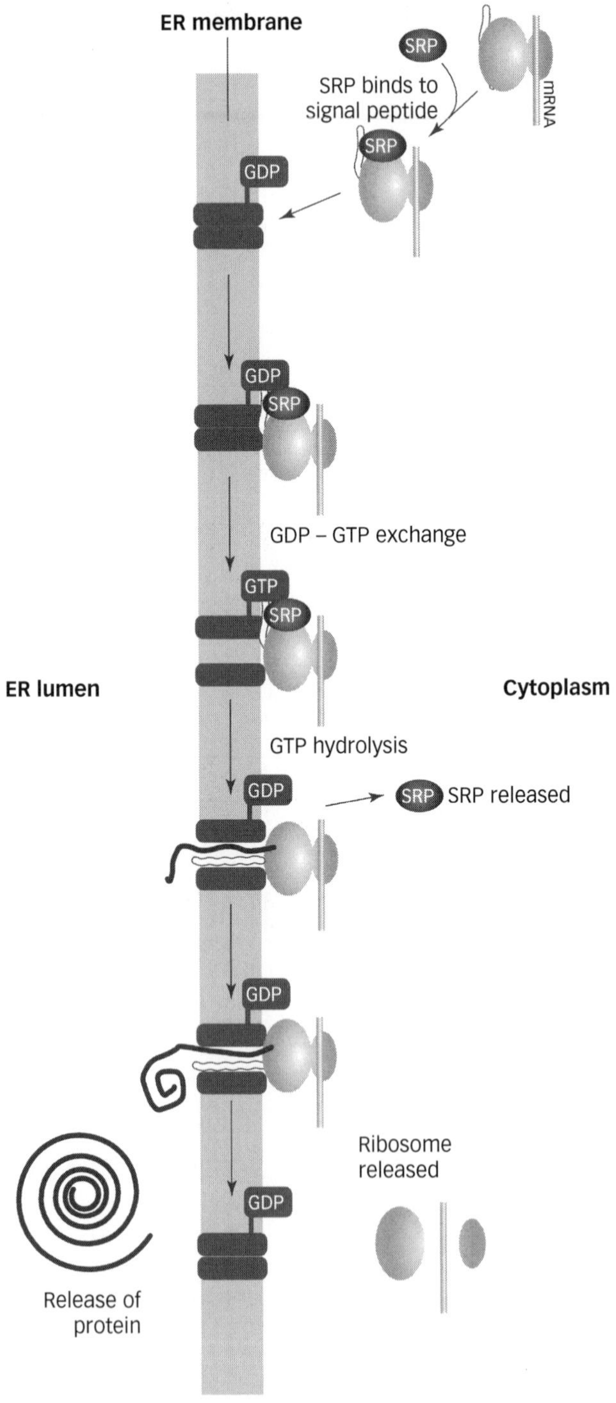

Figure 8.6 Sequence of events involved in the co-translational translocation of a polypeptide across the membrane of the rough endoplasmic reticulum. See relevant text for a description of each stage. The ribosome receptor has been omitted for clarity.

- The SRP is a complex structure composed of six different polypeptides and a 300 bp long ribonucleotide molecule of RNA. Three major binding sites are associated with the protein components:
  - One recognizes and binds to the amino terminal signal peptide
  - One associates with a portion of the ribosome to stall translation
  - One binds to a receptor on the ER membrane.
- The binding of the SRP to its ER membrane receptor directs the ribosome complex to an ER membrane **translocator**.
- Translocators are integral ER membrane proteins that form *hydrophilic* pores through the membrane. Translocators and their associated macromolecules are often called **translocons**.
- Translocators are closed until a ribosome binds to them, ensuring the ER membrane separates the ER lumen and cytosol as distinct compartments.
- Once a ribosome has been delivered to a translocator, the receptor is released to bind another complex. The SRP is also released back to the cytosol.
- Both the SRP and the ER translocator are **G proteins**, possessing GTP/GDP domains. Successive cycles of GTP binding by these proteins and its hydrolysis induce conformational changes that ensure the SRP is only released when the ribosome has appropriately bound to a translocator.

Looking for extra marks?

Blobel was awarded the 1999 Nobel Prize in Physiology or Medicine for his studies on the roles of signal peptides and the signal recognition particle in protein translocation across the endoplasmic reticulum membrane.

➔ *G proteins are discussed in other chapters (such as Chapters 3, 8, 15) and in the companion volume,* Thrive in Biochemistry and Molecular Biology.

- The ribosome remains attached to the cytosolic side of the membrane, giving rise to RER; the release of the SRP allows it to resume translation of its mRNA.
- The translocator inserts the growing polypeptide into its hydrophilic pore allowing it to cross the ER membrane into the ER lumen.
- A specific **signal peptidase** on the luminal side of the ER membrane catalyses the removal of the signal sequence.
- Translation then proceeds until elongation of the polypeptide is completed.
- Termination of translation allows the:
  - ribosome to dissociate into its subunits, which are released back into the cytosolic pool
  - finished polypeptide to be released into the lumen of the ER.
- Within the lumen, the polypeptide is appropriately folded and subjected to some post-translational modifications to form a protein, which is then transferred to the Golgi apparatus (Chapter 9).

Check your understanding

8.3 Globins are cytosolic proteins. (a) Where would you expect them to be produced? (b) Where would chimeric mRNA encoding a globin polypeptide within an amino terminal signal sequence be translated?

Integral membrane and ER proteins

Integral membrane proteins (IPs) are produced by ribosomes attached to the RER in a similar manner to those released as soluble proteins into its lumen.

- Unlike soluble proteins, IPs are retained within the ER membrane by transmembrane domains that typically consist of 20–30 hydrophobic amino acid residues (Chapter 3).
- Two main mechanisms are thought to be involved in the integration of the polypeptide into the ER membrane:
  1. one involves an amino terminal sequence and SRP
  2. the second is independent of an amino terminal ER signal sequence.
- In the former, the synthesis of the IP begins like that of a soluble ER-destined protein: an SP recognizes the amino terminal signal sequence and delivers it to a receptor and then a translocator as described earlier.
- Translation proceeds, with elongation of the polypeptide until the hydrophobic transmembrane sequence is formed.
- Figure 8.7 illustrates how this sequence operates as a **stop transfer sequence**, which halts *translocation* but allows *translation* to proceed.
- Thus, the remainder of the polypeptide is formed on the cytosolic surface of the ER membrane.
- The stop-transfer sequence is moved through a lateral side exit in the translocon into the lipid bilayer where it forms a permanent anchorage for the IP.
- The second mechanism of integrating proteins relies on the transmembrane sequence operating as a **start-transfer** sequence in the growing polypeptide.
- Translation proceeds in the cytosol until the start transfer sequence has been formed (Figure 8.8).
- The start transfer sequence is associated with two roles:
  1. It functions as an ER signal sequence allowing the SRP to deliver the ribosome complex to the receptor from which it is transferred to a translocator.
  2. The hydrophobic stretch of the start transfer sequence can be moved through the lateral opening of the translocon to form a membrane anchor.
- Figures 8.7 and 8.8 show the insertion of only a single transmembrane sequence; membrane proteins with multiple membrane-spanning regions have appropriate numbers of alternating start and stop transfer sequences to allow the polypeptide to pass back and forth across the membrane.

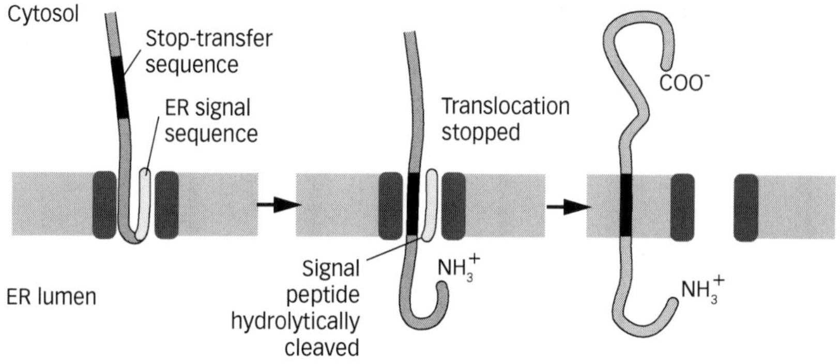

Figure 8.7 Schematic illustration of the role of the stop transfer sequence in the formation of an integral membrane protein. See text for details.

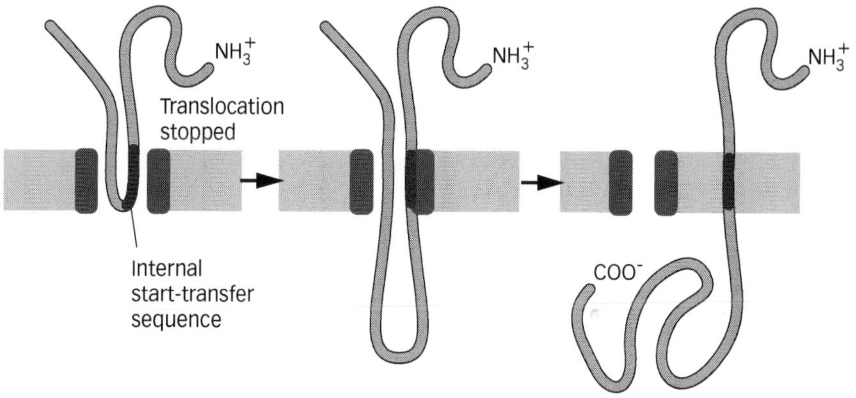

Figure 8.8 Schematic illustration of the role of the start transfer sequences in the formation of integral membrane proteins. For clarity, only one start transfer sequence has been shown. See text for details.

- The orientation of the start sequence at the point of insertion into the translocon determines which direction (N→C or C→N) the polypeptide crosses the bilayer and therefore whether the amino or carboxyl termini end up in the cytosol or lumen.
- Whatever method is used to form the IP, depending on the site of its function it can be:
  - retained in the ER membrane
  - transported to the nuclear envelope (Chapter 6)
  - incorporated into vesicles and then transported to the Golgi apparatus and from there to lysosomes or the plasma membrane (Chapter 9).

Key activities of the rough endoplasmic reticulum

Endoplasmic reticulum proteins

- A protein whose final destination is the ER itself possesses **ER retention signals**:
  - the retention signal of soluble proteins of the ER lumen is the carboxyl terminal sequence **Lys–Asp–Glu–Leu** (often called the **KDEL** sequence from the one letter abbreviations)
  - that of integral proteins of the ER membrane consists of the carboxyl terminal sequence, **Lys–Lys–X–X**.
- The ER retention signals ensure the proteins are eventually retained within the ER, even if they are transported to the Golgi apparatus, as described in Chapter 9.

Protein folding and post-translational activities of the ER

Polypeptides that will eventually become secreted proteins or proteins of the endomembrane or plasma membranes are folded and subjected to some post-translational modifications in the ER lumen to produce molecules with the necessary conformations.

- Protein folding begins as soon as the polypeptide has entered the lumen and may go through several stages until the final structure is reached.
- The lumen of the ER contains high concentrations of proteins called **chaperones** and enzymes that assist in the processes that fold proteins into their correct tertiary structures and allow quaternary proteins to be assembled from their subunits.

➔ *You can read about the structure of proteins in the companion volume,* Thrive in Biochemistry and Molecular Biology.

Chaperones and enzymes

- The major chaperone of the ER lumen is an **hsp** (**heat shock protein**) **70** called **BiP** (**Bi**nding **p**rotein). This chaperone has *slow* autocatalytic ATPase activity.
- BiP-ATP has a high affinity for exposed hydrophobic surfaces on partially folded proteins and binds to them. This masks the hydrophobic surfaces and prevents misfolding and aggregation of the polypeptide.
- However, over time, the BiP-ATP slowly hydrolyses:

$$\text{BiP-ATP} + H_2O \leftrightarrow \text{BiP-ADP} + P_i$$

- The slow hydrolysis changes the conformation of BiP to one that has a reduced affinity for the folded polypeptide.
- The BiP-ADP then dissociates from the protein, which can fold appropriately.
- An enzyme particularly prevalent in the ER lumen is **disulphide bond isomerase (PDI)**, which oxidizes proteins and assists in their folding by breaking and reforming disulphide bonds.

- This activity begins before the synthesis of the polypeptide has finished, which allows PDI to shuffle the positions of disulphide bonds in the polypeptide until they are formed between pairs of appropriate cysteine residues and can stabilize the protein's structure.
- The oxidizing environment within the ER lumen promotes disulphide bond formation.

Check your understanding

8.4 The chaperone BiP is found only in high concentrations in the ER lumen and not in other compartments of the endomembrane system or in the cytosol. Explain how these differential concentrations are achieved by the cell.

➔ *The formation of disulphide bonds in the periplasmic space of bacteria is briefly discussed in Chapter 5.*

- Most of the proteins that enter the ER lumen are also **glycosylated** by **glycosyl transferases** found on the luminal face: sugar residues are added to the growing polypeptides to eventually form **glycoproteins**. Glycosylation is continued in the Golgi apparatus (Chapter 9).

Quality control of folding

- Quality control of protein folding ensures that inappropriately folded proteins within the lumen are eventually degraded.
- Sensor molecules of the ER membrane are able to detect misfolded proteins and activate signalling pathways that lead to a reduction in the synthesis of most proteins but stimulate the production of those involved in protein folding and degradation.
- Incorrectly folded proteins do not accumulate within the lumen but are transferred to the cytosol where they are degraded by **proteasomes**, which allows their amino acids to be recycled.

➔ *More about proteasomes can be obtained in Chapter 16.*

- Correctly folded proteins are then allowed to leave the ER in vesicles from transitional elements and be transported to the Golgi apparatus before reaching their final destination.

Looking for extra marks?

Studies on subcellular fractions of ER are performed on the portion of the homogenate that sediments at 100,000*g* and is generally called the microsomal fraction (Chapter 2).

Check your understanding

8.5 Which of the following statements is/are true or is/are false?

a. The ER generally has the largest volume of all the intracellular compartments.
b. Smooth ER is involved in the synthesis of a wide variety of lipids and fatty acids.
c. Fatty acids are synthesized in the cytosol.
d. Transitional ER lacking ribosomes is involved in detoxification reactions.
e. Rough endoplasmic reticulum is simply smooth ER from which the ribosomes have detached.
f. Sarcoplasmic reticulum is a highly evolved form of smooth ER.
g. Any particular ribosome can translate any particular mRNA.
h. Only polypeptides possessing an appropriate amino terminal signal can cross the ER membrane.
i. The highly oxidizing environment within the RER promotes the formation of disulphide bonds in proteins.
j. Transitional vesicles carry materials from the ER to the Golgi apparatus.

Revision tip

Do browse through the further reading references given for this chapter online. Go to http://www.oxfordtextbooks.co.uk/orc/thrive/

9 Golgi apparatus, the endosomal-lysosomal system, and vacuoles

Key features of the Golgi apparatus

- The Golgi apparatus consists of stacks of flattened membranous and associated vesicles and tubules.
- The stacks are sometimes called dictyosomes and are polarized: they have *cis* and *trans* faces.
- Golgi sacs are usually called cisternae.
- Material exported from the endoplasmic reticulum, such as lipids and proteins, enter the *cis* Golgi apparatus, are sorted, and then sent to a variety of destinations from the *trans* face.
- The Golgi apparatus chemically modifies the structures of the proteins it receives by a series of post-translational modifications and also completes the synthesis of glycolipids.
- Various vesicles from the plasma membrane also fuse with the Golgi apparatus, allowing their intravesicular components to be recycled.

9.1 STRUCTURE OF THE GOLGI APPARATUS

The **Golgi apparatus** has a distinctive appearance when viewed by electron microscopy (Figure 9.1), although details of its morphology differ from cell type to cell type and species to species.

Structure of the Golgi apparatus

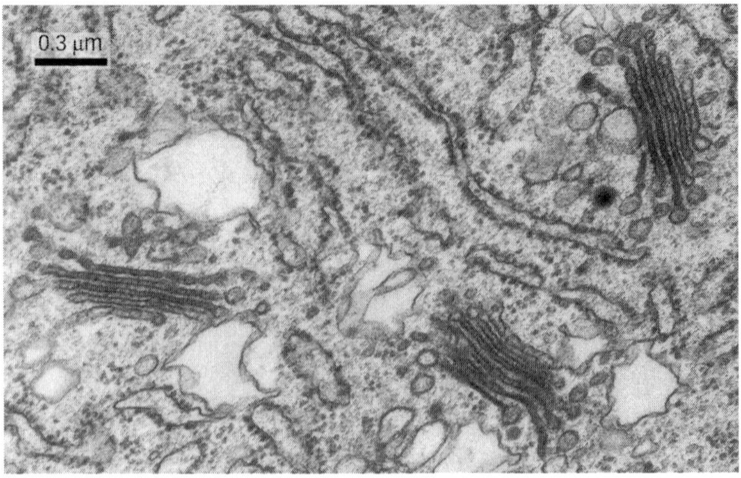

Figure 9.1 Electron micrograph of the Golgi apparatus.

- The Golgi apparatus always possesses flattened membranous stacks that are associated with numerous vesicles.
- In its most characteristic form, the Golgi apparatus shows three levels of organization:
  - **cisternae** or **saccula**
  - stacks or **dictyosomes**
  - the Golgi apparatus/complex/body.
- Golgi cisternae are typically flattened sacs of smooth membranes, with a central plate-like area of 0.5–1 μm diameter that is often continuous with peripheral tubules and associated with vesicles.
- Stacks of cisternae (sometimes called dictyosomes, although this term is often used to refer to the Golgi apparatus of plants) are the most obvious feature of the Golgi apparatus (Figure 9.1).
- Animal cells usually have five to eight cisternae per stack but it is not uncommon to find numbers in excess of 30 in so-called lower organisms and several 100s in some types of plant cells.
- The Golgi apparatus consists of an association of stacks of cisternae (Figure 9.1).
- In mammalian cells the stacks of the Golgi apparatus are generally linked together by tubules in a compact arrangement but in plants and invertebrates they are normally dispersed throughout the cytoplasm.

Looking for extra marks?

Read a short biography of the discoverer of the Golgi apparatus and Nobel Prize winner, Camillo Golgi at: http://www.nobelprize.org/nobel_prizes/medicine/laureates/1906/golgi-bio.html

9.2 FORMATION OF THE GOLGI APPARATUS

Cisternae are ***polarized*** (Figures 9.1 and 9.2) in having two opposing faces.

- One face is associated with the endoplasmic reticulum and is called the ***cis*** or **forming** face; the other face is the ***trans*** or (less commonly) **releasing** face.
- Cisterna between the *cis* and *trans* faces are referred to as ***medial.***
- The *cis* face is associated with a network of interconnected tubule and partial cisternal structures called the ***cis*** **Golgi network** (CGN)
- The *trans* face is associated with a complex of partial cisternal structures and budding vesicles called the ***trans*** **Golgi network** (TGN).
- Membrane-bounded transport vesicles called **coatomer-coated** (COP) **II transition vesicles** bud off from transitional endoplasmic reticulum and eventually fuse with the CGN.
- These coated vesicles differ from clathrin-coated vesicles reviewed earlier (see Chapter 3, Figure 3.23b–d and Figure 9.3).
- A small G protein called **ADP ribosylation factor** (ARF) mediates their formation.

➔ *The formation of clathrin-coated vesicles was described in Chapter 3. G proteins have numerous roles, many of which have already been revised in earlier chapters.*

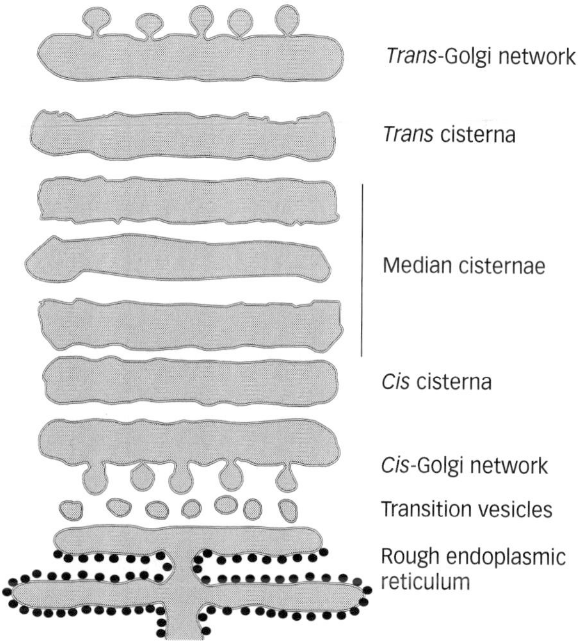

Figure 9.2 Schematic illustration of a stack of Golgi cisternae.

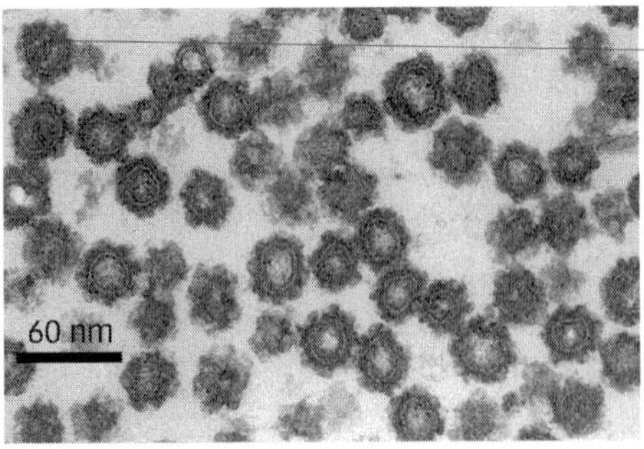

Figure 9.3 Electron micrograph of purified coatomer-coated vesicles from yeast endoplasmic reticulum. Courtesy of Professor L. Orci, Département de Morphologie, Faculté de Médicine, Université de Genèva, Switzerland.

- ADP ribosylation factor occurs in the cytosl in its ARF-GDP form. When it associates with a membrane bearing a specific **guanine nucleotide exchange factor** its GDP is exchanged for a GTP.
- The active ARF-GTP is able to bind to the membrane (the ER membrane in this case) and recruit coatomer proteins to the site of vesicle formation.
- These vesicles carry newly synthesized proteins and lipids produced in the lumen of the ER and by the smooth endoplasmic reticulum respectively to the CGN.
- These materials move through Golgi stacks and normally leave the Golgi apparatus through the TGN.

➔ *You can revise the synthesis of proteins and lipids and the role of the endoplasmic reticulum in Chapters 5, 7, and 8. Biochemical details of the biosynthesis of lipids are revised in the companion volume,* Thrive in Biochemistry and Molecular Biology.

Check your understanding

9.1 Which of the following statements is/are true?

a. The Golgi apparatus is formed from transition vesicles that are produced by the endoplasmic reticulum.

b. The *trans* or forming face of the Golgi apparatus is associated with the endoplasmic reticulum; the other face is called the *cis* or (less commonly) releasing face, while cisterna found between the *cis* and *trans* faces are called *medial* types.

c. The Golgi apparatus usually has five to eight cisternae per stack.

d. Coatomer-coated (COP) I transition vesicles bud off from transitional endoplasmic reticulum and eventually fuse with the *cis* Golgi network.

e. Adenosine diphosphate ribosylation factor (ARF) is a G protein that mediates the formation of ER-derived COP vesicles.

9.3 MOVEMENT THROUGH A GOLGI STACK

Two major models have been proposed to explain the movement of material through the Golgi stack:

1. **vesicular transport model**
2. **cisternal maturation model**.

- In vesicular transport (Figure 9.4a), vesicles are thought to bud from one cisterna and fuse with the next, carrying proteins, lipids, and carbohydrates within them.
- Thus, a cisterna would be a relatively stable structure, with molecules destined for sites post-TGN being moved in a *cis* to *trans* direction; usually called **antegrade** flow.
- In the cisternal maturation model (Figure 9.4b), cisternae progressively move along the stack: the Golgi apparatus is viewed as a *dynamic* structure with vesicles forming a cisterna at the *cis* face, which is then modified and matures as it progressively becomes a *medial* and then a *trans* cisterna before it is dissipated into vesicles at the TGN.

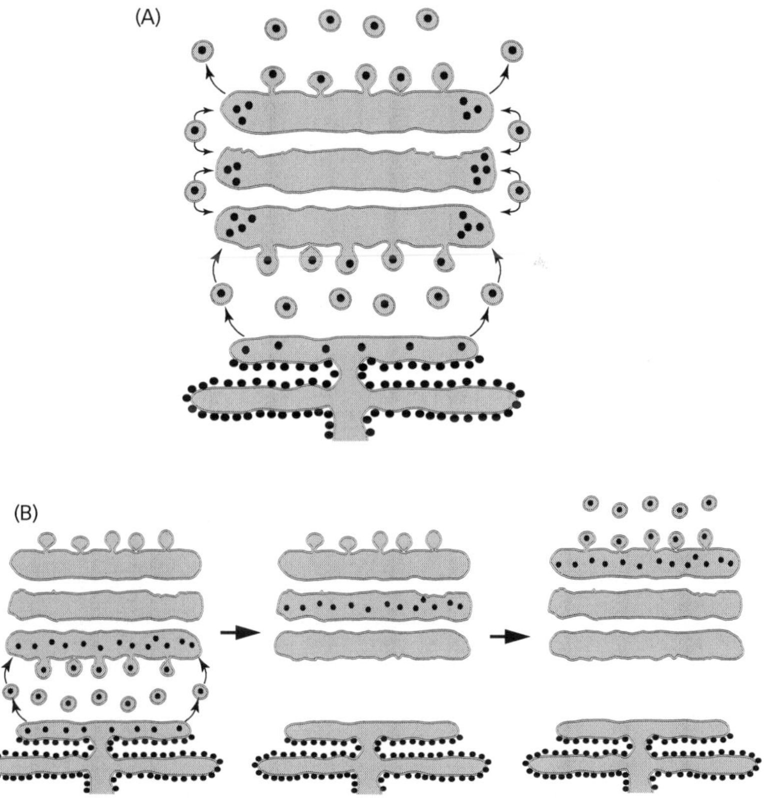

Figure 9.4 Schematics illustrating (A) the vesicular transport and (B) the cisternal maturation models of the movement of material through a Golgi stack. See text for details.

- This model requires modifications to cisternae as they move through the stack, such as the **retrograde** transport of enzymes in vesicles to preceding cisterna, for which there is some evidence.
- *Cis* Golgi membranes are 5–7 nm thick (similar to the ER membrane), while membranes of the *trans* face are 7–9 nm thick (more akin to the plasma membrane). Thus, maturation through the stack involves a gradual transformation from an intracellular-type membrane to one similar to the plasma membrane.
- Many of the enzymes of the Golgi apparatus are integral membrane proteins (Chapter 3). It is possible these enzymes are segregated to specific cisterna because their migration to the next is not possible if their membrane-spanning α helices are too short to span its membrane.
- Much evidence favours the cisternal maturation model but it is possible that neither model is mutually exclusive; vesicular transport may be used in some cells in some circumstances.

9.4 KEY ROLES OF THE GOLGI APPARATUS

Both the CGN and TGN package their contents into vesicles to which various 'address labels' are attached, which allows the vesicles to be transported to specific intracellular sites. Within the Golgi cisterna, contents are not only packaged but subjected to a number of biosynthetic activities.

- The Golgi apparatus has essential roles in:
  - a number of biosynthetic activities
  - the distribution of materials to sites within the cytoplasm
  - the secretion of materials from the cell
  - the distribution of integral membrane proteins
  - the production of some of the polysaccharides and their deposition during the formation of plant and fungal cell walls.

Biosynthetic activities of the Golgi apparatus

- A variety of biosynthetic events are performed within the Golgi apparatus. For example, it contains enzymes that complete the glycosylation of glycoproteins and glycolipids.
- Proteases complete the hydrolytic events that convert inactive proteins to their active conformations. For example, insulin is produced and secreted by β cells of the islets of Langerhans in the pancreas (Figure 9.5a).
  - Insulin is synthesized as an inactive precursor called **preproinsulin** by ribosomes attached to the rough endoplasmic reticulum (Chapter 8).
  - Removal of the signal peptide by signal peptidase activity in the lumen of the RER converts preproinsulin to another inactive form, **proinsulin**.

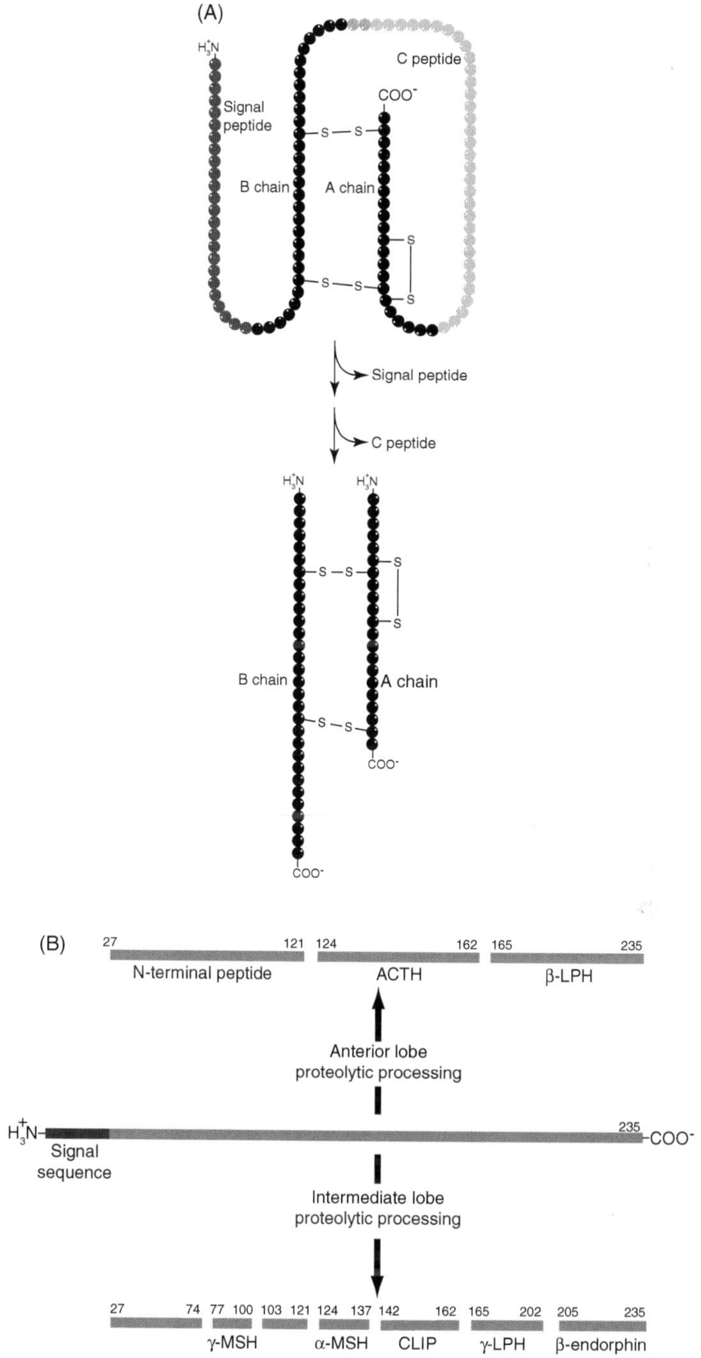

Figure 9.5 (A) Conversion of preproinsulin to active insulin. (B) Production of peptide hormones in the pituitary. ACTH, adrenocorticotropic hormone; CLIP, corticotrophin-like intermediate lobe protein; LPH, lipotrophin.

  - Proinsulin is delivered to the Golgi apparatus where further specific proteolyses in the TGN convert it to active insulin, which is secreted into the blood stream.
- A similar mechanism is sometimes used to produce bioactive peptides. For example, an extended polypeptide is differentially degraded by a series of specific proteolyses in the anterior and intermediate lobes of the pituitary gland to produce a whole series of different peptide hormones (Figure 9.5b).

Intracellular distribution from the Golgi apparatus

The Golgi apparatus is a major sorting station that packages and transfers the products of the ER to the appropriate cellular site including the ER itself, the plasma membrane and the endosomal system (Figure 9.6).

Transfer to the ER

- Proteins and enzymes that function in the ER but are transferred to the Golgi apparatus are returned to it from the CGN (Figure 9.6).
- Receptors in the CGN recognize the endoplasmic reticulum retrieval signals Lys–Lys–X–X and KDEL on membrane and soluble proteins respectively (Chapter 8).
- The proteins are concentrated into **COP I-coated** vesicles by the CGN for return to the endoplasmic reticulum.

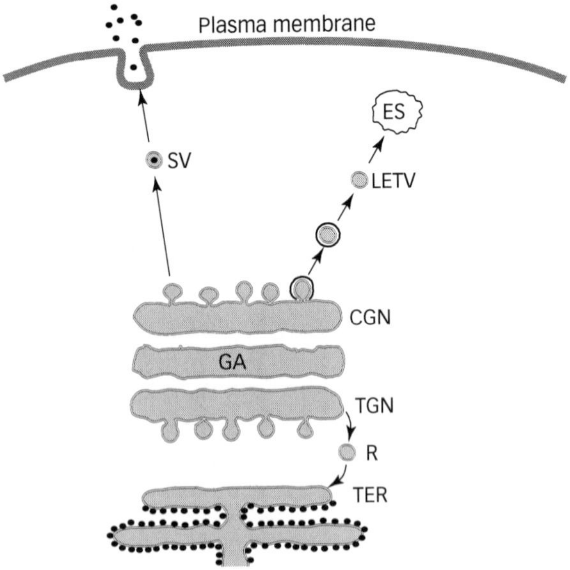

Figure 9.6 Overview of some of the distribution routes of materials from the Golgi apparatus. CGN; *cis* Golgi network; ES, endosomal system; GA, Golgi apparatus; LETV, lysosome enzyme transport vesicles; R, recycling of ER proteins; SV, secretory vesicle; TER, transitional endoplasmic reticulum; TGN, *trans* Golgi network. See text for details.

- These vesicles have a different coatomer coat to the COP II-coated vesicles produced by the ER membrane.
- The CGN then releases these vesicles to return the proteins to the ER.

Secretion

- The TGN packages materials for secretion from the cell into appropriate secretory vesicles (Figure 9.6). These materials include proteolytic enzymes and hormones from a variety of secretory cells.
- The transport vesicles move to the plasma membrane along fibres of the cytoskeleton (Chapter 7) and fuse with it in a process that involves the loss of the COP coating.
- The contents of the vesicle can then be released from the cell by **exocytosis** (Chapter 3).
- Secretion can be sporadic or continuous, depending on the cell type and the material secreted.
- For example, the release of insulin by pancreatic β cells or of digestive enzymes from pancreatic acinar cells is through **regulated secretory pathways**. They are initially stored in secretory vesicles before being released from the cell in sporadic events in response to an increase in the concentration of glucose in the blood or the ingestion of a meal respectively.
- In contrast, blood proteins are formed and secreted into blood plasma continuously by **constitutive secretory pathways** in hepatocytes of the liver.
- The Golgi apparatus also has an essential role in delivering enzymes to the endosomal-lysosomal system (Figure 9.6).

Distribution of integral membrane proteins

- Integral membrane proteins are synthesized by ribosomes of the RER, and are inserted into the membrane of the RER after synthesis, as described in Chapter 8.
- Given the ER and nuclear envelope are continuous, integral proteins of the outer and inner nuclear membranes can be transferred directly to them.
- Membranes of COP II-coated vesicles that deliver materials from the ER to the CGN therefore contain integral membrane proteins destined for such places as the Golgi membranes themselves or the plasma membrane or lysosomes.
- The Golgi apparatus sorts the proteins into appropriate COP I- and clathrin-coated vesicles, which transport the membrane proteins to their destinations.
- Fusion of the vesicles with the target membrane integrates the protein into that membrane.

9.5 LYSOSOMES

Lysosomes are eukaryotic organelles containing digestive enzymes that are most active at acidic pH.

Key features of lysosomes

Lysosomes are spherical organelles, limited by a single membrane (Figure 9.7).

- Lysosomes are formed when Golgi-derived clathrin-coated vesicles (Chapter 3) called **lysosomal enzyme transport vesicles,** which contain lysosomal enzymes, fuse with a type of membrane-bound vesicle called a **late endosome.**
- An ATP-driven H^+ pump in the membranes of these organelles lowers their luminal pH by transporting H^+ from the cytosol across their membranes.
- The lowering of the pH activates the 40–60 acid optimum hydrolytic enzymes that were present in the Golgi-derived vesicles.
- The digestive activities of these enzymes allow the original contents of the endosomes, and other vesicles called **autophagosomes** and **phagosomes,** to be partially or fully digested and the material they contain recycled.
- These enzyme activities digest unwanted cellular material and allow cellular components to be recycled; they also release material absorbed into the cell by endocytosis.
- The proteins of the lysosomal membrane are heavily glycosylated and their oligosaccharides form a dense layer lining the inner surface of the organelle, which protects the membrane from digestion.

Lysosomes and their discovery

Lysosomes were discovered by de Duve and co-workers who observed that the activity of an acid phosphatase in homogenates of animal tissues increased with time and storage in the cold.

- The 40–60 acid hydrolases of lysosomes are able to degrade all biological materials.

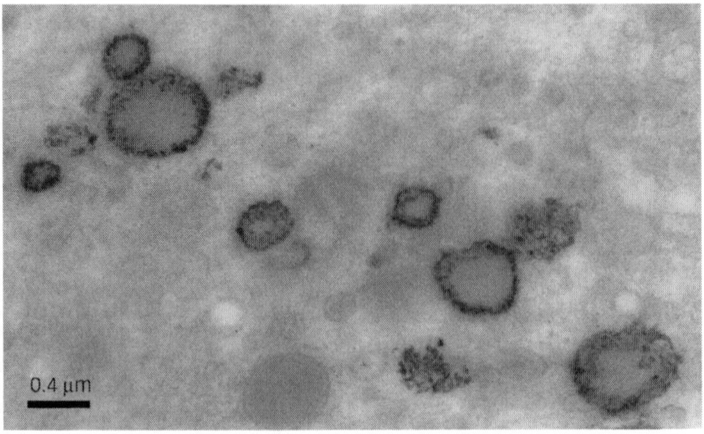

Figure 9.7 Electron micrograph showing lysosomes of a synovial cell visualized by aryl sulphatase activity. See text for explanations. Courtesy Dr C.J.P. Jones, Department of Pathology, University of Manchester, UK.

Lysosomal enzymes

Lysosomal enzymes are synthesized by the ribosomes of the rough endoplasmic reticulum (see Chapter 8) and enzymes within the Golgi apparatus.

- In the Golgi apparatus, biosynthetic enzymes recognize sites on the surface of the prelysosomal enzymes and tag them with mannose 6-phosphate (M6P) residues.
- The M6P residues act as molecular addresses, ensuring that the prelysosomal enzymes are recognized by M6P receptors (MPRs) on the inner face of TGN membrane.
- This allows the enzymes to be integrated into lysosomal enzyme transport vesicles (Figure 9.6), which can fuse with endosomes and eventually be delivered to the lysosomal system.

Check your understanding

9.2 The rupture of a *single* lysosome and release of its contents into the cytosol would most likely result in which of the following?

a. A substantial increase in the hydrolysis of cytosolic macromolecules.
b. Apoptosis of the cell.
c. Increased degradation of chloroplast DNA.
d. Stimulation of chloroplast division.
e. Relatively little damage to the cell.

Key features of the endosomal-lysosomal system

Among its many functions, the endosomal-lysosomal system:

- Mediates the internalization of extracellular material.
- Promotes the dissociation of ligands from their plasma membrane receptors.
- Facilitates the recycling of membrane receptors.
- Catalyses the digestion of materials of extracellular and cellular origin.

- Figure 9.8 gives an overview of the main organelles and vesicles involved in these processes.
- Lysosomes are discrete compartments within the endosomal-lysosomal system.
- They are formed when lysosomal enzyme transport vesicles shed their clathrin coats and fuse with and deliver their hydrolases to an endosome.
- Phagosomes and autophagosomes are membrane-bound vesicles that can fuse with late endosomes to form lysosomes or fuse with pre-existing lysosomes.
- These different organelles therefore deliver materials taken up by the cell by receptor-mediated endocytosis or phagocytosis or unwanted cytosolic structures to the lysosomal system for recycling.

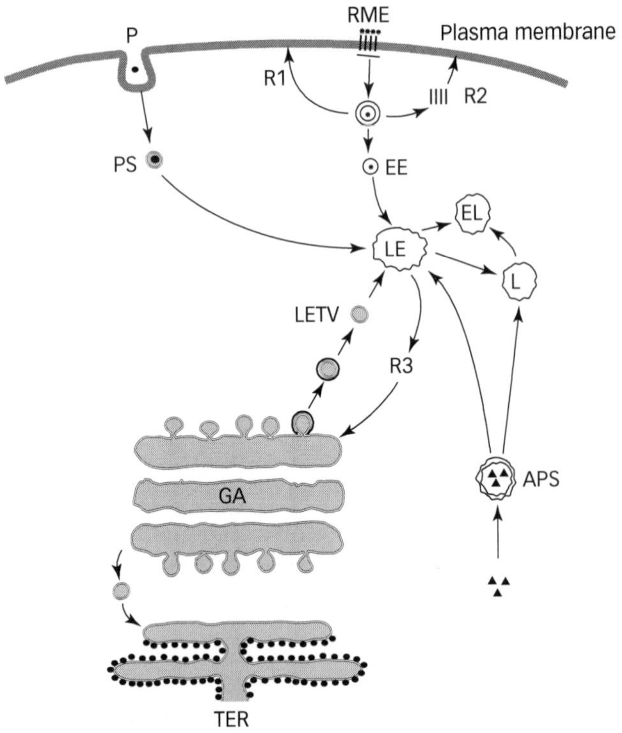

Figure 9.8 Overview of the activities of the endosomal-lysosomal system. APS, autophagosome; GA, Golgi apparatus; EE, early endosome; EL, endolysosome; L, lysosome; LE, late endosome; LETV, lysosome enzyme transport vesicles; P, phagocytosis; PS, phagosome; R1, recycling of EE coat components; R2, recycling of plasma membrane receptors; R3, recycling of MPRs; RME, receptor-mediated endocytosis. See text for explanations.

Receptor-mediated endocytosis

- Endosomes are formed when a particle or molecule (a ligand) binds to its specific receptor on the outer surface of the plasma membrane.
- The receptors also have cytoplasmic domains that can be recognized by an **adaptin** protein, which allows the ligand to be internalized into the cell in a clathrin-coated pit as described in Chapter 3.

➔ *The formation of clathrin-coated vesicles was described in Chapter 3.*

- Removal of the clathrin coating releases the vesicle and allows the adaptins and clathrin molecules to recycle back to the plasma membrane (Figure 3.23), and releases the vesicle into the cytoplasm.
- The fusion of endocytic vesicles in animal cells forms an **endosome**.
- Newly formed endosomes are generally referred to as *early* endosomes and mature ones as *late* endosomes.

- Early endosomes are part of the main plasma membrane receptor-recycling pathway. The low pH of early endosomes (5.9–6.5) causes the ligand-plasma membrane receptors complexes to dissociate.
- In many cases, the receptors are often recycled back to the plasma membrane, and the ligand retained within the endosome (Figure 9.8).
- Late endosomes are involved in intracellular digestion and the formation of vesicles that return the MPRs to the TGN (Figure 9.8).

Autophagy and phagocytosis

- Unwanted intracellular components are degraded in the endosomal-lysosomal system by a process called **autophagy** (Figure 9.8), during which regions of the cytoplasm, which may be cytosolic proteins or even organelles, are surrounded by a double membrane, possibly derived from the ER, and form an **autophagosome.**
- A number of cell types are specialized **phagocytes**, including macrophages and neutrophils of the immune system.
- Phagocytic cells can engulf particles and bacteria and internalize them within the cytoplasm in a membrane-bound **phagosome** (Figure 9.8).
- Degradation of the materials within a phagosome or an autophagosome occurs by essentially the same cellular process.
- The autophagosome or phagosome fuses with late endosomes giving rise to a lysosome or they fuse directly with a pre-existing lysosome (Figure 9.8).

Check your understanding

9.3 In terms of membranes, which of the following is the odd one out?

a. Golgi apparatus
b. Smooth endoplasmic reticulum
c. Transitional endoplasmic reticulum
d. Autophagosomes
e. Late endosomes.

- A major role of lysosomes is in the complete breakdown of the material within the autophagosome or phagosome.
- The acidification of late endosomes releases the hydrolases from the MPRs as described earlier. Continued H^+-ATPase activity in the lysosome reduces their internal pH to 4.5 to 5, which activates their acid hydrolases.
- It was previously thought that the different macromolecules of the material within the lysosome were catabolized to simple residues that were then translocated through the lysosomal membrane to the cytosol for recycling (Figure 9.9).

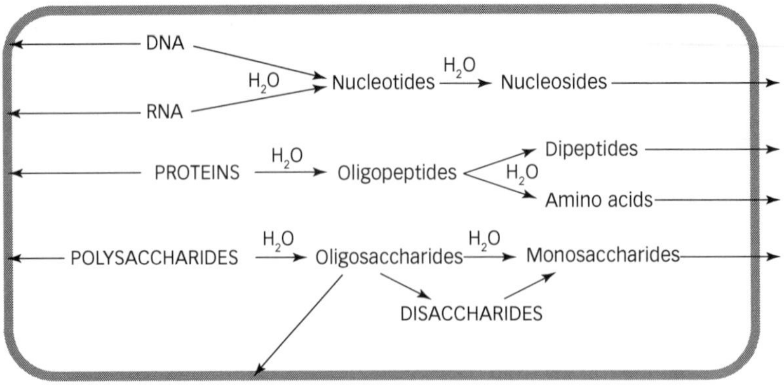

Figure 9.9 Schematic overview of the hydrolytic activities of lysosomes.

- However, this model of lysosome activity is now known to be too simple.
  - Late endosomes are also able to fuse with pre-existing lysosomes to form membranous structures, sometimes called **endolysosomes**, which are able to fuse with one another.
  - A pH gradient occurs across the endosomal-lysosomal system, with endososmes being the least acidic (about pH 6), lysosomes being the most (pH 4.5 to 5), and endolysosomes somewhere in between.
  - Since acid hydrolases are also present in endosomes, some digestion also occurs in them and in endolysosomes.
  - Also, evidence indicates that macromolecules are not invariably hydrolysed to their simplest components but that catabolic intermediates may exit the lysosome and be recycled, saving the cell wasting energy on synthesizing some molecules *de novo*.
- Following digestion of its internal material, the lysosome can begin the digestive cycle again by fusing with late endosomes or endolysosomes.
- Thus, the endosomal-lysosomal system is extremely diverse in terms of the sizes and morphologies of its vesicular components.
- Any material that cannot be degraded by the hydrolases of a lysosome is retained within the organelle (Figure 9.10), which is sometimes called a **secondary lysosome** or **residual body**.
- In protozoa, residual bodies eventually fuse with the plasma membrane and expel their contents.
- The residual bodies of vertebrate cells, however, can accumulate in the cytoplasm and may only be released when the cell is stressed.

Check your understanding

9.4 Arrange the following in terms of *increasing* acidity: late endosome, lysosome, early endosome, and cytosol.

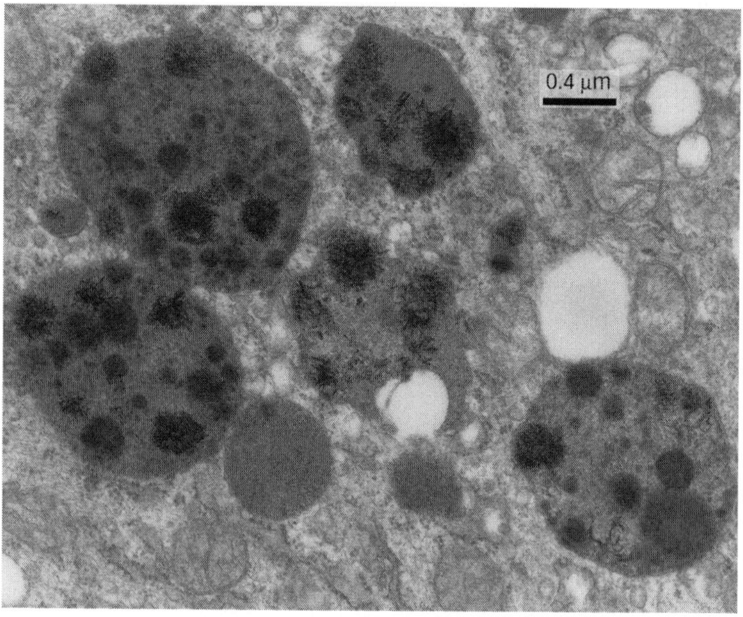

Figure 9.10 Electron micrograph of lysosomes containing indigestible colloidal gold particles in a macrophage of a rheumatoid arthritis patient who had been treated with gold injections. Courtesy of Dr C.J.P. Jones, Department of Pathology, University of Manchester, UK.

Melanosomes

Skin **melanocytes** produce and store the pigment **melanin** in specialized large lysosomes, diameter approximately 500 nm, called **melanosomes**.

- Melanosomes are transported along cytoskeletal fibres to the ends of extended cell processes where, unlike normal lysosomes, they are able to leave the cell.
- Once extruded from the melanocyte, the melanosome is taken up by keratinocytes of the epidermis resulting in pigmentation of the skin.

Lysosomal storage diseases

A genetic lesion leading to defects in the complement of lysosomal enzymes results in clinical defects known as **lysosomal storage diseases** because the substrate for the missing or defective enzyme accumulates in the lysosomal system.

- This usually impairs the functions of major organs such as the brain and liver with grave clinical consequences.
- More than 40 lysosomal storage diseases are known (Table 9.1). All are relatively uncommon in the general population although several are more prevalent among Ashkenazi Jews.

Disease	*Enzyme deficiency*	*Material (substrate) accumulated*
Gaucher's disease	β-glucocerebrosidase	Glucosylceramide
Inclusion cell disease (I-cell disease or mucolipidosis II)	Numerous lysosomal enzymes absent	Glycolipids, glycoproteins, sialyloligosaccharides
Metachromatic leucodystrophy	Arylsulphatase A	Sulphatides
Niemann–Pick disease	Sphingomyelinase	Sphingomyelin
Sphingolipidosis GM_1 gangliosidosis	β-galactosidase	GM_1 gangliosides
Tay–Sachs disease	Hexaminidase A	GM_2 gangliosides

Table 9.1 Examples of lysosomal storage diseases

Looking for extra marks?

Lysosomal storage diseases cannot be cured. However, adult Gaucher's disease is treated by enzyme replacement therapy involving intravenous injections of purified β-glucocerebrosidase. The enzyme is bound by receptors on macrophage surfaces and absorbed by the cells and delivered to the lysosomes where it catalyses the breakdown of the accumulated glucosylceramide.

9.6 VACUOLES

Vacuoles are the large vesicles that can occupy most of the cytoplasm of many eukaryotic unicellular organisms, fungi including yeasts, and mature plant cells, which lack lysosomes.

Revision tip

View beautiful electron micrographs and some animations of the Golgi apparatus, lysosomes and plant cells at The Cell: An Image Library™ of The American Society for Cell Biology at: http://www.cellimagelibrary.org/

Key features of vacuoles

The vacuole is the acidic subcellular compartment of many unicellular organisms, fungal and mature plant cells and is their equivalent to the lysosomes of animal cells.

- Vacuoles of mature plant cells are sometimes referred to as the **central vacuole** (Figure 9.11).
- Vacuoles are limited by a membrane called the tonoplast and contain a fluid called the cell sap.
- Constituents of vacuoles are synthesized by the ER and Golgi apparatus.
- Coated vesicles released from the Golgi apparatus transfer these materials to a provacuole, which can be regarded as an organelle analogous to the endosome.

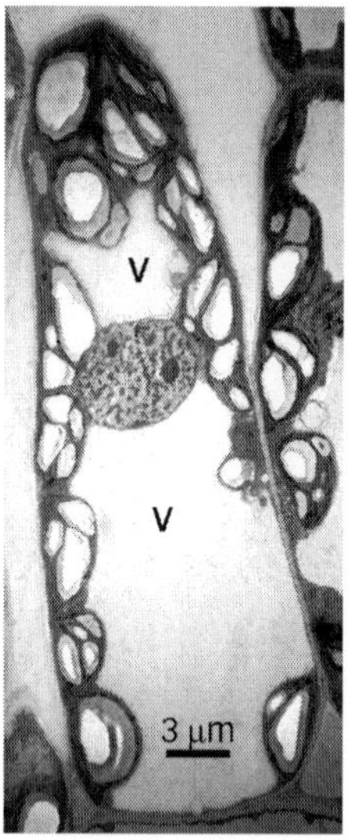

Figure 9.11 Electron micrograph of a mature plant cell. Note how most of the cytoplasm is occupied by the large central vacuole (V). Courtesy of Dr E. Sheffield, Department of Cell and Structural Biology, University of Manchester, UK.

- Provacuoles grow and mature to form the central vacuole, which can occupy as much as 30–90% of the cell volume.
- Vacuoles are similar to animal lysosomes in being acidic compartments containing acid hydrolases but they have a number of specific roles not associated with the lysosomes of animal cells, including maintaining turgor pressure and the storage of a variety of materials.
- The acidity of the cell sap is maintained by an H^+-ATPase in the tonoplast that is able to pump H^+ into the cell sap at the expense of ATP and is typically about two pH divisions lower than that of the cytosol.
- In many cases, the major functions of vacuoles are in maintaining turgor pressure and to store a variety of materials.

Check your understanding

9.5 In terms of $[H^+]$, how much more acidic are the contents of a vacuole compared to the cytosol?

Key roles of vacuoles

The biological role of vacuoles is much broader than that of lysosomes but in most plants it probably plays a limited role in intracellular digestion.

- A key function of the central vacuole is in providing **turgor pressure**.

- Turgor pressure arises from the osmotic pressure due to solutes in the vacuole forcing the tonoplast outwards until it is counteracted by the cell wall pressure.

  ➔ *Cell walls of eukaryotic cells are revised in Chapter 13.*

- This hydrostatic tension prevents the cytoplasm collapsing and contributes mechanical support to non-woody tissues; it is also necessary for the opening and closing of stomata.
- The H^+-ATPase of the tonoplast is able to contribute to buffering the cytosol by counteracting any acidification of the cytosol by pumping H^+ into the vacuole.
- A variety of materials are stored by plants in the central vacuole. These include:
  - the temporary storage of ions, sugars, and amino acids
  - anthocyanin pigments that impart colour to the petals and the sepals of some plants
  - toxic compounds such as phenols and glycoside whose release deters predation by a number of pests and herbivores
  - in crassulacean acid metabolism (CAM) plants (Chapter 10), approximately 90% of the malate formed is stored in the vacuole, which prevents acidification of the cytosol
  - waste materials are often stored by plants in their vacuoles
  - in some seeds, proteins are often stored in the vacuole until needed at germination.

Check your understanding

9.6 Which of the following statements is/are true or is/are false?

a. Membranes of the CGN have a thickness of about 8 nm and are thinner than those of the TGN.

b. Both COP and clathrin-coated vesicles are formed by the Golgi apparatus.

c. Golgi transport vesicles for secretion and those involved in transferring lysosomal enzymes are COP and clathrin-coated respectively.

d. Late endosomes are formed from lysosomal enzyme transport vesicles.

e. Lysosomal enzyme transport vesicles can fuse with endosomes, phagosomes, and autophagosomes.

f. The presence of a carbohydrate lining protects lysosomes from autodigestion.

g. In receptor-mediated endocytosis, membrane receptor molecules are recycled from lysosomes.

h. Autophagy is the only method for intracellular digestion.

i. Residual bodies invariably discharge their contents to the exterior of the cell.
j. Vacuolar functions mirror those of lysosomes.

Revision tip

Do browse through the further reading references given for this chapter online. Go to http://www.oxfordtextbooks.co.uk/orc/thrive/

10 Chloroplasts and photosynthesis

Key features of chloroplasts

- In plants, chloroplasts are rounded oval structures approximately 4–10 μm long and 1–5 μm in diameter (Figure 10.1).
- Chloroplasts are surrounded by a chloroplast envelope that encloses a stroma containing a third type of membrane called the thylakoid system (Figure 10.1).
- Variations in structure occur between the chloroplasts of plants, algae, and unicellular eukaryotic photosynthetic organisms.
- The envelope regulates the exchange of materials between the cytosol and the chloroplast.
- The main photosynthetic tissues in most plant cells are mesophyll cells of the leaf. Mesophyll cells of the palisade contain up to several hundred chloroplasts, those of the spongy mesophyll about 40.
- Chloroplasts, like mitochondria and hydrogenosomes (Chapter 11), are energy transduction organelles, which convert energy from one form to another.
- Chloroplasts capture light energy in the light-dependent reactions of the thylakoid system and use that energy to produce NADPH and ATP.
- Carbohydrates are synthesized from CO_2 and H_2O in the light-independent or dark reactions in the stroma using the products of the light reactions.
- Water is the source of the reducing hydrogen. The dioxygen or molecular oxygen (O_2) released is excreted to the atmosphere.

- Starch is formed and stored in chloroplasts, where it can also be converted to transportable soluble sugars.
- Other biosynthetic reductive reactions occur in chloroplasts.
- Chloroplasts most probably evolved from cyanobacterial endosymbionts of early eukaryotic cells.

Revision tip

Revise the structures of chloroplasts in different organisms using the excellent electron micrographs in The Cell: An Image Library™ of The American Society for Cell Biology at: http://www.cellimagelibrary.org/

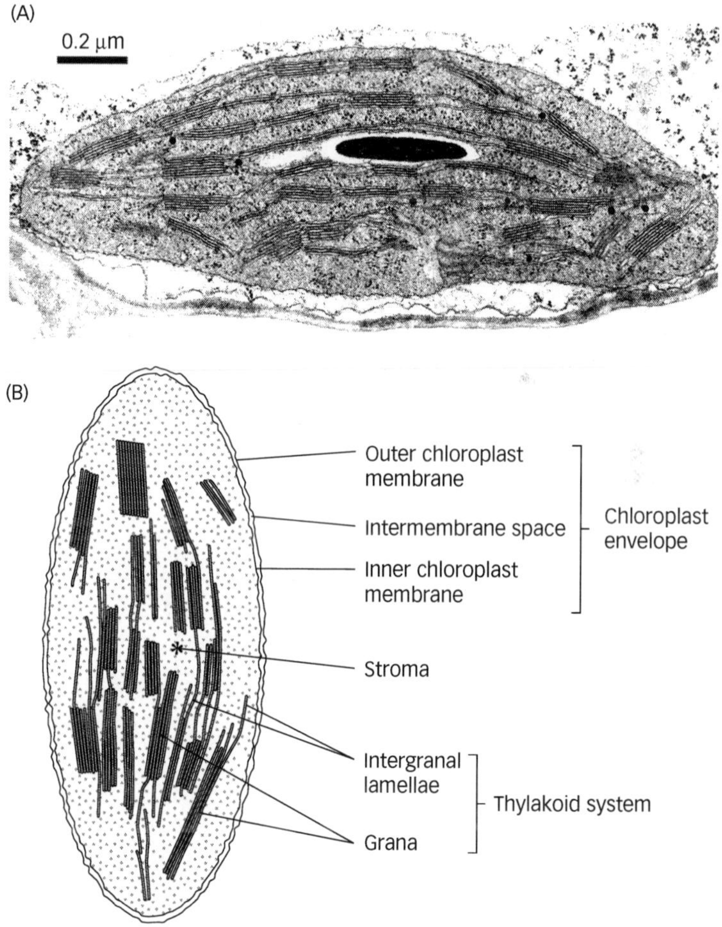

Figure 10.1 (A) Electron micrograph of a chloroplast of a plant cell. Courtesy of Professor M.W. Steer, University College Dublin, Republic of Ireland. (B) Schematic illustration of a typical plant chloroplast.

10.1 CHLOROPLAST STRUCTURE

The chloroplasts of green plants (Figure 10.1) are relatively uniform in structure; those of unicellular organisms and algae show considerably more variations (Figure 10.2). This section will concentrate on the better studied plant chloroplasts.

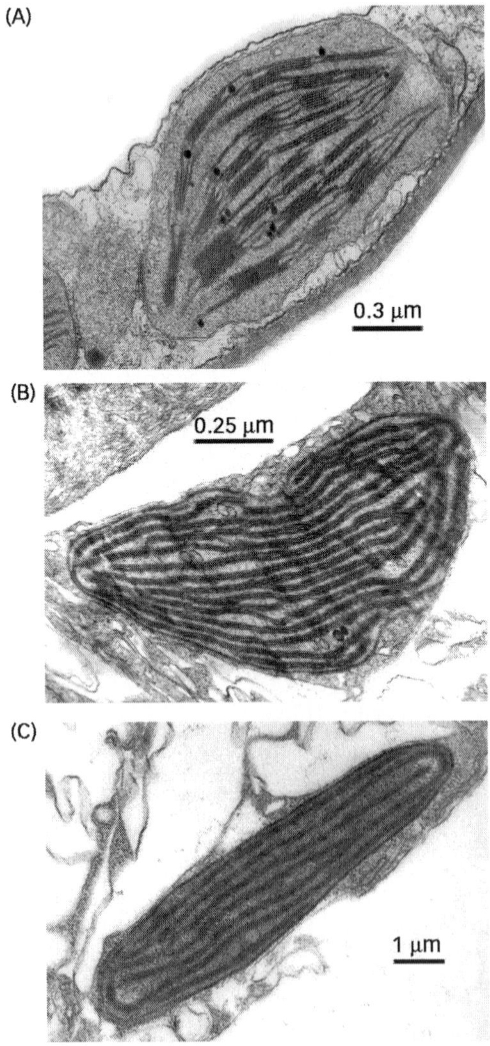

Figure 10.2 Electron micrographs of chloroplasts from (A) pea (*Pisum sativum*), (B) a red seaweed of the Porphyra group, and (C) a brown seaweed (*Fucus serratus*). Note the absence of grana in the seaweeds. (A) Courtesy of Dr G. Beakes, School of Biology, Newcastle University, UK.

Chloroplast envelope

The **chloroplast envelope** consists of **outer** and **inner membranes** (**OCM** and **ICM**, respectively) separated by an **intermembranous space**.

- The OCM contains **porins** (Chapters 4 and 11).
- Porins are proteins that form pores in membranes. Their presence in the OCM means it is relatively freely permeable to numerous ions and metabolites of low M_r (up to approximately 5000).
- The ICM is selectively permeable to most metabolites, and controls what enters and leaves the chloroplast.
- The ICM also contains the protein machinery that regulates the entry of proteins into the chloroplast.

Thylakoid membranes

Thylakoid membranes of plants form an internal system of membranes within the stroma and are the site of the **light-dependent reactions** of photosynthesis.

- In plants the thylakoid system is divided into two structural domains:
  - cylindrical stacks of flattened sacs called **grana**
  - interconnecting single membrane regions called **stroma lamellae** or **intergranal frets** (Figures 10.1 and 10.3).
- The thylakoid system forms a highly folded, continuous network of membranes that encloses a space called the **lumen**, which is distinct from the stroma.
- Chloroplasts of algae lack grana and their thylakoid system consists of extended appressed membranes.

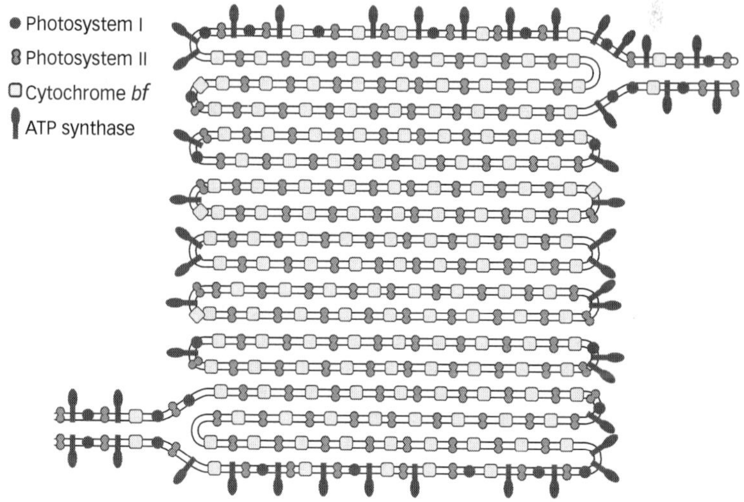

Figure 10.3 Simplified schematic indicating the separation of grana and interconnecting stroma lamellae.

Chloroplast stroma

The stroma is the site of the **light-independent** or **dark reactions** of photosynthesis.

- The stroma is a protein-rich aqueous gel that appears granular in electron micrographs (Figure 10.1).
- The proteins include enzymes that catalyse the light-independent reactions of photosynthesis.
- The stroma also contains chloroplast DNA and chloroplast tRNA molecules, and chloroplast ribosomes.

Check your understanding

10.1 Which of the following statements is/are true?

a. Animal cells contain chloroplasts.
b. Plant cells generally contain mitochondria.
c. All plant cells contain chloroplasts.
d. Animal cells do not contain mitochondria.
e. All plant cells contain mitochondria and chloroplasts.

10.2 MECHANISM OF PHOTOSYNTHESIS IN PLANTS

The various forms of bacterial photosynthesis are reviewed in Chapter 4.

Overview of photosynthesis

Photosynthesis consists of *two* linked processes or sets of reactions, called the light-dependent and light-independent (or dark) reactions respectively. Their interdependence is summarized in Figure 10.4.

- In the light reaction, light is absorbed and the acquired energy used to split water:

$$2\,H_2O \rightarrow 4H^+ + 4e^- + O_2$$

- Electrons (e^-) enter an electron transport chain and are eventually used to reduce $NADP^+$ to NADPH. The O_2 formed is excreted to the atmosphere.
- During electron transport, H^+ are pumped from the stroma into the lumen. These are returned to the stroma via an ATP synthase, a process that forms ATP.
- The formation of ATP during photosynthesis is generally called **photophosphorylation**.
- The useful products of the light-dependent reactions are therefore NADPH and ATP.

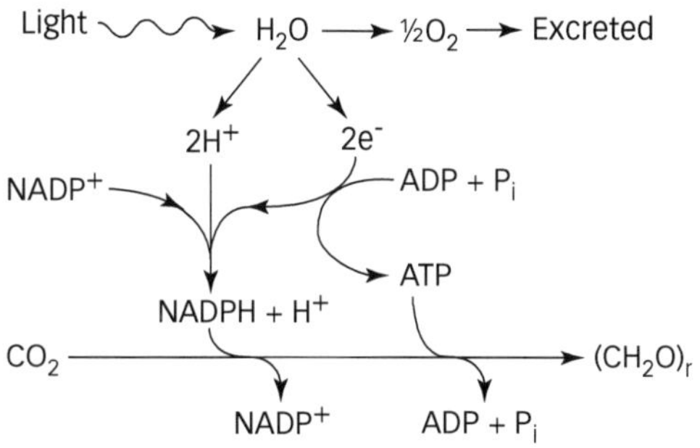

Figure 10.4 Overview of events of the light-dependent and light-independent (or dark) reactions of photosynthesis.

- In the light-independent reactions, NADPH and ATP are used to reduce CO_2 to carbohydrate according to the (greatly simplified) equation:

$$6CO_2 + 6\,H_2O \xrightarrow{\text{NADPH, ATP}} C_6H_{12}O_6 + 6O_2$$

- The light-dependent reactions occur in the thylakoid system, the independent reactions in the stroma.

Photosynthetic pigments and light absorption

The thylakoid membranes of plants and green algae contain **chlorophylls *a*** and ***b***. A number of other accessory light absorbing pigments, including carotenoids, xanthophylls, and phycobilins, are distributed among photosynthetic organisms.

- Photosynthetic pigments are **conjugated** molecules (contain alternate single and double bonds) as shown for the chlorophylls *a* and *b* in Figure 10.5.
- Conjugated molecules, and therefore photosynthetic pigments, are extremely efficient photoreceptors of visible light.
- Chlorophylls have maximum absorption at wavelengths of approximately 420 and 660 nm.
- The maximum absorption of light by accessory pigments is between approximately 450 and 650 nm and complements that of the chlorophyll pigments (Figure 10.6).
- Thus, photosynthesis is driven by light of wavelengths approximately 400–700 nm, which correspond to the peak intensity of visible radiation from the sun reaching the earth's surface (Figure 10.6).

Photosystems

- Chlorophylls, accessory pigments, and associated proteins are organized within the thylakoid membrane into functional units called **photosystems** (**PSs**).

Mechanism of photosynthesis in plants

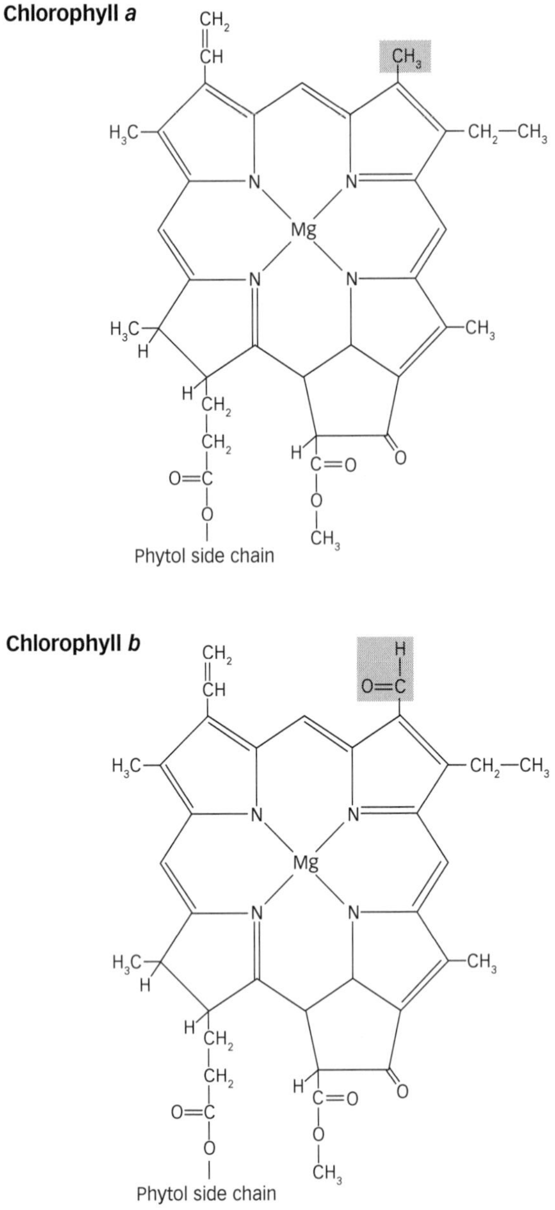

Figure 10.5 Structures of chlorophylls *a* and *b*.

- Two photosystems, called **I** and **II** respectively, occur in plants and algae.
- Photosystem I is mainly found in stromal lamellae granal membranes, as is the **chloroplast ATP synthase** (Figure 10.3).
- Photosystem II is mainly found in granal membranes along with the **cytochrome** b_6f **complex** (Figure 10.3).

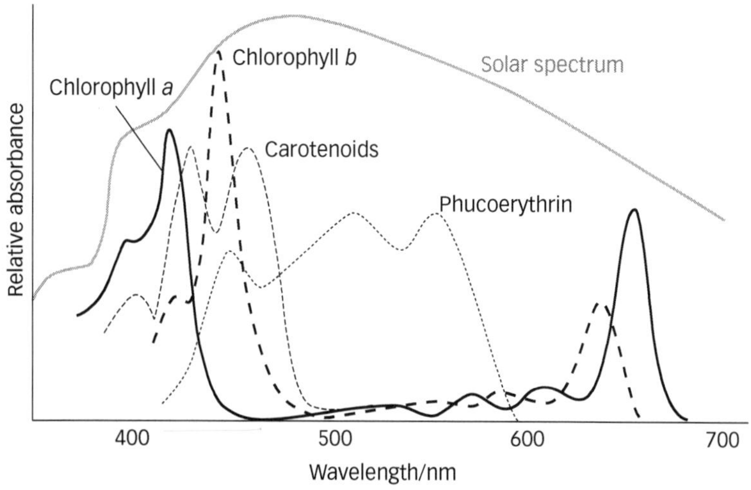

Figure 10.6 Absorption spectra of chlorophylls *a* and *b* and a number of accessory pigments used in photosynthesis compared with that of the visible solar spectrum.

- The proteins distribute and orientate the pigments to maximize the efficiency of light collection (harvesting).
- Most photosynthetic pigments within a photosystem act as **antennae** molecules to maximize the area available for light absorption.

Reaction centres

- Each of the PSs contains a **reaction centre** (**RC**).
- Reaction centres are dimers of chlorophyll molecules (often called '**special pairs**') located near the lumen side of the thylakoid membrane.
- The chlorophyll molecules of RCs are chemically identical to those of other chlorophylls but absorb light maximally at different wavelengths because they occur in a different protein environment.
- These special pairs differ slightly between PS I and PS II.
- In PS II, the RC absorbs photons with a wavelength of 680 nm, and it is therefore called **P680**. In PS I it absorbs photons at 700 nm, and is called **P700**.

Light-harvesting complexes

- Photosystems are associated with **light-harvesting complexes** (**LHCs**), which also contain photosynthetic pigments. PS I is associated with **LHC I**, PS II with **LHC II**.
- In plants and algae these LHCs consist of 80–250 chlorophyll *a* and *b*, and carotenoid molecules, together with pigment-binding proteins, but they do *not* possess RCs.
- Photosystems and LHCs are mobile in the membrane and therefore able to transfer the energy they absorb to other pigment molecules including RCs.

Light-dependent reactions

Figure 10.7 summarizes the light-dependent reactions of photosynthesis.

- Light energy absorbed by chlorophylls and accessory pigments of antennae and LHCs of PS II is passed from molecule to molecule until it reaches the P680 RC.
- The accepted means of transferring energy between chlorophylls is by **resonance energy transfer** that is in a step-wise manner.
- However, **quantum mechanical effects** are also thought to be involved, in which energy is transferred in a wave form and so follows several paths simultaneously. If this seems confusing do *not* worry: quantum mechanics is confusing!
- The energy is trapped at the RC because its special pair of chlorophylls has excited states with lower energies than other energized chlorophyll molecules.
- Thus, in PS II, the light energy funnelled to its RC (P680) results in the centre losing e^-, which are captured by an electron transport system (Figure 10.7).
- The electron-deficient or oxidized form of the RC is often abbreviated to **P680*** or **P680**$^+$.

Electron transport in the thylakoid membrane

- The *photoexcited* e^- are first passed from the RC to other pigments within the PS to eventually reach and reduce the lipid-soluble membrane e^- carrier, **plastoquinone** (*PQ*) to a plastoquinol (PQH_2).
- The next step is the transfer of e^- from PQH_2 to the cytochrome b_6f complex. This complex consists of seven membrane proteins, including two cytochromes and an iron-sulphur protein.

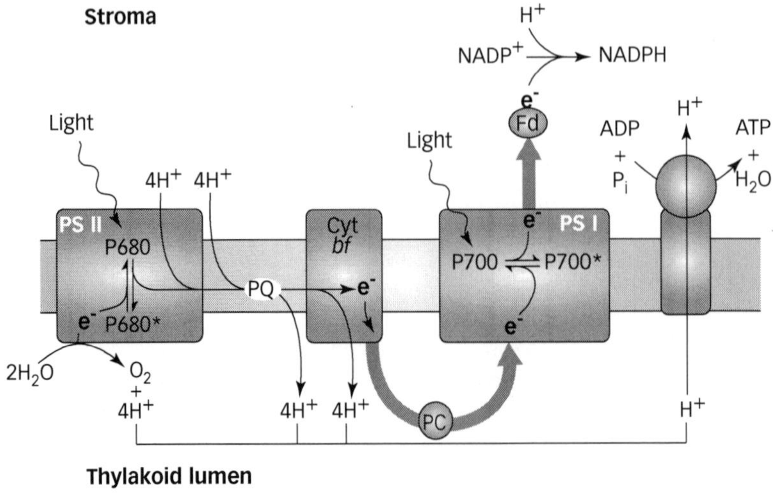

Figure 10.7 Summary of the light-dependent reactions of photosynthesis. Refer to the text for an explanation of the events. Cyt*bf*, cytochrome b_6f; Fd, ferredoxin; PQ plastoquinone; PS, photosystem.

- Reduced cytochrome *f* transfers e^- to a mobile peripheral protein situated on the luminal surface of the thylakoid membrane called **plastocyanin** (**PC**, Figure 10.7).
- The photoexcitation of PS I and LHI results in energy being funnelled to its RC, P700. Its special pair then lose e^-, forming the oxidized or electron-deficient, P700*/$P700^+$ form.
- Thus, P700* can be reduced by accepting e^- from reduced PC and then transferring them through pigment molecules in PS I to the soluble protein **ferredoxin** (**Fd**, Figures 10.7 and 10.8) in the stroma.
- Reduced Fd can be used to reduce the final e^- acceptor, $NADP^+$, to NADPH, in a reaction catalysed by **ferredoxin-$NADP^+$ reductase:**

$$Fd_{reduced} + NADP^+ + H^+ \xrightarrow{\text{Ferredoxin-NADP}^+\text{ reductase}} Fd_{oxidized} + NADPH$$

➔ *More information on cytochromes and iron-sulphur proteins is given in Chapter 11.*

- A linear transfer of e^- from P680 to $NADP^+$ has been used to reduce the latter to NADPH.
- This linear transport is usually called **non-cyclic electron transport** (*cyclic* electron transport is described in the 'Cyclic electron transport' section).
- However, the electron-deficient P680* must be reduced to allow further electron transport.
- Electrons to reduce P680* are supplied from the **photolysis** of water.

Photolysis of water

- Electrons lost from P680 to PQ are replaced by ones from H_2O as shown in Figure 10.7.

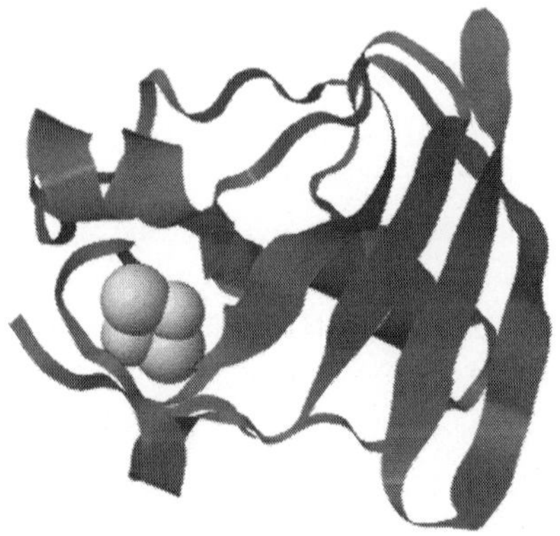

Figure 10.8 Molecular model of ferredoxin from spinach (*Spinacia oleracea*). The iron-sulphur centre involved in electron transfer is shown in the lighter tone. PDB 1A7O.

- Oxidized P680* is a strong oxidizing agent, with a greater affinity for e^- than that of O_2 and so it can extract them from water.
- Electrons are obtained from water by a complex of Mn^{2+}-containing proteins found on the luminal side of PS II called the **water-splitting centre** or **oxygen-evolving complex** (OEC).
- The photolysis of two molecules of H_2O releases four H^+ and four e^- (Figure 10.9).
- The four Mn^{2+} of the OEC capture the four e^- extracted from water and donate them sequentially to P680* (Figures 10.7 and 10.9), which prevents the formation of dangerous reactive oxygen species such as hydrogen peroxide or superoxide anions.
- The H^+ formed are released into the thylakoid lumen.

➔ *More information about reactive oxygen species is given in Chapter 12.*

- The O_2 produced from water is released into the lumen of the thylakoid system and is eventually excreted from the plant into the atmosphere.
- The other useful product of the light-dependent reactions, ATP, is formed by photophosphorylation.

Formation of ATP by photophosphorylation

- Photophosphorylation is the light-dependent formation of ATP.
- The mechanism of ATP generation is called **chemiosmosis** and depends on the formation of a transmembrane gradient of H^+ across the thylakoid membrane.

➔ *Chemiosmosis is introduced in Chapter 4 and is discussed in greater detail in Chapter 11.*

- The synthesis of ATP is performed by a H^+-dependent ATP synthase, commonly called CF_0CF_1-ATPase (C for chloroplast; F for factor). This enzyme strongly resembles the F_0F_1-ATP synthase of mitochondria (Chapter 11).
- CF_0CF_1-ATPase is a complex enzyme and consists of two multisubunit components:
  1. The F_0 component is a transmembrane complex that forms an H^+-specific channel through the thylakoid membrane.
  2. The F_1 component is anchored in the F_0 portion and has a knob-like structure with ATP synthase activity that projects into the stroma.
- In general, the thylakoid membrane is impermeable to H^+; however, the movement of e^- is associated with the transport of H^+ across the membrane into the lumen.
- This movement is largely associated with the electron transport activities of PQ and the cytochrome b_6f complex, which transfer eight H^+ across the membrane into the lumen (Figure 10.7).
- The photolysis of water contributes 4 H^+ to the lumen (Figure 10.7).
- The concentration of H^+ in the lumen is therefore higher than in the stroma such that its pH is lower. Also, the lumen is more positively charged.

(A)

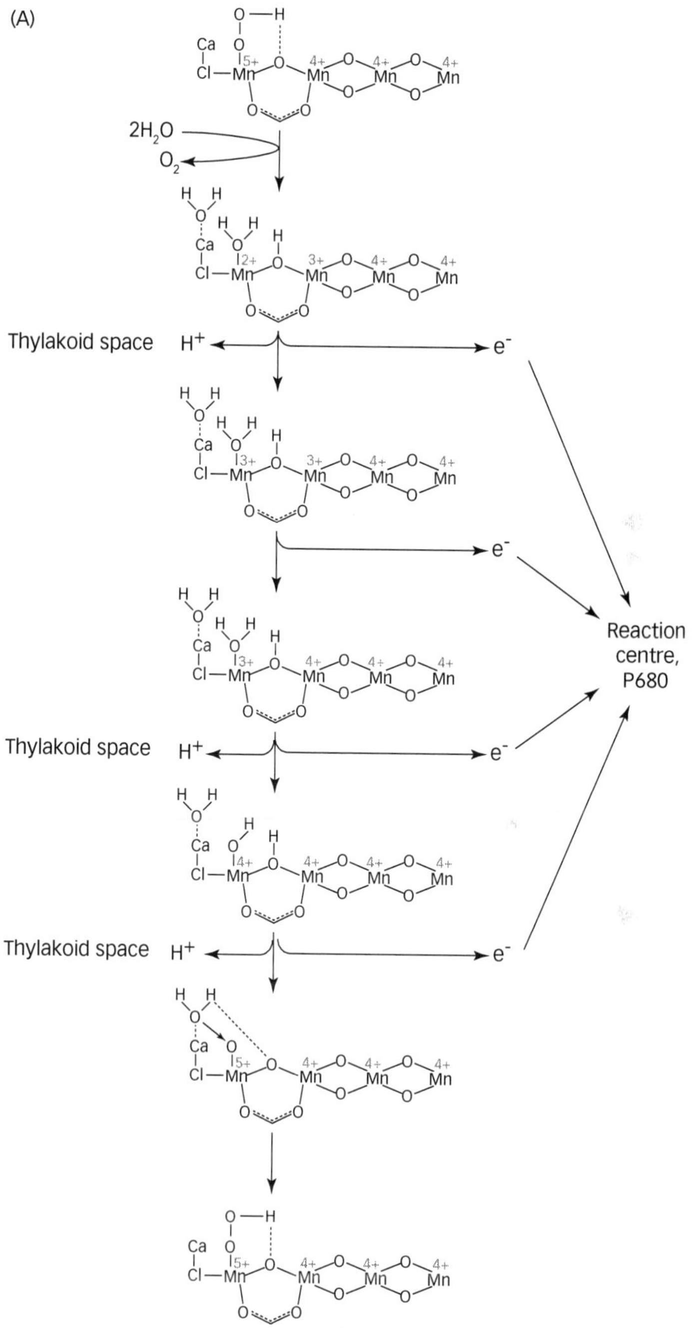

Figure 10.9 (A) Proposed scheme for the photolysis of water by the four Mn^{2+} cluster of the water-splitting centre or oxygen-evolving complex (OEC) of photosystem II. Note how the binding of *two* molecules of water is followed by successive releases of *single* electrons and H^+. (B) Molecular model of photosystem II of the cyanobacterium, *Thermosynechococcus elongates*, which includes the OEC. Model constructed using PDB file 1S5 L.

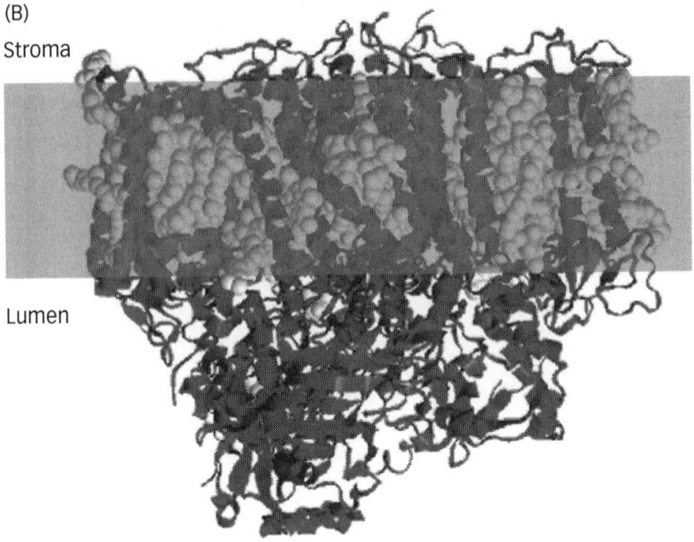

Figure 10.9 Continued

Proton motive force

- The differences in pH (ΔpH) and charge ($\Delta u_H{}^+$) across the thylakoid membrane form an electrochemical gradient or **proton motive force** (**pmf**) between the lumen and the stroma.
- The removal of H^+ from the stroma by the activity of ferredoxin-$NADP^+$ reductase also contributes to the electrochemical gradient.
- The pmf often has a value of approximately 0.15 V generated by a pH difference of 2 and a membrane potential of 0.03 V (lower and positively charged in the lumen respectively).
- Proton motive forces are sources of free energy, which can be coupled to the synthesis of ATP in the chloroplast.
- In normal circumstances, H^+ can only move in accordance with their electrochemical gradient back to the stroma through the CF_0CF_1-ATPase in the thylakoid membrane.
- This movement of H^+ through the synthase stimulates the synthesis of ATP from ADP and P_i as described in greater detail for the more widely studied mitochondrial form of the enzyme in Chapter 11.
- It is thought the synthesis of one ATP molecule requires the translocation of four H^+ through the CF_0CF_1-ATPase.

  ➔ *See Chapter 11 for a more detailed description of ATP synthesis by chemiosmosis.*

Biochemical yields of the light dependent reaction

- The useful products of the light-dependent reaction are NADPH and ATP.
- If reduced ferredoxin is used only to reduce $NADP^+$, then two molecules of NADPH are produced for every eight photons (four by PS I and eight by the cytochrome b_6f system and PQ respectively) by non-cyclic electron transport, as indicated in Figure 10.7.

- Electron transport and the photolysis of water produces 12 H^+ for the eight photons absorbed (Figure 10.7).
- Given the CF_0CF_1-ATPase requires the translocation of four H^+ for every ATP molecule formed, the yield is three ATP for every two NADPH formed.

Cyclic electron transport

- It is unlikely that ATP and NADPH will be used to the same extent within the chloroplast; typically ATP is used to support a greater range of activities (for example, active transport across membranes), but the concentration of NADPH is not lowered to the same extent. Hence, the ratio of NADPH:ATP increases.
- High values of NADPH:ATP promote *cyclic* electron transport, which is shown in Figure 10.10.
- When the NADPH:ATP is high, reduced ferredoxin uses the *cytochrome* b_6f *complex*, rather than $NADP^+$, as an e^- acceptor.
- Hence e^- flow from reduced ferredoxin to the cytochrome b_6f complex and then to PC.
- Electron transport by the cytochrome b_6f complex is associated with H^+ translocation into the lumen and therefore ATP synthesis.
- The PC is reoxidized by P700*, which completes the cycle of electron transport.
- Photosystem II is *not* involved in non-cyclic electron transport, hence:
  - photolysis of H_2O does not occur
  - H^+ pumping is only associated with the cytochrome b_6f complex
  - $NADP^+$ is not reduced.
- In contrast to non-cyclic phosphorylation, cyclic photophosphorylation forms only ATP, which helps re-establish the NADPH:ATP ratio to physiological values that support the light-independent reactions.

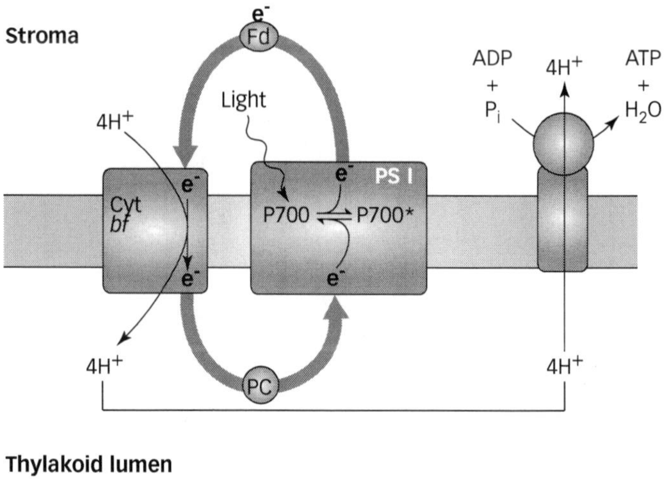

Figure 10.10 Outline of cyclic electron transport. See accompanying text for an explanation of the process. Cyt*bf*, cytochrome b_6f; Fd, ferredoxin; PC, plastocyanin; PS, photosystem.

Revision tip

An animation showing electron transport and ATP synthesis in the light reactions of photosynthesis can be viewed at: http://www.youtube.com/watch?v=hj_WKgnL6MI

Check your understanding

10.2 In photosynthesis, water is used in which of the following?

a. To excite electrons in chlorophyll *a*.
b. The oxidation of chlorophyll *a*.
c. It supplies electrons to the electron transport chain.
d. The oxidation of NADPH.
e. To supply electrons in cyclic photophosphorylation.

Light-independent or dark reaction

The light-independent or dark reaction of photosynthesis is the reduction of CO_2 to form carbohydrate.

- Although light is not required for this process, it does require reducing power (NADPH) and chemical potential (ATP) and therefore depends on the products of the light-dependent reactions.
- The formation of carbohydrate from CO_2 occurs in the stroma in a series of reactions called the **Calvin cycle** (after the famous biochemist), which is illustrated in Figure 10.11.
- Carbon dioxide (C_1 molecule) is fixed by its covalent bonding to the acceptor **ribulose 1,5-bisphosphate** (C_5) to form an unstable C_6-intermediate that is immediately hydrolysed to two molecules of **glycerate 3-phosphate** ($2 \times C_3$).
- This reaction is catalysed by **ribulose-1,5-bisphosphate carboxylase/oxygenase** (**rubisco**), generally regarded as the most abundant protein in the biosphere.

➔ *Photosynthesis is introduced in Chapter 4. More information about the activities of ribulose-1,5-bisphosphate carboxylase/oxygenase is given in Chapter 12.*

- Glycerate 3-phosphate is next phosphorylated using ATP (light-dependent product) to form glycerate 1,3-bisphosphate, which, in turn, is reduced using NADPH (light-dependent product) and dephosphorylated to give the key intermediate, **glyceraldehyde 3-phosphate** (C_3).
- If the cycle was considered to start with *three* CO_2 and *three* ribulose 1,5-bisphosphate molecules ($3 \times C_1 + 3 \times C_5$) then *six* glyceraldehyde 3-phosphate ($6 \times C_3$) would be formed.
- Five of the six glyceraldehyde 3-phosphate are used in the rest of the reactions of the cycle to regenerate the acceptor molecule, ribulose 1,5-bisphosphate and allow the cycle of reactions to continue.

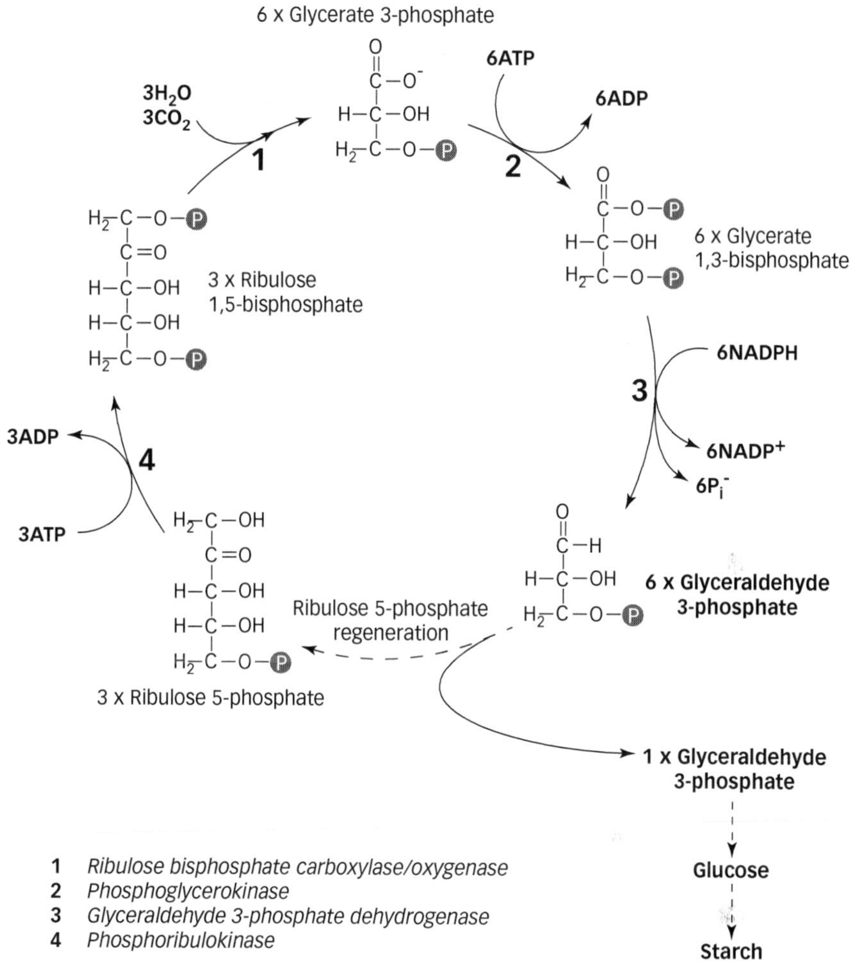

Figure 10.11 Outline of the Calvin cycle reactions. See main text for an accompanying description.

- The 'remaining' glyceraldehyde 3-phosphate is equivalent to the three fixed CO_2 molecules.
- The ICM contains a glyceraldehyde 3-phosphate/dihydroxyacetone phosphate: P_i antiporter that can exchange these triosephosphates for P_i allowing their export to the cytosol. (Remember the intermembrane space is equivalent to the cytosol with respect to most small molecules).

➔ *Transport mechanisms across membranes are revised in Chapter 3.*

- Glyceraldehyde 3-phosphate can be transported to the cytosol and enter the glycolytic pathway (Chapter 7) providing ATP and NADH for the rest of the cell. It can also be used as a precursor in the synthesis of the carbohydrates **starch** and **sucrose**, and other organic molecules.

Looking for extra marks?

A short biography of Melvin Calvin can be read at: http://www.nobelprize.org/nobel_prizes/chemistry/laureates/1961/calvin-bio.html

Formation of starch and sucrose

- Starch and sucrose are the major carbohydrate energy store and transport sugar of plants respectively. Both are formed from glyceraldehyde 3-phosphate, which can be interconverted to dihydroxyacetone phosphate:

  Glyceraldehyde 3-phosphate $\rightleftharpoons$ dihydroxyacetone phosphate

- Within the stroma, these two compounds can be used for the synthesis of starch as shown in Figure 10.12.
- Once formed, the starch may be packaged in grains that often form prominent structures in chloroplasts observed by electron microscopy (Figure 10.1).
- Both glyceraldehyde 3-phosphate and dihydroxyacetone phosphate can be exported from the stroma using the triosephosphate: P_i antiporter mentioned earlier.
- Within the cytosol, sucrose is formed as illustrated in Figure 10.12.
- Many enzyme activities are common to both the synthesis of starch and sucrose, although the stroma and cytosol contain different molecular forms of the enzymes; they are therefore **isoenzymes**.

Revision tip

An animation showing how Nobel Prize winner Calvin and coworkers elucidated the light-independent reactions of photosynthesis—the Calvin cycle—is available at: http://bcs.whfreeman.com/thelifewire/content/chp08/0802003.html

Check your understanding

10.3 Which of the following statements is/are false?

a. Glyceraldehyde 3-phosphate and dihydroxyacetone phosphate are used in the cytosol for the synthesis of starch.

b. In the light-dependent reactions carbon dioxide is fixed by covalent bonding to ribulose 1,5-bisphosphate.

c. This first reaction of the Calvin cycle is catalysed by ribulose 1,5-bisphosphate carboxylase/oxygenase.

d. In light-independent reactions, glycerate 1,3-bisphosphate is reduced using NADPH and dephosphorylated to form glyceraldehyde 3-phosphate.

e. Glyceraldehyde 3-phosphate can be transported from the chloroplast to the cytosol.

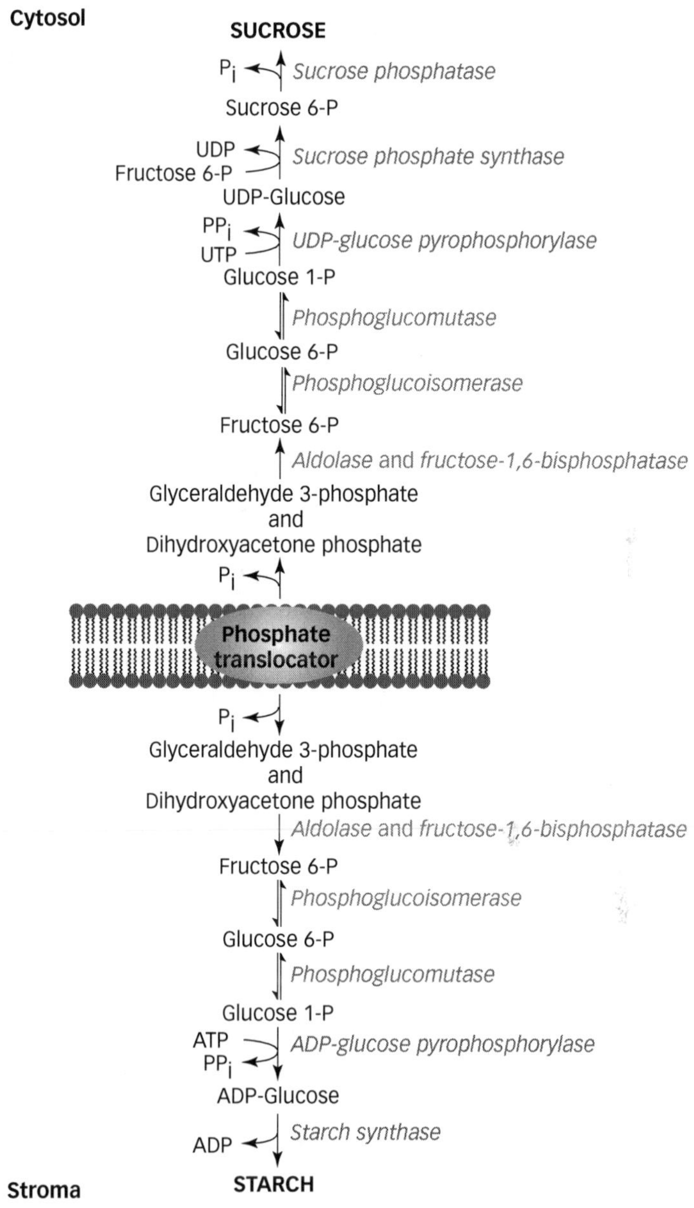

Figure 10.12 The biosynthesis of starch and sucrose from glyceraldehyde 3-phosphate and dihydroxyacetone phosphate.

Rubisco has two catalytic activities

- The major enzyme of carbon fixation in the Calvin cycle, ribulose 1,5-bisphosphate carboxylase-oxygenase (rubisco), catalyses two distinct reactions:

Mechanism of photosynthesis in plants

1. It is a *carboxylase* and catalyses the fixing of CO_2:

$$\text{Ribulose 1,5-bisphosphate} + CO_2 \xrightarrow{\text{Rubisco}} 2 \times \text{3-phosphoglycerate}$$

2. It also acts as an *oxygenase* catalysing the formation of 2-phosphoglycolate:

$$\text{Ribulose 1,5-bisphosphate} + O_2 + CO_2 \xrightarrow{\text{Rubisco}} \text{2-phosphoglycolate} + \text{3-phosphoglycerate}$$

Photorespiration

- The two catalytic activities of rubisco are in opposition, making rubisco essentially an *inefficient* enzyme for fixing CO_2.
- Only carboxylase activity produces the required 3-phosphoglycerate.
- Oxygenase activity adds an O_2 to ribulose 1,5-bisphosphate to form both 3-phosphoglycerate and 2-phosphoglycolate in a process called **photorespiration**.
- Photorespiration is dependent on the metabolic activities of peroxisomes (Chapter 12).
- Two mechanisms that have evolved in plants and at least partially overcome the inefficiency of rubisco are:
  1. C_4 metabolism or the Hatch–Slack pathway
  2. crassulacean acid metabolism (CAM).

C_4 metabolism or the Hatch–Slack pathway

- C_4 metabolism or the **Hatch–Slack pathway** is seen in agriculturally important grasses—for example, maize, sugarcane, and sorghum.
- The hot and dry environment inhabited by these types of plants has several consequences:
  - Higher temperatures decrease the solubility of CO_2 to a greater extent than O_2 in plant fluids.
  - Dry conditions stimulate stomal closure, which prevents CO_2 entering leaves, while photolysis of water continues to produce O_2.
- All these features tend to increase the concentration of O_2 in the plant relative to that of CO_2 and so promote the oxygenase activity (photorespiration) of rubisco.
- In C_4 plants, this inefficiency is overcome by having *two* separate enzymes to fix CO_2 and having them located in two different types of cells.

Kranz anatomy

- The mesophyll cells of C_4 plants possess a distinctive anatomy that is often referred to as **Kranz anatomy** (Figure 10.13).
- Leaves have a single layer of dark green-coloured cells called the **bundle sheath** that surrounds the vascular bundles.
- Bundle sheaths are, in turn, surrounded by mesophyll cells.
- Bundle sheath cells are dark green in colour because they contain large numbers of chloroplasts (Figure 10.13): they are specialized photosynthetic cells.

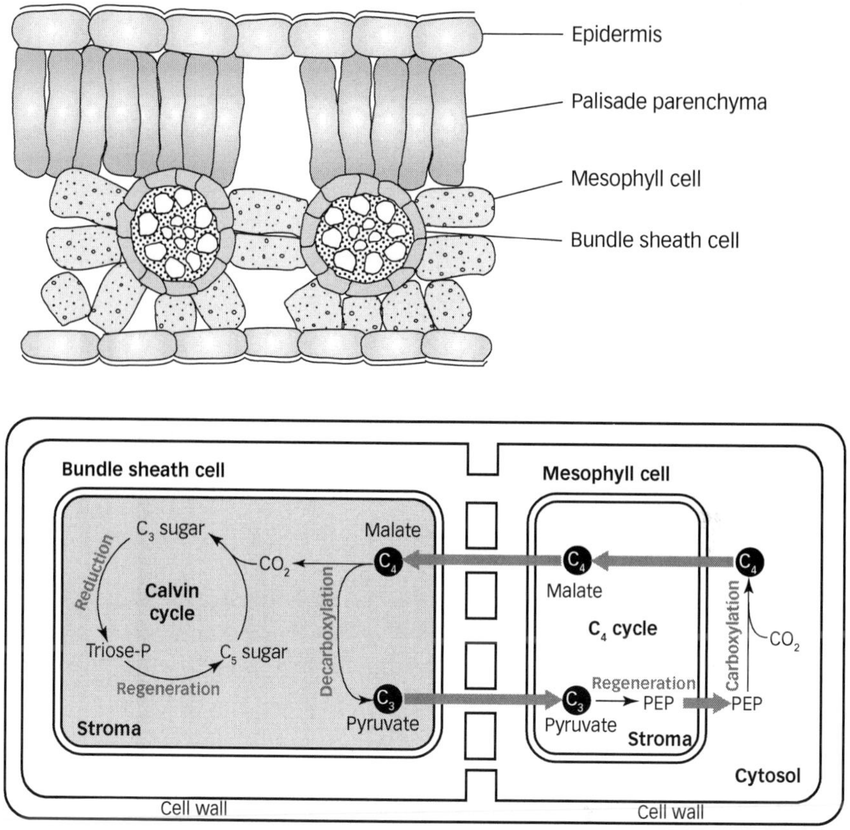

Figure 10.13 Schematic illustration of Krantz leaf anatomy and metabolic pathway of C_4 metabolism or the Hatch–Slack pathway. See the main text for explanation.

Hatch–Slack pathway

- The Hatch–Slack pathway is essentially a carboxylation and decarboxylation activity that increases the concentration of CO_2 in bundle sheath cells.
- In C_4 metabolism, mesophyll cells do *not* contain rubisco but are rich in **phosphoenolpyruvate (PEP) carboxylase** activity, which catalyses the carboxylation of PEP producing oxaloacetate (OAA, Figure 10.13):

$$CO_2 + H_2O + PEP + NADP^+ \xrightarrow{\text{Phosphoenolpyruvate carboxylase}} OAA + NADPH + H^+$$

- Phosphoenolpyruvate carboxylase has a high affinity for CO_2 but none for O_2 and is therefore able to fix the former effectively.
- The OAA produced is converted to malate, which is transferred to the neighbouring bundle sheath cell (Figure 10.13) and is decarboxylated to pyruvate, releasing CO_2. This increases the concentration of CO_2 in the bundle sheath cell up to 60-fold.

- Bundle sheath cells *do* contain rubisco, and the increased amounts of CO_2 formed by the Hatch–Slack pathway allow it to operate efficiently despite the hot arid external environment.
- The pyruvate is returned to the mesophyll cell where it is converted to PEP (Figure 10.13), which can be carboxylated to form OAA by PEP carboxylase activity.

Crassulacean acid metabolism

- Crassulacean acid metabolism was first observed in plants belonging to the *Crassulaceae*, hence its name. It has since been found in a number of species including many cacti and succulents.
- These types of plants cope with hot dry conditions by only opening their stomata at night when the environment is more moist and cooler.
- Carbon dioxide enters the mesophyll cells of CAM plants where it is fixed in malate in reactions that resemble a partial Hatch–Slack pathway (previously described).
- During the night, the malate is transported into the central vacuole of the mesophyll cell, which prevents any acidification of the cytosol.
- During the day, the vacuole exports the malate back to the cytosol, where it is decarboxylated by the delayed Hatch–Slack pathway reactions to release CO_2. However, O_2 concentrations within the chloroplast are low because stomata are closed.
- The CO_2 is able to diffuse into the chloroplast stroma and reach the high concentrations that stimulate the carboxylase activity of rubisco and so promote the Calvin cycle.
- Given both the carboxylation of PEP and the decarboxylation of malate occur in the cytosol, PEP carboxylase activity of CAM plants is strongly inhibited during daylight.

Check your understanding

10.4 Which of the following statements is/are true?

a. ATP produced in the chloroplast is exported to the rest of the cell.

b. Respiration in plants only occurs at night.

c. Photorespiration only occurs in C_4 plants.

d. Useful products of the light dependent reactions of photosynthesis are O_2, ATP, and NADPH.

e. Photolysis occurs in photosystem II.

10.3 OTHER CHLOROPLAST FUNCTIONS

In addition to reducing CO_2 to carbohydrates, chloroplasts are involved in several other biosynthetic reductive processes.

- Nitrate, obtained from the soil, is reduced to nitrite, which is further reduced to form ammonia.

$$NO_3^{2-} + 2\,H^+ + 2\,e^- \xrightarrow{\text{Nitrite reductase}} NO_2^- + H_2O$$

$$NO_2^- + 8\,H^+ + 6\,e^- \xrightarrow{\text{Nitrite reductase}} NH_4^+ + 2\,H_2O$$

- Ammonia is the 'fixed' form of nitrogen required by the plant for synthesizing organic compounds such as amino acids and nucleotides.
- Hydrogen sulphide is produced in the chloroplast by reducing soil-derived sulphate, and is used in the production of sulphur-containing amino acids (for example, cysteine).

➔ *Details of amino acid synthesis are revised in the companion volume,* Thrive in Biochemistry and Molecular Biology.

Revision tip

Construct a simple table to summarize the activities of chloroplasts and the sites where they occur.

10.4 EVOLUTION AND DEVELOPMENT OF CHLOROPLASTS

Chloroplasts, like mitochondria (Chapters 1 and 11), are thought to have evolved from primitive bacteria, in the case of chloroplasts from cyanobacterial-types that were endosymbionts of early eukaryotic cells.

- Various evidence exists to support this ancient origin:
    - A number of contemporary endosymbiotic relationships involve photosynthetic bacteria which occupy niches within the cytoplasm of unicellular eukaryotes and cells of invertebrate organisms. This supports the supposition that primitive cyanobacteria could be the ancestors of chloroplasts.
    - The **chloroplast chromosome** resembles bacterial types (Chapter 3) in being circular and lacking histones (Chapter 5).
    - Each contains a single circular DNA molecule that lacks repetitive sequences and **introns** (Chapters 5 and 6), and therefore have a high proportion of coding DNA.
    - **Chloroplast ribosomes** more closely resemble bacterial types (Chapter 3) than the 80 S ribosomes of the cytosol (Chapter 7).
    - The nucleotide sequences of **chloroplast rRNA** molecules are similar to those found in cyanobacteria.
    - Protein synthesis in chloroplasts and bacteria are similar in the types of protein factors used and in their sensitivities to antibiotics.

Looking for extra marks?

The different types of chloroplasts found in plants and seaweeds imply more than one endosymbiont relationship. Some evidence supports the theory that the chloroplasts of plants and green algae arose from a single evolutionary event, while the chloroplasts of red and brown algae originated in at least one more endosymbiotic association.

Chloroplast DNA

Chloroplast genomes encode only a restricted number of genes, and **chloroplast DNA** forms only a relatively small fraction of the total cellular DNA, most being found in the nucleus (Chapter 5).

- Chloroplast DNA molecules typically have circumferences of approximately 50 μm and contain about 120,000–160,000 base pairs encoding roughly 120 genes.
- The chloroplasts genome encodes:
  - chloroplast transfer and ribosomal RNA molecules
  - a total of around 100 proteins, which include protein factors involved in protein synthesis in the stroma, and several protein components of both PSs and two subunits of rubisco.
- Chloroplast DNA is transcribed within the stroma to form chloroplast mRNA molecules that are translated by chloroplast ribosomes in protein synthesis processes that are much like those that occur in bacteria (Chapter 5).

  ➔ *Overviews and details of protein synthesis respectively are given in Chapters 5 and 7 and are revised in the companion volume,* Thrive in Biochemistry and Molecular Biology.
- The vast majority of genes of the original endosymbiont have been transferred to the nucleus (Chapter 6) or lost as redundant genes.

Development of chloroplasts

Chloroplasts in newly differentiated plant cells develop from **proplastids**, which are smaller than mature chloroplasts and lack their complex internal membranes. They do contain DNA and can replicate by division.

- Given that the zygote receives most of its cytoplasm from the *female* parent, all proplastids in the zygote are inherited in a maternal fashion. Note how this resembles the mitochondrial inheritance described in Chapter 11.
- Proplastids are found in plant embryonic and young root, shoot and leaf tissues, and in the meristems of mature plants.
- In normal growing conditions (for example, adequate light intensities), thylakoid membranes develop from vesicles that bud from the ICM into the interior of the organelle and the numerous components of their photosynthetic apparatus are synthesized and assembled.

- The development of chloroplasts depends in part on the production of new proteins. Chloroplasts, however, can synthesize only approximately 100 of their proteins internally.
- The 2000-or-so chloroplast proteins encoded by nuclear DNA and synthesized in the cytosol must be imported into the chloroplast by specific **chloroplast protein targeting** mechanisms.

Chloroplast protein targeting

Given that more than 95% of chloroplast proteins are encoded by nuclear DNA and produced in the cytosol (Chapter 7), an efficient method of targeting them to the chloroplast and then importing them to the appropriate chloroplast compartment is required.

- Protein import into the chloroplast is by a rather complex transportation system.
- Proteins for import into chloroplasts are maintained as unfolded polypeptides by the binding of **chaperone proteins** (**Hsp70**) during transportation (Figure 10.14). Folded proteins could not traverse the membrane.

➔ *Protein translocation across the nuclear envelope, the endoplasmic reticulum membrane, and into mitochondria are revised in Chapters 6, 8, and 11 respectively.*

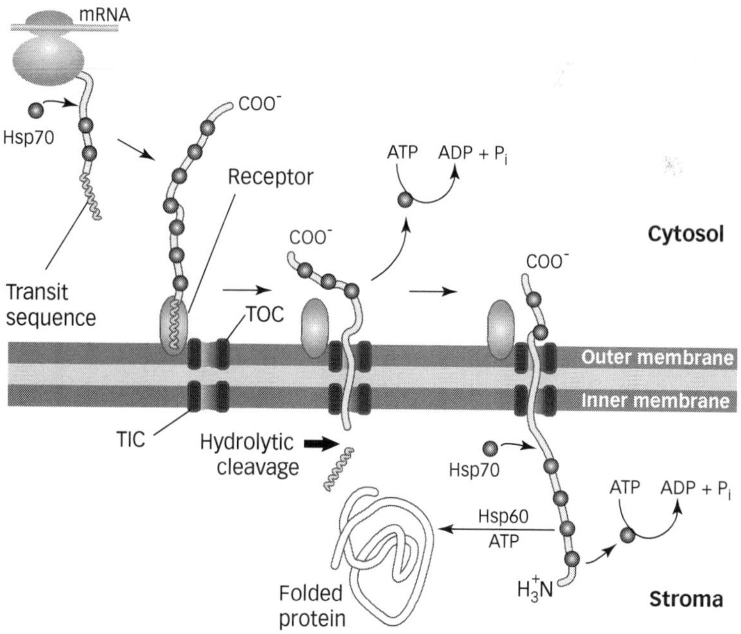

Figure 10.14 Overview of the import of polypeptides from the cytosol to the interior of the chloroplast. See text for details. Compare with Figure 11.8.

- Such polypeptides are directed into chloroplast by sequences of amino acid residues called **transit** or **chloroplast targeting sequences**.
- Transit sequences are found at the amino terminus of the polypeptide and typically contain hydrophobic and amphiphilic amino acid residues but other types of chloroplast targeting sequences occur as internal sections.
- The import of cytosolic proteins to the chloroplast occurs at specialized sites where the OCM and ICM are in close contact.
- At these sites, **import translocases** of the OCM and ICM, called TOCs and TICs respectively, can form a continuous pore through the envelope (Figure 10.14).
- Transit signals are recognized by **receptor proteins** on the OCM which directs the polypeptide towards import translocases.
- Once polypeptides enter the pore of a translocase their transit signals are normally quickly removed by a specific hydrolase.
- Other chaperones, such as Hsp60, then assist in their folding to give active conformations.
- Chloroplast transit signals differ from those concerned with import into mitochondria.

Translocation to different chloroplast compartments

- Polypeptides destined for different compartments of the chloroplast (OCM, intermembrane space, ICM, stroma, thylakoid membrane, and the thylakoid lumen) have additional signals to direct their transport to the appropriate site.
- For example, polypeptides destined for the thylakoid membrane enter the stroma where the transit signal is removed, unmasking a **thylakoid signal**, a sequence of hydrophobic amino acid residues that directs the polypeptide to the thylakoid system.
- If the polypeptide is to form an integral protein of the thylakoid membrane, the stretch of hydrophobic residues may spontaneously allow its insertion into the membrane and form its transmembrane segments (Chapter 3).
- Some proteins of the thylakoid membrane require the involvement of a GTP-dependent protein (compare the signal recognition particle used in the insertion of polypeptides into the endoplasmic reticulum membrane; Chapter 8).
- Polypeptides destined for the lumen are translocated fully through the thylakoid membrane and their thylakoid signal sequences are hydrolytically removed during entry.
- All these transport activities require metabolic energy, which is provided by the hydrolysis of ATP.

Check your understanding

10.5 Which of the following statements is/are false?

a. Chloroplasts in newly differentiated plant cells develop from proplastids.

b. Translocases of the OCM and ICM are called TICs and TOCs respectively.

c. Chloroplast proteins are kept unfolded by the binding of Hsp70 in the cytosol.
d. Chloroplast proteins are directed to chloroplasts by transit or chloroplast targeting sequences.
e. Once a protein has entered the chloroplast, its folding is assisted by the action of Hsp60.

Looking for extra marks?

Note there are many similarities between the import of polypeptides into chloroplasts and into mitochondria.

Revision tip

Do browse through the further reading references given for this chapter online. Go to: http://www.oxfordtextbooks.co.uk/orc/thrive/

Check your understanding

10.6 Which of the following statements is/are true or is/are false?

a. Photosynthesis occurs in the cytoplasm.
b. The light-independent reactions of photosynthesis can only occur in the dark.
c. Photosystem I and the chloroplast ATP synthase are found mainly in membranes forming the stromal lamellae.
d. Photosystem II and the cytochrome b_6f complex are mainly found in granal membranes.
e. Cyclic photophosphorylation uses both P680 and P700 reaction centres.
f. Reaction centres consist of special pairs of chlorophyll molecules situated on the stromal side of the thylakoid membrane.
g. The reaction centre of PS II is called P680, while that of PS I is P700.
h. Non-cyclic electron transport involves the transfer of electrons from P680 to $NADP^+$.
i. Cyclic photophosphorylation forms ATP and NADPH.
j. High temperatures and dry conditions tend to increase the concentration of CO_2 in plants relative to that of O_2.

11 Mitochondria, hydrogenosomes, and mitosomes

Key features of mitochondria

- Figure 11.1 shows the typical structure of mitochondria found in the cells of organisms that live in aerobic environments.
- The number of mitochondria varies between cell types; in many cases, the number of mitochondria present reflects the metabolic requirements of the cell.
- A human liver cell typically possesses approximately 1000 mitochondria but in other types, such as spermatozoa and some yeasts, the mitochondrion is a single continuous network.
- When viewed by electron microscopy (Chapter 2), mitochondria are cylindrical and approximately 1.5–10 μm long and 0.25–1.0 μm in diameter. The detailed morphology of mitochondria varies between those isolated from different tissues.
- Mitochondria are limited by an envelope consisting of outer and inner mitochondrial membranes (OMM and IMM respectively) separated by an intermembranous space, which encloses a matrix (Figure 11.1).
- The presence of pore-forming proteins called porins in the OMM make it freely permeable to solutes with M_r up to approximately 5000.

- Similar porins are found in the outer membranes of Gram-negative bacteria (see Chapter 4, Figure 4.4) and chloroplasts (Chapter 10).
- The OMM is therefore relatively 'leaky' with respect to small molecules and ions, and can be regarded as a continuation of the cytosol (Chapter 7).
- The IMM is folded into cristae, which greatly increases its surface area compared to that of the OMM (Figure 11.1).
- Cristae have been regarded as broad flattened extensions of the IMM. However, microscopical studies indicate that, in at least some cells, cristae are tubular structures that associate together to form lamellar cristae, the morphology of which resembles that of the grana of chloroplasts (Chapter 10) in some respects.
- Mitochondria of kidney tubules, which are heavily dependent on biological energy to power active transport functions, or those of cardiac muscle that need ATP to power contraction, contain a much higher proportion of cristae than liver mitochondria.
- The relatively larger volume of matrix found in liver cell mitochondria is a reflection of the role of the liver in biosyntheses.
- The IMM is concerned with active transport, electron transport, and the synthesis of ATP by oxidative phosphorylation. It is relatively impermeable to many solutes but contains specific transporters for a number of molecular species such as ADP-ATP and Ca^{2+}-Na^{+} and various carboxylate antiporters and a P_i-H^{+} symporter (Chapter 3).
- Many of the components of the electron transport chain (ETC) are embedded within the IMM and it is also generally rich in the F_1F_0-ATP synthase.
- Sites where the two membranes of the envelope juxtapose are thought to be regions where proteins produced in the cytosol are imported into the mitochondrion.
- The envelope encloses an interior gel-like aqueous matrix.
- The matrix is a protein-rich environment containing numerous enzymes that catalyse activities such as the breakdown of pyruvate, the TCA cycle and β oxidation.
- It produces numerous small molecules for biosynthetic reactions
- The matrix also contains mitochondrial DNA (mtDNA), mtRNA molecules, and mt ribosomes. It is therefore the site of mtDNA replication and transcription, and translation of the mt mRNA (mitochondrial protein synthesis).
- Mutations in mitochondrial DNA lead to a variety of clinical conditions.
- The application of mitochondrial DNA sequence studies has proved useful in evolutionary studies.
- Mitochondria have evolved from endosymbiotic bacteria.
- Hydrogenosomes and mitosomes occur in some organisms and appear to have evolved from mitochondria as a response to anaerobic environments.

continued

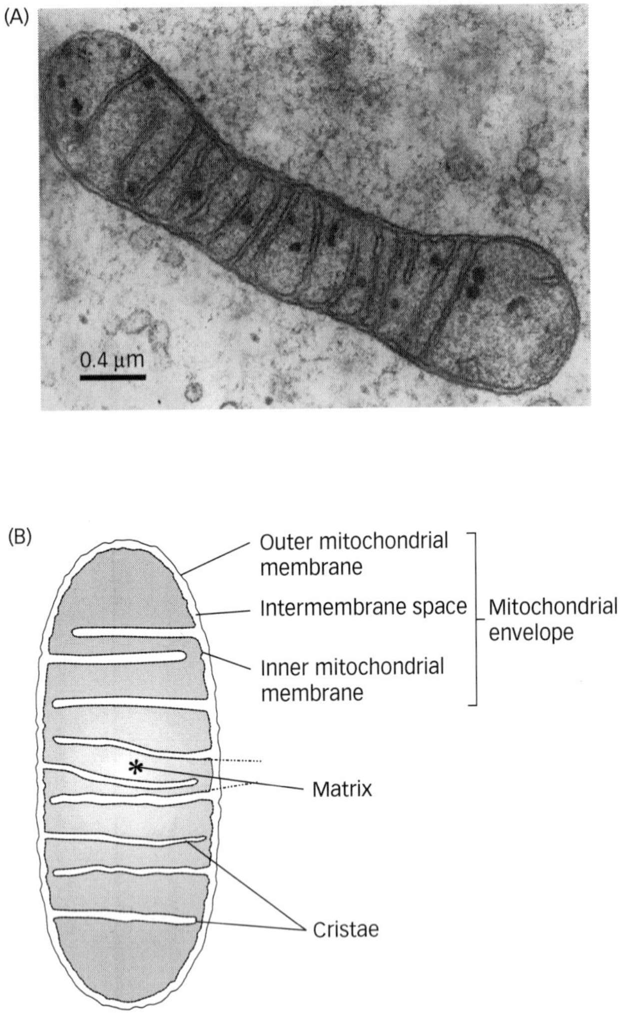

Figure 11.1 (A) Electron micrograph of a mitochondrion from a hen oocyte. Courtesy of Emeritus Professor Ruth Bellairs, University College London, UK.
(B) Schematic outline illustrating the structure of a typical mitochondrion.

- Hydrogenosomes produce ATP using protons (H^+) as electron acceptors, forming dihydrogen (H_2).
- Mitosomes do not produce ATP but retain the mitochondrial capacity to synthesize iron-sulphur proteins.

Check your understanding

11.1 How many cylindrical mitochondria of length 5 μm and diameter 0.5 μm would be found in a hepatocyte of mean volume 5000 μm^3 if they occupy 20% of the cell volume? Give your answer to the nearest 100.

Revision tip

Revise the structures of mitochondria in different tissues and organisms using the beautiful electron micrographs in The Cell: An Image Library™ of The American Society for Cell Biology at: http://www.cellimagelibrary.org/

Electron transport and oxidative phosphorylation in prokaryotes is revised in Chapter 4 and in eukaryotic photosynthesis in Chapter 10.

11.1 KEY FUNCTIONS OF MITOCHONDRIA

Mitochondria have numerous functions.

- The matrix is the site of:
  - the oxidations (decarboxylation) of pyruvate and fatty acids
  - the **TCA cycle**
  - the production of heat in some thermogenic tissues
  - the production of precursor molecules for biosyntheses
  - the formation of **iron-sulphur** (Fe-S) **proteins**
  - storage of Ca^{2+}
  - some of the reactions of the urea cycle.
- The **inner mitochondrial membrane** (IMM) is the site:
  - of the production of ATP by oxidative phosphorylation
  - for initiating apoptosis.

Oxidation of pyruvate

A specific pyruvate symporter (Chapter 3) of the IMM cotransports a pyruvate along with an H^+ into the matrix.

- Within the matrix, pyruvate is converted to acetyl CoA by its oxidative decarboxylation, in a reaction catalysed by pyruvate dehydrogenase:

$$\text{Pyruvate} + \text{NAD}^+ + \text{CoASH} \xrightarrow{\text{Pyruvate dehydrogenase}} \text{Acetyl CoA} + \text{NADH} + \text{H}^+ + \text{CO}_2$$

- Pyruvate dehydrogenase comprises many subunits (24 in the *Escherichia coli* version) and requires thiamine pyrophosphate, a cofactor derived from vitamin B_1 or thiamine, for activity.

Oxidation of fatty acids

Fatty acids are oxidized by **β-oxidation**, a cyclic series of four reactions that reduce the length of the fatty acid chain by two carbon atoms, which are released as acetyl CoA.

- Each shortening in length also forms the reduced coenzymes NADH and $FADH_2$.

The TCA cycle

The TCA cycle is the central means in the cell for oxidizing fuel molecules to form the reduced coenzymes NADH and $FADH_2$.

Reactions of the TCA cycle

- The TCA cycle consists of eight reactions, which are outlined in Figure 11.2.
- Fuel substrates of the cycle include, for example, acetyl CoA, which is formed from pyruvate and fatty acid oxidations.

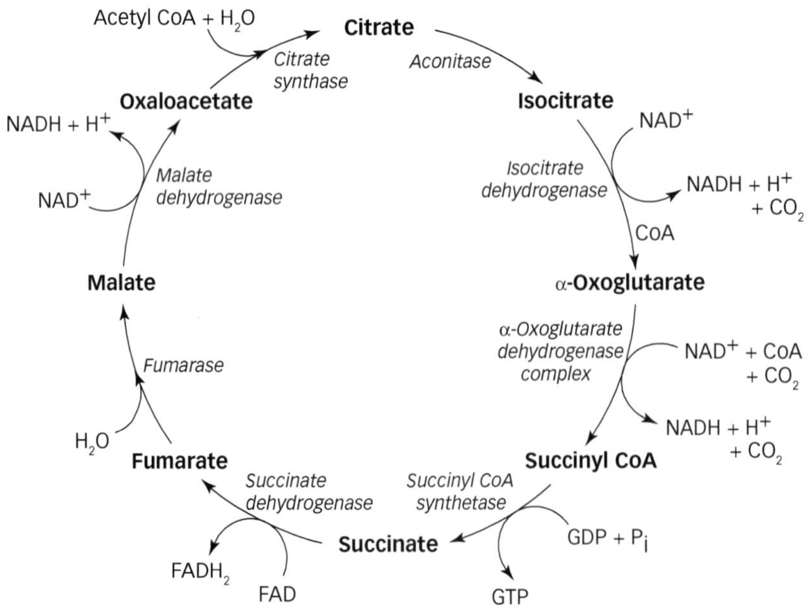

Figure 11.2 Reactions of the tricarboxylic acid (TCA) cycle. This is also known as the Krebs or citric acid cycle.

- Amino acids derived from endogenous proteins or obtained in the diet can also be converted to intermediates of the cycle and so be oxidized. The intermediates formed may be acetyl CoA (directly or via pyruvate), oxaloacetate, succinyl CoA, fumarate, or α oxoglutarate, depending upon the identity of the amino acid.
- In the first reaction, acetyl CoA reacts with oxaloacetate (OAA) to form the tricarboxylic acid, citrate (hence the name of the cycle).
- The remaining reactions of the cycle regenerate OAA and form one molecule of GTP (equivalent to an ATP), three of NADH + H^+, and one of $FADH_2$ for each acetyl CoA oxidized.
- The cycle may be summarized as:

$$\text{Acetyl CoA} + H_2O + 3NAD^+ + FAD + GDP + P_i \rightarrow \text{CoASH}$$
$$+ 2CO_2 + 3NADH + 3H^+ + FADH_2 + GTP$$

Biochemical details of the TCA cycle can be found in the companion volume, Thrive in Biochemistry and Molecular Biology.

Check your understanding

11.2 Arrange the following in an appropriate order: fumarate, acetyl CoA/ oxaloacetate, malate, isocitrate, succinyl CoA, α oxoglutarate, succinate, citrate.

Oxidative phosphorylation

Mitochondria are the principal sites of ATP synthesis by **oxidative phosphorylation**, the major way in which a cell makes ATP in aerobic conditions.

- During oxidative phosphorylation the oxidation of reduced coenzymes (NADH and $FADH_2$) is tightly **coupled** to the phosphorylation of ADP to give ATP (Figure 11.3).

Electron transport

- Electron transport requires an electron transport chain (ETC) and an H^+-dependent ATP synthase, which are associated with the IMM (see Figure 11.3).
- The ETC consists of flavo- and **iron-sulphur** (**Fe-S**) proteins, cytochromes, copper-containing cytochromes, and a quinine called **ubiquinone** or **coenzyme Q**. Most of the proteins are arranged in four large multiprotein membrane **complexes** referred to as **I**, **II**, **III**, and **IV**.
- Cytochrome *c* is a small protein loosely attached to the matrix surface of the IMM.
- Electrons from NADH are fed into the chain at complex I, those from $FADH_2$ at complex II.

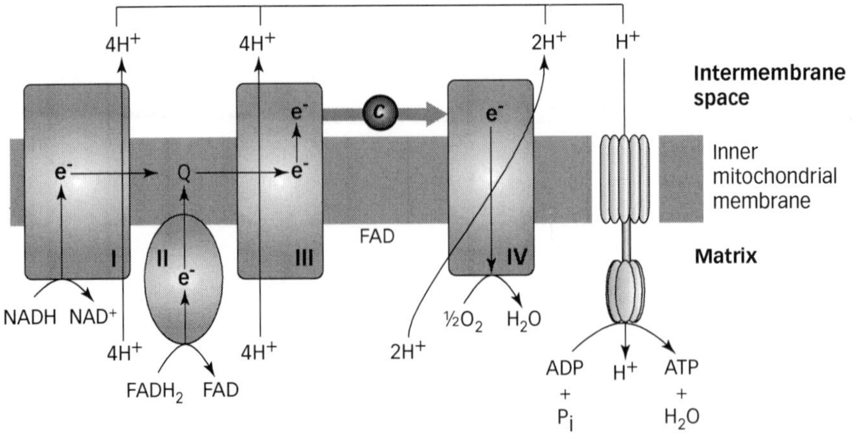

Figure 11.3 Schematic outline of the electron transport chain and ATP synthase of the inner mitochondrial membrane. Compare this with Figure 10.1 showing the similar structures of the chloroplast. I, II, III and IV, protein complexes I, II, III and IV respectively; c, cytochrome c.

- The final electron acceptor in aerobic organisms is molecular or **dioxygen** (O_2) and results in the formation of water, H_2O:

  $$O_2 + 4\,e^- + 4\,H^+ \rightarrow 2\,H_2O$$

- In some circumstances, mitochondria can produce dangerous **reactive oxygen species.** These are compounds that possess an unpaired electron such as superoxide anions (O_2^-) and hydroxyl radicals ($OH^\bullet$).
- Superoxide dismutase activity can convert O_2^- to the damaging oxidizing agent hydrogen peroxide.
- Vitamins (for example, vitamins C and E) help protect the cell by functioning, at least partially, as free radical traps. The activities of some microbodies in protecting the cell are described in Chapter 12.

  ➔ *Revise the protective functions of microbodies in Chapter 12.*

Looking for extra marks?

The mitochondria of eukaryotic facultative *anaerobic* organisms use a variety of organic substances, for example, fumarate, or inorganic materials, such as nitrate, as final electron acceptors.

The synthesis of ATP

- The process by which ATP is formed in oxidative phosphorylation is often referred to as **chemiosmosis.**

- Chemiosmosis is the process by which ATP is synthesized by membrane-based molecular complexes.
- Electron transport along the ETC promotes the active transport of H^+ from the mitochondrial matrix across the IMM into the intermembrane space by complexes I, III, and IV, as indicated in Figure 11.3.
- The concentration of H^+ in the intermembrane space is therefore higher than in the matrix (that is, its pH is lower), and it is also a more positively-charged compartment.

Check your understanding

11.3 Which of the following statements is/are false?

a. The role of dioxygen in cells is as an electron acceptor.

b. The electron transport chain of the mitochondrion is situated in the inner mitochondrial membrane.

c. Protein complexes I, III, and IV pump H^+ across the inner mitochondrial membrane.

d. Water is used in glycolysis to split glucose.

e. Cytochrome *c* and ubiquinone are relatively mobile electron carriers.

Proton motive force

- The differences in pH (ΔpH) and charge ($\Delta\mu_{H^+}$) constitute an **electrochemical gradient** or **proton motive force** (**pmf**) between the matrix and the intermembrane space.
- The pmf often has a value of 0.16V and a pH difference of 1 (positive and lower in the intermembrane space respectively).
- Proton motive forces are sources of **free energy**. Most of the free energy is associated with the Δu_{H^+} component.
- The free energy of the pmf is linked to the synthesis of ATP in the mitochondrion.

F_0F_1-ATPase

- The IMM is impermeable to H^+, which, in normal circumstances, can only move in accordance with their electrochemical gradient back to the matrix through the H^+-dependent ATP synthase of the IMM.
- This synthase is commonly referred to as an **F_0F_1-ATPase** (F for factor) or **complex V** (Figure 11.4).
- The F_0F_1-ATPase is a complex multisubunit enzyme but its structure can be described relatively simply.
  - The F_0 component is a transmembrane complex forming a pore that allows H^+ to flow through the IMM down their pmf to the matrix.

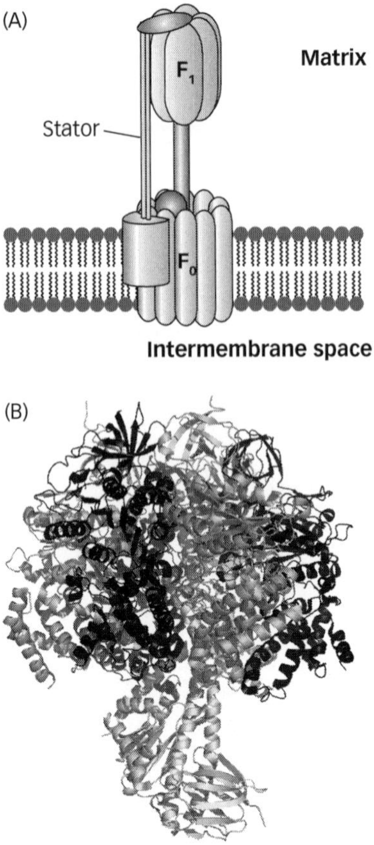

Figure 11.4 (A) Schematic showing the structure of the F_1F_0-ATPase. (B) Molecular model of an F_1F_0-ATPase. Courtesy J. Crowe, Oxford University Press, UK.

- It also forms a stator, which is an extended rod that extends into the matrix connecting F_0 and F_1 complexes. The stator prevents the F_1 complex from rotating.
- The F_1 component is associated with ATP synthase activity. It forms a knob-like structure that is anchored in the F_0 portion and projects into the matrix.
- The protein subunits of an F_0F_1-ATPase are shown in Table 11.1.
- Their actions in forming ATP are listed in Table 11.1.

ATP formation and release

- F_0 provides a channel for the flow of H^+ from the intermembrane space to the matrix down their electrochemical gradient (Figure 11.3).
- The flow of H^+ through the channel causes the c_{10} ring (Table 11.1 and Figure 11.4) to rotate. The rotation of the c_{10} is imparted to the γ subunit of the F_1 complex.
- The movement of H^+ provides free energy that results in the binding of ADP and P_i and the release of ATP from the F_1 complex in a process called the **binding charge model**.

Factor	Polypeptide component			Role
	Designation	Approximate M_r	Number present	
F_0	*a*	30,000	1	Forms H^+ channel
	b	17,000	2	Forms stator connecting F_0 and F_1 complexes
	c	8000	10	Forms rotating ring and transmits the rotary movement to the γ subunit of F_1
F_1	α	52,000	3	Activates the β subunit
	β	55,000	3	Contains the catalytic sites for ATP synthesis
	δ	19,000	1	Anchors the $\alpha_3\beta_3$ ring to the stator of the F_0 *b* subunit
	γ	31,000	1	Its rotary movement transmits the energy of the rotating c_{10} ring of F_0 to the $\alpha_3\beta_3$
	ε	15,000	1	Attaches the F_1 subunit to the c_{10} ring of F_0

Table 11.1 Composition of the F_0F_1-ATPase of *Escherichia coli*

- The $\alpha_3\beta_3$ ring of the F_1 subunit *cannot* be rotated by the γ subunit because of the anchoring stator.
- The rotating γ subunit causes each of the catalytic β subunits to progress sequentially through three distinct conformational states called **open** (O), **loose** (L) and **tight** (T) that result in the formation and release of an ATP.
- Initially the active sites of the three catalytic β subunits will be in O, L, and T conformations respectively. Consider the sequence of events for only a *single* substrate (Figure 11.5).
- Initially, the β subunit is in its O conformation and its active site is empty and thus able to be occupied by the substrates ADP and P_i.
  - A 120° rotation of the γ subunit converts the β subunit to its L state where the ADP and P_i are loosely bound in the active site but catalytic activity does not occur.
  - A further 120° turn by the γ subunit induces the T state, which favours the condensation of ADP and P_i to form ATP and water.
  - Another 120° turn of the γ subunit opens the site, allowing the release of the ATP molecule and another ATP-forming cycle to begin.
- During this cycle, the other two β subunits experience similar conformational changes resulting in the formation of ATP.
- Two points are noteworthy:
  - Unlike the aqueous conditions of the general cytoplasm, conditions in the active site of a β subunit in its T state are favourable for formation of ATP.
  - The free energy of the pmf has not been used to drive the formation of ATP but rather its release from the F_1 complex.

➔ *More details of the biochemistry of oxidative phosphorylation can be found in the companion volume,* Thrive in Biochemistry and Molecular Biology.

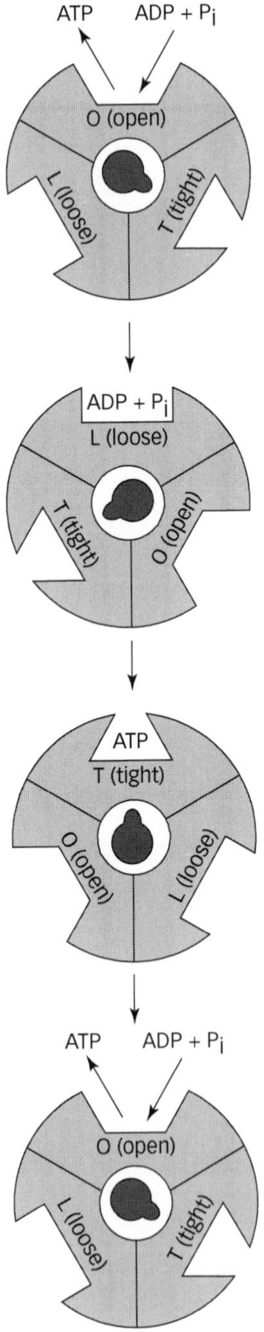

Figure 11.5 Formation and release of ATP by the F_1F_0-ATPase. For the sake of clarity, only the events at one of the three α subunits have been highlighted.

Revision tip

Two excellent animations that concentrate on different aspects of the formation of ATP by the mitochondrial ATP synthase can be seen at: http://vcell.ndsu.edu/animations/atpgradient/movie-flash.htm and http://www.sigmaaldrich.com/life-science/metabolomics/learning-center/metabolic-pathways/atp-synthase.html

Stochiometry of ATP synthesis

- It is generally accepted that three H^+ must pass through the F_0 complex to result in the synthesis of one ATP molecule by the F_1 synthase (Figure 11.3).
- For every NADH oxidized by the ETC, 10 H^+ are transferred across the IMM (four by complex I, four by complex III, and two by complex IV).
- The oxidation of $FADH_2$ results in the transfer of six H^+ (four by complex III and two by complex IV).
- Thus, the oxidations of NADH and $FADH_2$ will yield approximately 2.5 and 1.5 molecules of ATP respectively.

Transport of adenosine phosphates

- The IMM is impermeable to most materials. However, a specialized antiporter transfers an ATP molecule out of the mitochondrion in exchange for the import of an ADP.
- Energy to drive this exchange comes from the membrane potential of the pmf, since the export of ATP carries four negative charges (ATP^{4-}) out of the matrix, while the import of ADP only transfers three in (ADP^{3-}).
- Phosphate (P_i) crosses the IMM to the matrix through a symporter that cotransfers an H^+.
- Hence, ADP and P_i enter the matrix to support the synthesis of ATP, which is then exported to the rest of the cell.

Uncouplers and thermogenesis

Uncouplers are compounds that break the link (couple) between the oxidation of reduced coenzymes and the phosphorylation of ADP to form ATP.

- They associate with the IMM and increase the permeability of the membrane to ions despite allowing electron transport to proceed.
- The increased permeability of the IMM to H^+ allows their return through the membrane to the matrix, which dissipates the pmf.
- Since the electrons have by-passed the F_1F_0-ATPase, electron transport and the formation of ATP are uncoupled. Hence the free energy of electron transport is dissipated as *heat.*

Thermogenesis

- In some **thermogenic tissues**, such as **brown adipose tissue** (**BAT**), heat is generated by natural uncouplers. The means of generating heat is called **non-shivering thermogenesis** (NST).
- In humans, BAT is mainly found behind the shoulder blades and along the spinal column.
- Brown adipose tissue contains numerous mitochondria and has an excellent supply of blood capillaries, hence its brown colour.
- The IMM of mitochondria in BAT are relatively deficient in ATPase activity but do contain the protein **thermogenin**, which is a *natural* uncoupler of oxidative phosphorylation.
- The activity of thermogenin is tightly controlled by the sympathetic nervous system and is inhibited by GTP.
- Non-shivering thermogenesis is particularly significant in neonates, which can rapidly lose heat to their surroundings, and when hibernating animals first begin to wake.
- As children grow, the deposits of BAT atrophy and NST becomes insignificant as a means of maintaining body temperature.

Precursors for biosyntheses

Mitochondria produce many of the precursors for biosynthetic (anabolic) reaction pathways.

- For example, intermediates of the TCA cycle can be used for the synthesis of biological materials, such as sugars, lipids, nucleotides and nucleic acids, and porphyrins such as haem and chlorophylls (Figure 11.6).

Iron-sulphur proteins

Mitochondria produce Fe-S clusters for insertion into Fe-S proteins.

- Iron-sulphur proteins are ubiquitously distributed among eukaryotic (and, indeed, prokaryotic) organisms, and have essential roles in electron transfer reactions, the catalytic activities of a number of enzymes, and in regulating metabolism.
- An environment low in O_2 is necessary for their synthesis, making the mitochondrial matrix an optimal site because oxidative phosphorylation rapidly utilizes the O_2 present.

Ca^{2+} storage

Calcium ions are key ions in the cytoplasm of cells.

- Their roles include:
  - acting as secondary messengers in many signal transduction processes (Chapter 3 and 15)

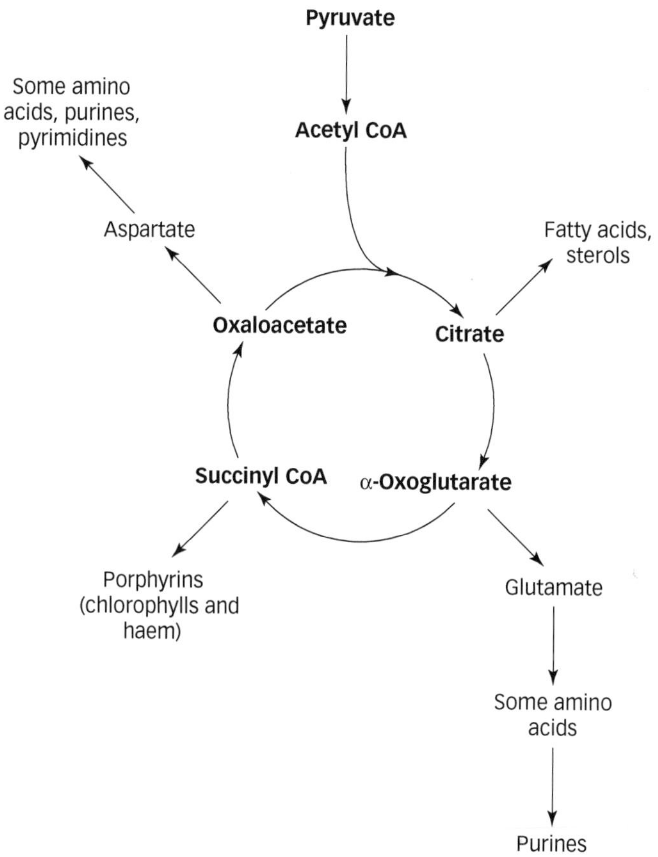

Figure 11.6 Partial list of biological compounds formed from intermediates of the TCA cycle.

- regulating a number of cellular processes, for example muscular contraction (Chapter 6) and the fusion of secretory vesicles with the plasma membrane (Chapters 3 and 9)
- activating or acting as cofactors for a number of enzymes.

- Mitochondria can accumulate Ca^{2+} from the cytosol using a uniporter in the IMM. The free energy driving uptake is provided by the pmf.
- Although the endoplasmic reticulum stores the bulk of cytoplasmic Ca^{2+} (Chapter 8), the Ca^{2+} transiently stored in mitochondria can be released *rapidly*.
- Thus, mitochondria have a key role in buffering the concentration of cytosolic Ca^{2+}.

Apoptosis

Apoptosis is the process associated with the genetically programmed death of a cell (Chapter 17).

- Mitochondria can begin the intrinsic pathway to initiate apoptosis by releasing cytochrome *c* into the cytosol.

Urea cycle

The urea cycle is the mechanism by which extremely soluble and toxic NH_3 is converted to the much less soluble and toxic urea.

- Several reactions of the urea cycle occur in liver mitochondria, others in the cytosol.

➔ *Biochemical details of the urea cycle can be found in the companion volume,* Thrive in Biochemistry and Molecular Biology.

11.2 EVOLUTIONARY ORIGINS OF MITOCHONDRIA

Mitochondria, like chloroplasts, are thought to have evolved from primitive bacteria which were endosymbionts of early eukaryotic cells, an event that occurred at least 2000 million years ago. Evidence for this ancient origin includes the following:

- The mt chromosome resembles bacterial types (Chapter 4) in being circular and lacking histones (Chapter 5).
- Each contains a single circular DNA molecule with an absence of repetitive sequences and therefore a high proportion of coding DNA.
- Mitochondrial genomes are encoded by the circular DNA molecules of the mt chromosome.
- Mitochondrial genes also resemble bacterial types in being relatively small and generally lacking **introns** (Chapter 5).
- The OMM is similar in composition to other eukaryotic membranes. In comparison, the IMM contains bacterial type lipids.
- Mitochondrial ribosomes are the 70 S (bacterial) types in contrast to the 80 S ribosomes found in the cytosol of the cell (Chapters 4 and 6).
- Mitochondrial rRNA molecules are similar to bacterial types in their nucleotide sequences, the types of protein factors they require for protein synthesis and in their sensitivities to antibiotics.

Check your understanding

11.4 Mitochondria are thought to have arisen from which domain of organisms?

a. Archaea
b. Bacteria
c. Cyanobacteria
d. Viruses
e. Eucarya.

Mitochondrial DNA

As mentioned earlier, mitochondria contain their own DNA. However, mtDNA encodes a restricted number of subunits that are essential polypeptides of ETC proteins and of the F_0F_1-ATPase.

- These are evolutionary descendants of the genome of the original endosymbiont that gave rise to mitochondria.
- In general, mtDNA accounts for only a fraction of the total cellular DNA, most being found in the nucleus (Chapter 6).
- Mitochondrial genomes are diverse, ranging from molecules of less than 10 kb to others nearly 70 kb in size, and encode only a few genes. For example, those of animals and fungi contain genes for mitochondrial tRNAs and rRNAs and a few protein genes.
- The protein coding genes include a small number of essential inner membrane proteins of respiratory complexes I to IV involved in electron transport and of complex V, the ATP-synthesizing enzyme (Figures 11.3 and 11.4)
- The human mitochondrial DNA molecule (Figure 11.7) is typical of the type:
  - It is a double-stranded molecule consisting of 16,569 base pairs and has a circumference of about 5 μm.
  - It contains 37 genes, comprising 13 protein genes encoding subunits of NADH dehydrogenase, cytochrome *b*, cytochrome *c* oxidase, and the mitochondrial ATPase; 22 genes for mt tRNA molecules; and two genes for mt rRNA molecules.
- Each mitochondrion contains two to ten mitochondrial chromosomes.

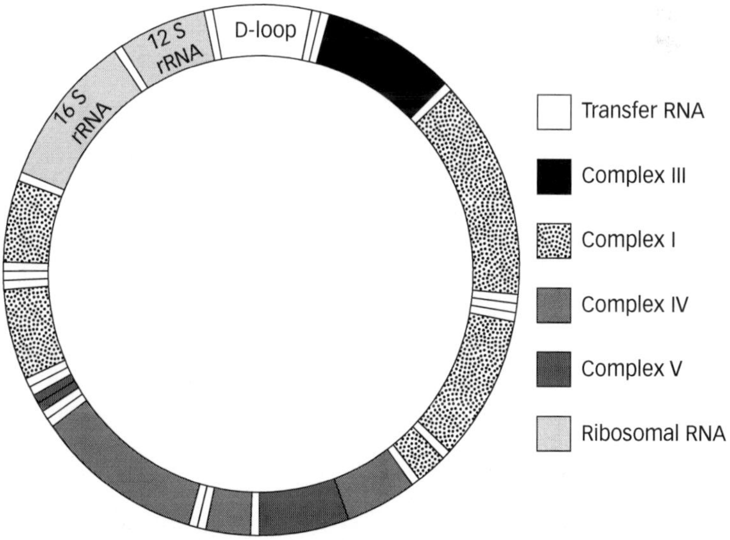

Figure 11.7 Structure of human mitochondrial DNA.

- Plant mtDNA often encodes more proteins than animal types (but still a greatly restricted number).
- Mitochondrial DNA is transcribed within the mitochondrion to form mt mRNA molecules that are translated by mt ribosomes in protein synthesis processes that are much like those that occur in bacteria.

➔ *Overviews and details of protein synthesis respectively can be found in Chapters 5–7 and the companion volume,* Thrive in Biochemistry and Molecular Biology.

- The vast majority of genes of the original endosymbiont have been transferred to the nucleus (Chapter 6) or lost as redundant genes.
- Thus, the hundred or more of mitochondrial proteins encoded by nuclear DNA must now be synthesized in the cytosol and imported into the mitochondria by **mitochondrial protein targeting.**

Mitochondrial protein targeting

Given that only a small number of proteins are encoded by mtDNA and produced in the matrix, most mitochondrial proteins are produced on cytosolic ribosomes.

- Proteins are imported into the mitochondrion as unfolded polypeptides by rather complex transportation systems.
- Folding in the cytosol is prevented by the binding of the **chaperone protein, Hsp70** to the polypeptide to maintain it in an extended linear conformation (which is essential since folded proteins could not traverse the membrane).
- These polypeptides are directed into mitochondria by the presence of sequences of amino acid residues called **transit** or **mitochondrial targeting sequences.**
- Transit sequences are found at the amino terminus of the polypeptide but other types of mitochondrial targeting sequences occur as internal sections.

➔ *See Chapters 5, 8, and 10 for signals associated with transport into the nucleus, endoplasmic reticulum and chloroplast respectively.*

- The import of cytosolic proteins to the mitochondrial matrix occurs at specialized sites where the OMM and IMM are in close contact.
- Transit signals are recognized by **receptor proteins** on the OMM which directs the polypeptide towards import translocases.
- At these sites, import translocases of the OMM and IMM called **TOM**s and **TIM**s respectively, can form a continuous pore through the envelope.
- A polypeptide destined to form an OMM protein possesses an internal signal sequence that allows the TOM to deliver it to the membrane.
- Imported polypeptides destined for insertion into the IMM possess a second leader signal that is exposed when the transit sequence is hydrolytically removed, and which directs them to the membrane.
- They also contain **topogenic** or **hydrophobic sorting signals** consisting of stretches of hydrophobic amino acid residues which allow the protein to be inserted into the IMM (Chapter 3).

- Following translocation, the transit signal is removed by hydrolysis.
- Those polypeptides destined to form matrix proteins are rapidly moved through both the TOM and TIM and delivered to the matrix (Figure 11.8).
- Once in the matrix, the transit signal is hydrolysed from the imported polypeptide by a matrix transit protease.
- Two mitochondrial chaperones, Hsp60 and Hsp70, supervise the folding of the imported polypeptide after it has entered the matrix.
- All these transport activities require metabolic energy, which is provided by the electrochemical gradient across the IMM and by the hydrolysis of ATP.
- The hydrolysis of ATP in the mitochondria is probably associated with the folding of the polypeptides to give active proteins.

Looking for extra marks?

Note there are many similarities between the import of polypeptides into mitochondria and into chloroplasts.

Mitochondrial replication and inheritance

Mitochondria arise from the growth and subsequent division of pre-existing mitochondria.

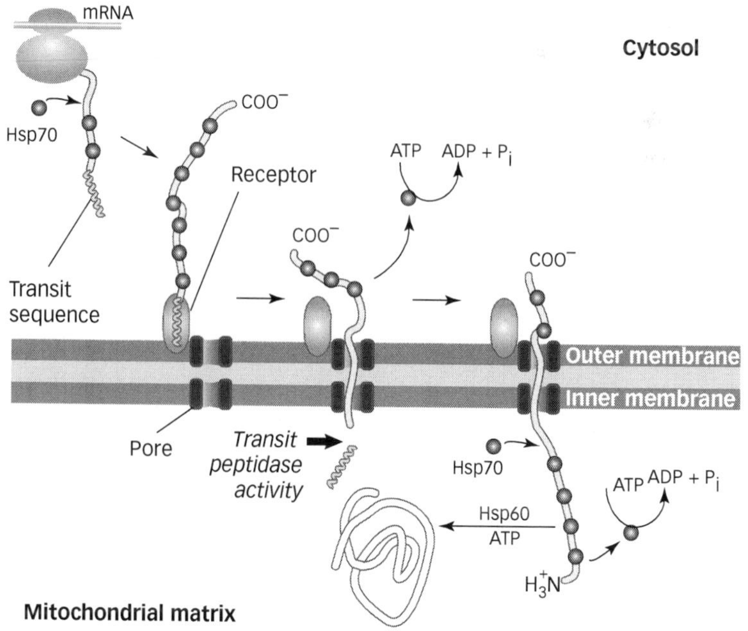

Figure 11.8 Overview of the import of a mitochondrial matrix polypeptide and its subsequent folding to form an active protein. Compare with Figure 10.4.

Evolutionary origins of mitochondria

- A growing mitochondrion eventually reaches a state that triggers its division to form two daughter mitochondria.
- This division maintains a number of mitochondria appropriate to the cell type.
- In some cases the total number of mitochondria per cell can change; for example, prolonged muscle activity can increase their number in muscle tissues.
- When cells divide by **mitosis** (Chapter 16), the mitochondria are segregated equally between the daughter cells.
- Prior to mitochondrial division, the mtDNA is replicated by **DNA dependent DNA polymerase** γ (**DNA polymerase** γ, **DNA pol** γ).
- Mitochondrial DNA replication introduces more mutations than does nuclear copying: DNA polymerase γ does not possess the **proofreading** capacity of nuclear DNA polymerases, which have error rates of approximately one mismatched base in 10^6–10^7 copied. Comparable rates for DNA polymerase γ are estimated to be about ten times greater (Chapters 5 and 6).
- In sexual reproduction, the inheritance of mitochondrial genes is **maternal** because spermatozoa possess little cytoplasm and all the mitochondria found in the zygote are donated from the egg.
- Thus, all offspring (male and female) inherit their mitochondria only from their female parent, and males *cannot* transmit their mitochondria to subsequent generations.
- Mutations in mtDNA replication are associated with a group of diseases called **mitochondrial myopathies.**

Mitochondrial myopathies

Mitochondrial myopathies are associated with mutations of mtDNA and also with mutations in nuclear genes that affect the import of mitochondrial proteins.

- Despite its small size, a relatively large number of mutations are associated with mtDNA and are significant contributors to human disease (Table 11.2). Their frequency in the UK is estimated at approximately seven in 100,000.
- Mutations in mtDNA result in inherited myopathies that can affect both sexes but which can only be passed on by affected females.
- Mitochondrial myopathies are a heterogeneous group of disorders. They generally impair mitochondrial oxidative phosphorylation systems; affecting energy demanding tissues, such as the central nervous system and sense organs, heart and skeletal muscles, kidneys and secretory tissues.
- They may also affect the morphology of the mitochondrion (Figure 11.9).
- The severity of any given mitochondrial myopathy varies between different patients depending upon the proportion of mutated to normal mtDNA molecules. When both mutated and normal are present, the condition is described as **heteroplasmy**; if all are mutated then **homoplasmy**.
- The higher rates of mutations and purely maternal inheritance associated with mitochondrial replication has proved useful in some genetic and evolutionary studies, such as establishing the evolution of modern humans in sub-Saharan Africa about 200,000 years ago.

Myopathy	*Abbreviation*	*Genes affected*	*Features*
Chronic progressive external ophthalmoplegia	CPEO	Deletion of several genes	Myopathy of eye muscle
Leber hereditary optic neuropathy	LHON	Those for subunits of NADH dehydrogenase	Blindness resulting from damage to optic nerve
Mitochondrial encephalomyopathy, lactic acidosis, and stroke-like episodes	MELAS	Those for $tRNA^{leu}$, $tRNA^{Glu}$ and NADH dehydrogenase	Myopathy, dementia, seizures, lactic acidosis
Myoneurogastrointestinal encephalopathy	MNGIE	Multiple deletions	Defects to eyes, leucoencephalopathy, gastrointestinal problems, myopathy
Myoclonic epilepsy and ragged-red fibre	MERRF	Those for $tRNA^{lys}$, $tRNA^{Leu}$	Myoclonic seizures, ataxia, lactic acidosis, 'ragged-red fibres' (an abnormality of the muscle tissue observable by microscopy)
Neurogenic muscle weakness, ataxia and retinitis pigmentosa	NARP	Those for subunits of F_1F_0-ATP synthase	Muscle wasting, ataxia, blindness

Table 11.2 Examples of mitochondrial myopathies

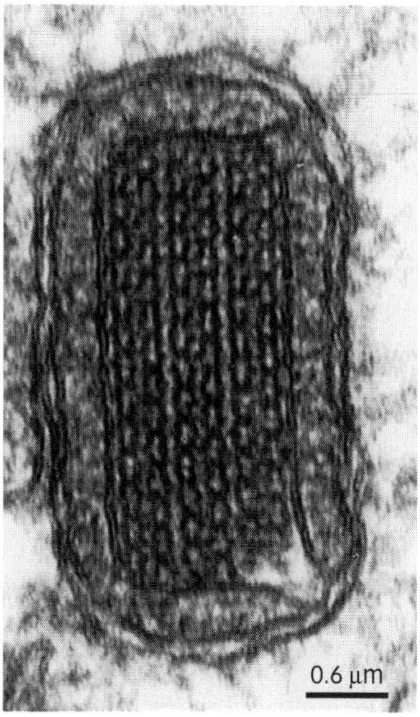

Figure 11.9 Electron micrograph of a mitochondrion displaying abnormal morphology (compare with Figure 11.1A) associated with a mitochondrial myopathy. Courtesy of M.J. Cullen, Newcastle General Hospital, UK.

Looking for extra marks?

Acquired mitochondrial DNA defects occur in HIV-1 sufferers treated with **highly active antiretroviral therapy (HAART)**, a treatment that uses inhibitors of the virus reverse transcriptase (Chapter 18) and the HIV protease. Zidovudine (AZT), the first inhibitor of reverse transcriptase used clinically, inhibits mtDNA replication and reduces the amount of mtDNA present. This leads to side effects that mimic the symptoms of *inherited* mitochondrial disorders. Patients treated with HAART must be monitored for mitochondrial functions.

Check your understanding

11.5 Which of the following statements is/are false?

a. The TCA cycle occurs in the cytoplasm.

b. Cristae of the OMM greatly increase its surface area compared to that of the IMM.

c. Electron transport promotes the active transport of H^+ into the intermembrane space by complexes I, II, and IV.

d. The free energy of the pmf is used to drive the release of ATP from the F_0 complex.

e. Glycolysis always occurs in the cytoplasm.

11.3 HYDROGENOSOMES

Hydrogenosomes are spherical-shaped organelles of approximately 1–2 μm diameter that are able to produce molecular or dihydrogen (H_2), hence their name.

- Hydrogenosomes occur *only* in eukaryotic organisms that live in anoxic environments and lack mitochondria. These include:
    - rumen-dwelling ciliates and fungi, as well as some free-living ciliates
    - the sexually transmitted parasite *Trichomonas vaginalis*, where some of the best-studied hydrogenosomes are found (Figure 11.10).
    - In 2010, organelles resembling hydrogenosomes were found in a group of multicellular organisms called loriciferans that live in anoxic conditions in a deep sea basin of the Atlantic Ocean.
- Hydrogenosomes lack the enzymes that catalyse the TCA cycle, and their inner membrane is not organized into cristae: they do not carry out oxidative phosphorylation.

Production of H_2 by hydrogenosomes

Within hydrogenosomes, the production of H_2 is coupled to the formation of ATP as shown in Figure 11.11.

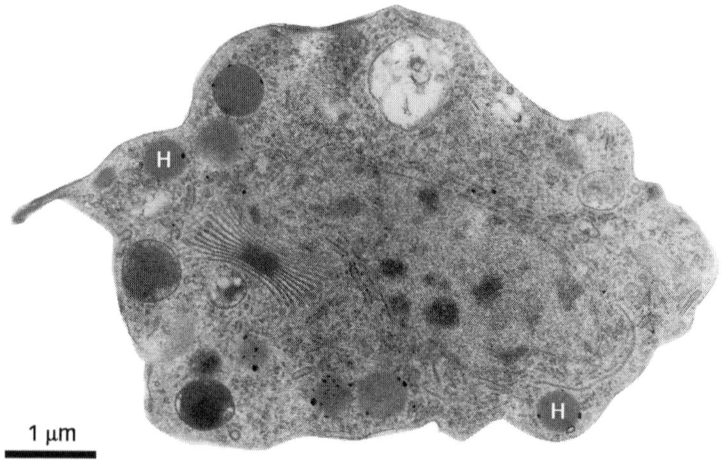

Figure 11.10 Electron micrograph of *Trichomonas vaginalis*. The hydrogenosomes (H) of this parasite are among the best studied. Courtesy of Dr A. Curry, Manchester Royal Infirmary, UK.

- Pyruvate and malate formed in the cytosol from glucose can enter the hydrogenosome where they are substrates for the enzymic actions of pyruvate:ferredoxin oxidoreductase and malic enzyme respectively.
- The pyruvate is converted to acetyl CoA and CO_2 in a reaction catalysed by acetate:succinate CoA-transferase.
- The acetyl CoA can be used to form ATP via succinyl CoA (Figure 11.11). Since the electron transport chain is not involved in this synthesis it is a substrate-level phosphorylation of ADP.

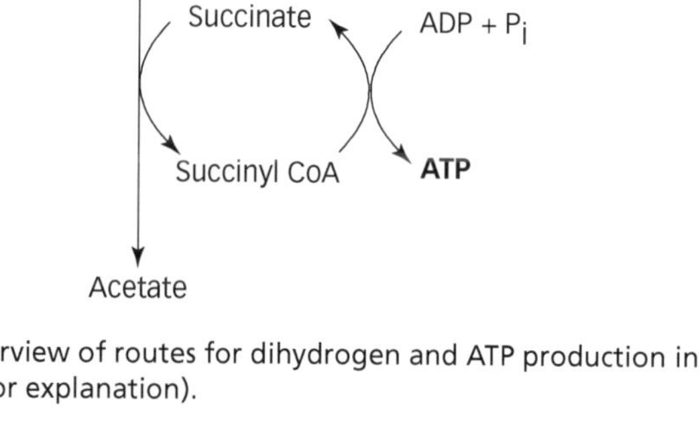

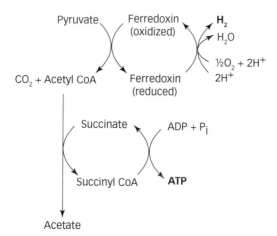

Figure 11.11 Overview of routes for dihydrogen and ATP production in hydrogenosomes (see text for explanation).

- The electrons released from pyruvate are transferred by ferredoxin to an hydrogenase (not shown in Figure 11.11 for simplicity), which uses H^+ as electron acceptors forming H_2.
- Malate can be used directly to reduce NAD^+ to NADH, from which electrons can be transferred to ferredoxin by the NADH:ferredoxin oxidoreductase activity of complex I (see 'Oxidative phosphorylation' section and Figure 11.3).
- The net result of these reactions is the conversion of pyruvate and malate to acetate, CO_2 and H_2.
- These reactions are coupled to the formation of ATP by the substrate level phosphorylation of ADP.

Evolutionary origins of hydrogenosomes

Hydrogenosomes appear to be have evolved from mitochondria in organisms that adapted to anaerobic environments.

- Hydrogenosomes do not possess electron transport chains or genomes. However, hydrogenosomes present in *Nyctotherus ovalis*, found in the hind gut of termites, have a partial electron transport chain and the remnants of a mitochondrial genome suggesting that *Nyctotherus* is an evolutionary link between organisms with true mitochondria (with electron transport chains and genomes) and fully evolved hydrogenosomes (lacking electron transport chains and genomes).
- The wide distribution of hydrogenosomes among diverse phyla, and the differences in the details of their biochemistries, suggests they have evolved independently from mitochondria in anaerobic-dwelling ancestors on several occasions.

11.4 MITOSOMES

Mitosomes, like hydrogenosomes, are degenerate mitochondria.

- Mitosomes are found in a number of unicellular eukaryotic species that occupy anaerobic environmental niches.
- Mitosomes resemble hydrogenosomes in appearance, being roughly spherical with diameters of about 1 μm, and limited by a double membrane.
- Their inner membrane in some species has a rudimentary system of cristae.
- Unlike mitochondria and hydrogenosomes, mitosomes are not involved in the formation of ATP, but produce Fe-S proteins, which requires anoxic conditions.

Revision tip

Do browse through the further reading references given for this chapter online. Go to http://www.oxfordtextbooks.co.uk/orc/thrive/

Check your understanding

11.6 Which of the following statements is/are true or is/are false?

a. Glycolysis supplies all the pyruvate used by the TCA cycle.

b. Porins are found in the outer membranes of Gram-negative bacteria (Chapter 4) and chloroplasts and mitochondria.

c. The relative volume of matrix to the surface area of cristae in mitochondria reflects their roles in the cell.

d. Within the matrix, pyruvate dehydrogenase catalyses the conversion of pyruvate to acetyl CoA with the release of CO_2 and formation of NAD^+.

e. Each cycle of β oxidation reduces the length of a fatty acid chain by two carbon atoms with the release of acetyl CoA and formation of NADH and $FADH_2$.

f. Hydrogenosomes are found in some eukaryotes of anoxic environments that lack mitochondria.

g. Evidence suggests that hydrogenosomes and mitosomes evolved from mitochondria in ancestors that lived in anaerobic environments.

h. Hydrogenosomes coupled the production of H_2 to the formation of ATP.

i. Mitosomes resemble mitochondria and hydrogenosomes in being limited by a double membrane.

j. Hydrogenosomes possess the enzymes that catalyse the TCA cycle and their inner membrane is organized into cristae that are capable of carrying out oxidative phosphorylation.

12 Microbodies

Key features of microbodies

- Microbodies are globular-shaped organelles surrounded by a single biological membrane.
- Three types of microbodies are recognizable in different types of cells:
  1. peroxisomes
  2. glyoxysomes
  3. glycosomes.
- All three perform different functions but all produce ATP by substrate-level phosphorylation.
- Plant peroxisomes are involved in photorespiration.
- Mutations in the genes for peroxisomal proteins or for their insertion in peroxisomes lead to a variety of clinical conditions.
- Glyoxysomes are found in plant cells and facilitate the glyoxylate cycle.
- Glycosomes are most notably associated with human pathogenic trypanosomes.

Check your understanding

12.1 Which of the following is the odd one out?

a. Glyoxysomes
b. Peroxisomes
c. Glycosomes
d. Mitochondria
e. None of these.

12.1 KEY FEATURES OF PEROXISOMES

Peroxisomes are roughly spherical organelles that are widely distributed in eukaryotes.

- They are approximately 0.2–2.0 μm in diameter (Figure 12.1).
- Like all microbodies they are limited by a single membrane.
- Peroxisomes are enriched in urate and D-amino acid oxidases, and catalase activities, whose presence, particularly catalase, is regarded as a defining feature.
- Peroxisomes contain about 40 enzymes involved in oxidation-reduction reactions.

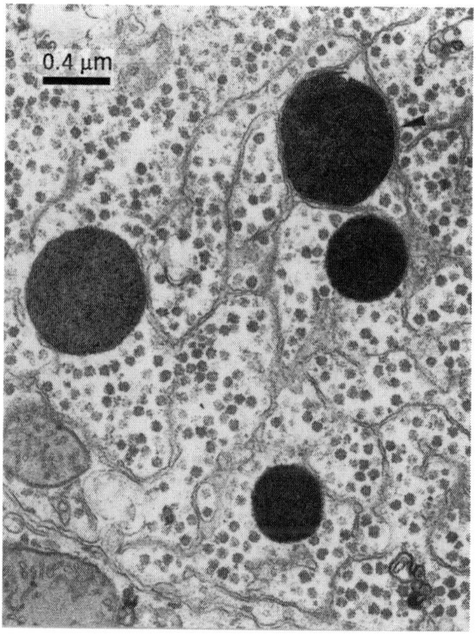

Figure 12.1 Electron micrograph of peroxisomes of liver visualized by their catalase activity. Courtesy of Drs M. Espeel and F. Roels, Department of Anatomy, Embryology and Histology, University of Ghent, Belgium.

Key activities of peroxisomes

- They participate in four major metabolic activities:
  1. the catabolism of certain types of long chain and very long chain fatty acids (LCFAs and VLCFAs respectively) to acetyl CoA
  2. certain detoxification reactions, such as the catabolism of ethanol
  3. the catalysis of hydrogen peroxide production and breakdown
  4. in plants, leaf peroxisomes are the sites of photorespiration.

> **Revision tip**
>
> Revise the structure of peroxisomes using the beautiful electron micrographs at The Cell: An Image Library™ of The American Society for Cell Biology at: http://www.cellimagelibrary.org/

12.2 KEY ACTIVITIES OF PEROXISOMES

General peroxisomal activities are summarized in Figure 12.2.

- Peroxisomes contain the enzymes that catalyse the β-oxidation of long chain (C_{16-22}), very long chain (C_{24-26}), and branched fatty acids to form acetyl CoA, which is exported to the cytosol where it can be used for the biosynthesis of fatty acids or subsequently enter mitochondria and be degraded by the TCA cycle (Chapter 11).
- Urate is formed by the catabolism of nucleic acids and certain proteins. Its oxidation to allantoin is catalysed by urate oxidase.
- The breakdown of a variety of other substances is also performed by peroxisomal enzymes. For example, D-amino acids, which cannot be metabolized by most amino acid using enzymes that generally recognize only L-amino acids, are absorbed by the gastrointestinal tract following the breakdown of the cell walls of gut bacteria.

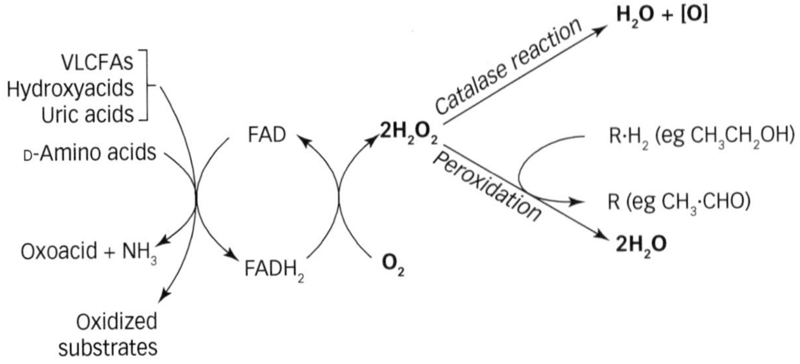

Figure 12.2 Summary of general peroxisomal activities. VLCFA, very long chain fatty acid.

- The oxidation reactions of the peroxisome produce the toxic oxidizing agent, hydrogen peroxide.
- Hydrogen peroxide is rendered harmless by one of two reactions, both catalysed by catalase.
  - *Catalase* activity degrades hydrogen peroxide to water and dioxygen.
  - As a *peroxidase*, catalase can use electrons from organic donors to reduce hydrogen peroxide to water, an activity that can be used to detoxify harmful substances such as ethanol.
- The actions of catalase, other peroxidases, and superoxide dismutase in peroxisomes protect the cell from **reactive oxygen species**, such as superoxide anions (O_2^-) and hydroxyl radicals ($OH^\bullet$).
- In some circumstances these may be produced as by-products of, for example, the oxidative activities of mitochondria (Chapter 11).

Photorespiration

Photorespiration is a light-dependent respiratory process that occurs in most photosynthesizing plants (see Chapter 11).

- Photorespiration occurs because, as explained in Chapter 11, the principal enzyme of carbon fixation in the **Calvin cycle** of **photosynthesis**, **ribulose 1,5-bisphosphate carboxylase-oxygenase** (**rubisco**), catalyses two distinct activities:
  - as a *carboxylase* it catalyses the fixation of CO_2:

$$\text{Ribulose 1,5-bisphosphate} + CO_2 \rightarrow 2 \times \text{3-phosphoglycerate}$$

  - as an *oxygenase* it promotes the formation of 2-phosphoglycolate:

$$\text{Ribulose 1,5-bisphosphate} + O_2 + CO_2 \rightarrow \text{2-phosphoglycolate} + \text{3-phosphoglycerate}$$

- An outline of photorespiration or glycolate cycle is given in Figure 12.3.

Key features of photorespiration

- Within the chloroplast, the combined carboxylase-oxygenase activities of rubisco cleaves ribulose 1,5-bisphosphate to 2-phosphoglycolate and 3-phosphoglycerate. The latter is hydrolysed to glycolate, which is exported to large leaf peroxisomes in the cell.
- In the peroxisome, glycolate is converted to the amino acid glycine, which, in turn, is transported to mitochondria, where it is used to form another amino acid, serine.
- The next stage of photorespiration is the export of the serine to the peroxisome and its conversion into glycerate. Glycerate is moved to the chloroplast and

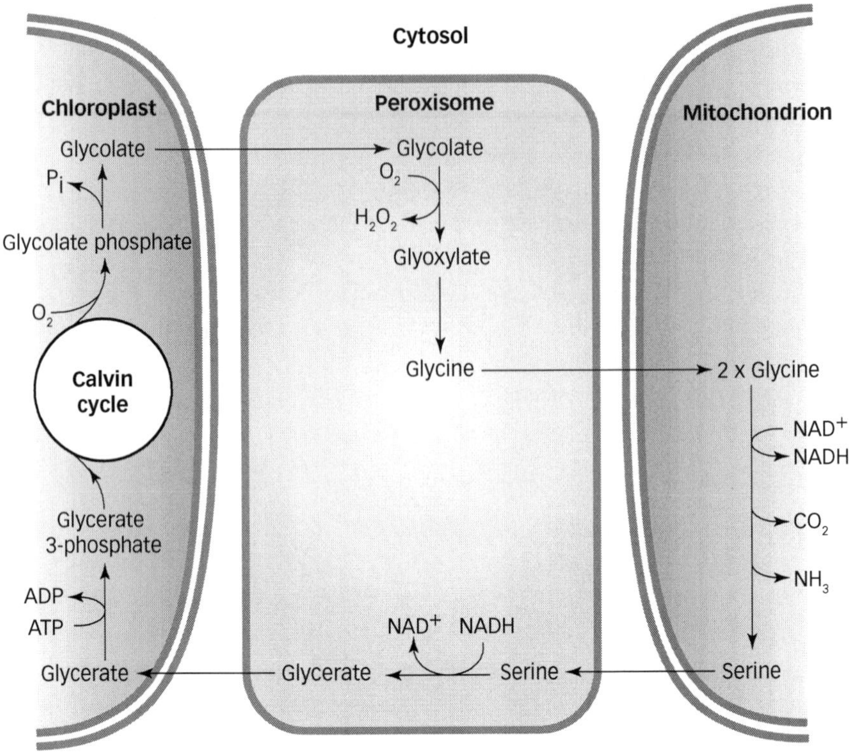

Figure 12.3 Simplified outline of photorespiration or glycolate cycle.

phosphorylated to give 3-phosphoglycerate, which can enter the Calvin cycle of photosynthesis.

- Photorespiration is therefore dependent on the interconnected activities of three organelles: chloroplasts, peroxisomes, and mitochondria.
- Photorespiration is promoted by conditions of high light intensity on hot, dry days when the rate of photosynthesis is high and stomatal closure is promoted. This increases the concentration of O_2 in leaf chloroplasts relative to that of CO_2, allowing it to compete against it for the common active site on rubisco.
- The recycling of the 2-phosphoglycolate to form 3-phosphoglycerate is wasteful of ATP and NADPH.
- Photorespiration is also wasteful because it returns only about 75% of the carbon fixed in 2-phosphoglycolate to the Calvin cycle, since one of the four carbons present in the two glycine molecules (derived from two 2-phosphoglycolate molecules) is lost as CO_2.
- The waste of ATP, NADPH, and fixed carbon are unavoidable consequences of the dual carboxylase-oxygenase activities of rubisco.

➔ *Biochemical details of photorespiration can be found in the companion volume,* Thrive in Biochemistry and Molecular Biology.

- Peroxisomes also have a number of diverse functions including the biosynthesis of plant hormones and reactions involved in the nitrogen metabolism in the root nodules of leguminous plants.

Check your understanding

12.2 Which of the following is the odd one out regarding photorespiration?

a. Peroxisomes

b. Chloroplasts

c. Mitochondria

d. Glycosomes

e. None of these.

Biogenesis of peroxisomes

Peroxisomes can form *de novo* as protoperoxisomes from vesicles derived from the endoplasmic reticulum, but appear to *increase* in number when *pre-existing* peroxisomes grow and divide.

- New materials for peroxisome growth are synthesized elsewhere in the cell.
  - **Lipids** required for peroxisomal membranes extension are formed in the smooth endoplasmic reticulum (Chapter 8) and delivered to the peroxisome by **phospholipid exchange proteins**.
  - **Peroxisomal proteins** are formed on *free* ribosomes in the cytosol (Chapter 7) and delivered into the organelles by specific peroxisomal proteins called **peroxins**, in a poorly understood ATP-dependent process.
- Peroxisomal proteins are synthesized in the cytosol and imported into the organelles by a complex process that is specific to the protein concerned.
- The possession of a **peroxisomal targeting signal** (**PTS**) by a putative peroxisomal protein is essential for its entry.
- The major signal (PTS1) consists of three carboxyl terminal amino acids residues with the consensus sequence –Ser–Lys–Leu–COO^-. Another type of PTS is also known, but this is situated near the amino terminus.

Looking for extra marks?

A number of rare terminal genetic diseases (approximately 1 in 50,000 live births) are caused by mutations associated with peroxisomal enzymes. The commonest are adrenoleucodystrophy (ALD), Zellweger syndrome, and Refsum's disease.

Check your understanding

12.3 Which of the following statements is the odd one out regarding peroxisomes?

a. They catabolize some long chain and very long chain fatty acids.
b. They are one of the possible sites where glycolysis occurs.
c. They participate in some detoxification reactions.
d. They catalyse the breakdown of hydrogen peroxide.
e. Leaf peroxisomes are the sites of photorespiration.

12.3 KEY FEATURES OF GLYOXYSOMES

Glyoxysomes are specialized peroxisomes found particularly in plant lipid storage structures such as the endosperm or the cotyledons of germinating seeds; they also occur in filamentous fungi.

- In addition to being found associated with lipid bodies (Figure 12.4), glyoxysomes are often situated near mitochondria.
- In addition to oxidative enzymes, glyoxysomes also contain those that catalyse the **glyoxylate cycle** or **shunt**, a series of reactions that allow the energy in insoluble lipid reserves to be transported to the **meristems** (growing regions) of the developing seedling as the soluble sugar, **sucrose** (Figure 12.5).

The glyoxylate cycle

An overview of the glyoxylate cycle is given in Figure 12.5.

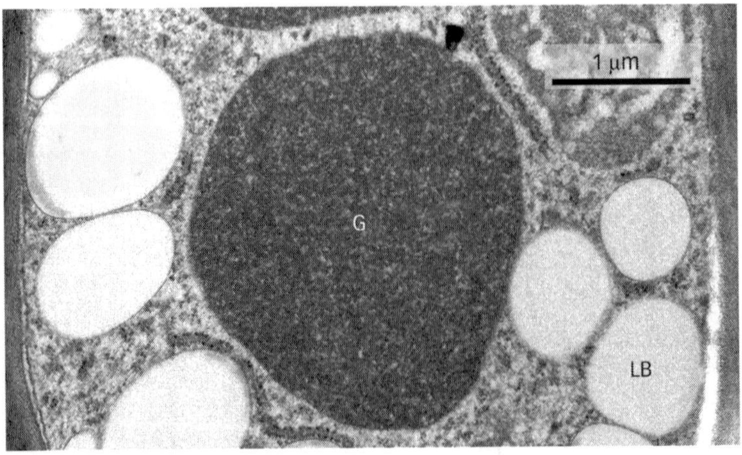

Figure 12.4 Electron micrograph showing the association of glyoxysomes (G) with lipid bodies (LB) in a developing seedling. Courtesy of Professor I.A. Graham, Department of Biology, University of York, UK.

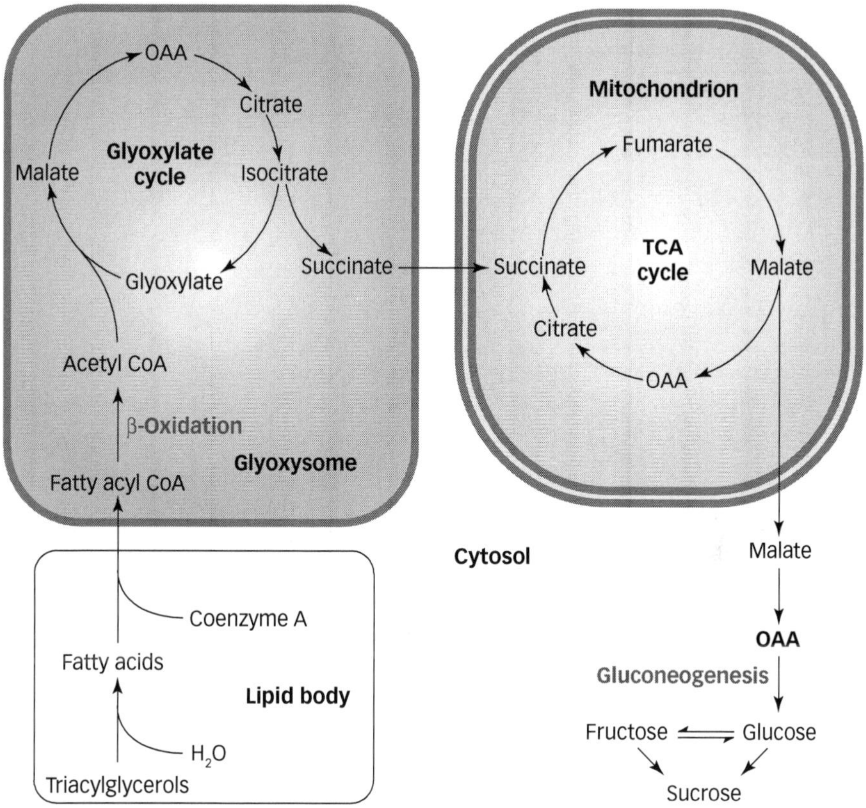

Figure 12.5 Outline of the glyoxylate cycle.

- The hydrolysis of triacylglycerol reserves in lipid bodies releases fatty acids, which enter the glyoxysome.
- In the glyoxysome, β-oxidation of fatty acids forms acetyl CoA, which enters the glyoxylate cycle.
- The glyoxylate cycle consists of five reactions that generate **succinate**, while forming and degrading glyoxylate.
- The net reaction of the cycle is:

$$2 \text{ Acetyl CoA} + NAD^+ \rightarrow \text{Succinate} + NADH + H^+$$

- Succinate is exported to the cytosol, from where it is taken up by mitochondria and enters the TCA cycle, and is converted to malate.
- Malate is then exported from the mitochondrion into the cytosol, where it is reduced to oxaloacetate.
- Oxaloacetate is a substrate for **gluconeogenesis** and can be converted to fructose and glucose 6-phosphates, from which sucrose, the major transport sugar of plants, can be synthesized (Figure 10.12).

- The glyoxylate cycle is therefore essential to allow energy to be transported to the shoot and root meristems to promote their growth.
- Many organisms, including humans, lack a glyoxylate cycle and thus cannot convert lipids to carbohydrates, although they can do the reverse—hence the ease with which an excessive dietary carbohydrate intake can lead to obesity.

➔ *Biochemical details of the glyoxylate cycle can be found in the companion volume,* Thrive in Biochemistry and Molecular Biology.

Interconversion of glyoxysomes and peroxisomes

As described earlier, geminating seeds contain glyoxysomes. However, as seedlings grow and actively photosynthesize, leaf peroxisomes participate in photorespiration (Figures 12.3 and 12.6).

- When leaves senesce, their peroxisomes are converted back to glyoxysomes.

12.4 GLYCOSOMES

Glycosomes are membrane-enclosed organelles of diameter 0.2–0.8 µm, which occur in a relatively few species of protozoa called trypanosomes (Figure 12.7).

- Human pathogenic trypanosomes are the causative agents of, for example, sleeping sickness, Chagas' disease, and leishmaniasis.

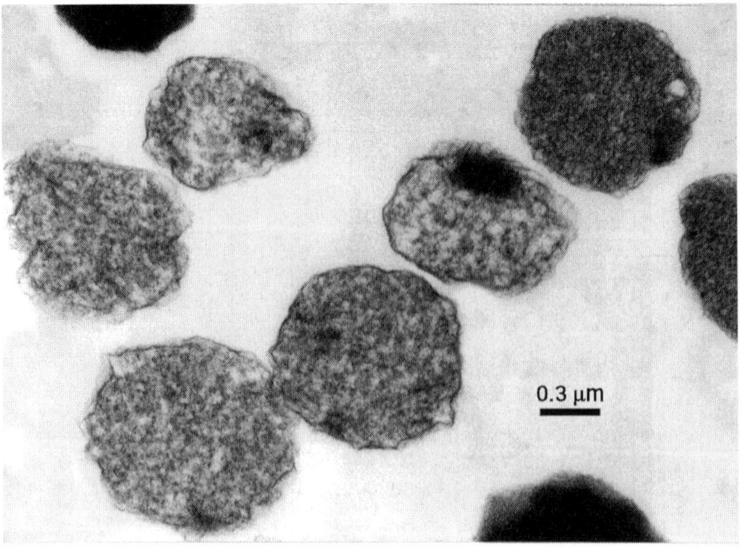

Figure 12.6 Electron micrograph of spinach leaf peroxisomes. Courtesy of Professor D.G. Robinson, Pflanzenphysiologisches Institut der Universität Göttingen, Germany.

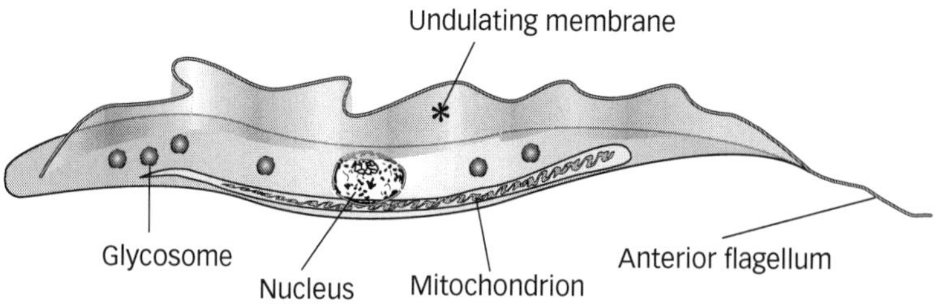

Figure 12.7 Schematic illustration of *Trypanosoma brucei*.

Revision tip

Look at two animations at the Wellcome Trust relating to stages in Chagas' disease at: http://www.wellcome.ac.uk/Education-resources/Teaching-and-education/Animations/Protozoans/WTDV027431.htm and http://www.wellcome.ac.uk/Education-resources/Teaching-and-education/Animations/Protozoans/WTDV027431.htm

Looking for extra marks?

Darwin, the greatest of all biologists, suffered ill health throughout the later stages of his life and it has been suggested he was infected with *Trypanosoma cruzi*, the causative agent of Chagas' disease during his 5-year round-the-world voyage on HMS *Beagle*.

- Although glycosomes appear morphologically similar to other microbodies, they contain enzymes that catalyse, at least in part, the reactions of **glycolysis** (Chapter 7 and Figures 7.2 and 12.8), hence their name. However, they also contain a number of other enzymes.
- The glycolytic enzymes of glycosomes differ in structure to their cytosolic counterparts of other organisms.
- The first seven enzymes of glycolysis and two enzymes of glycerol metabolism are confined to the glycosome (Figure 12.8).
- In the glycosome, glucose is metabolized to glyceraldehyde 3-phosphate, which is exported to the cytosol as the 2-phosphate derivative.
- Alternatively, dihydroxyacetone phosphate may be converted to glycerol 3-phosphate, which can be released to the cytosol or converted to glycerol in a glycosomal reaction that generates ATP.
- In aerobic conditions, a mitochondrial glycerol-1,3-phosphate oxidase (GPO) can reconvert the glycerol 3-phosphate back to DHAP, which is able to enter the glycosome and be metabolized further.

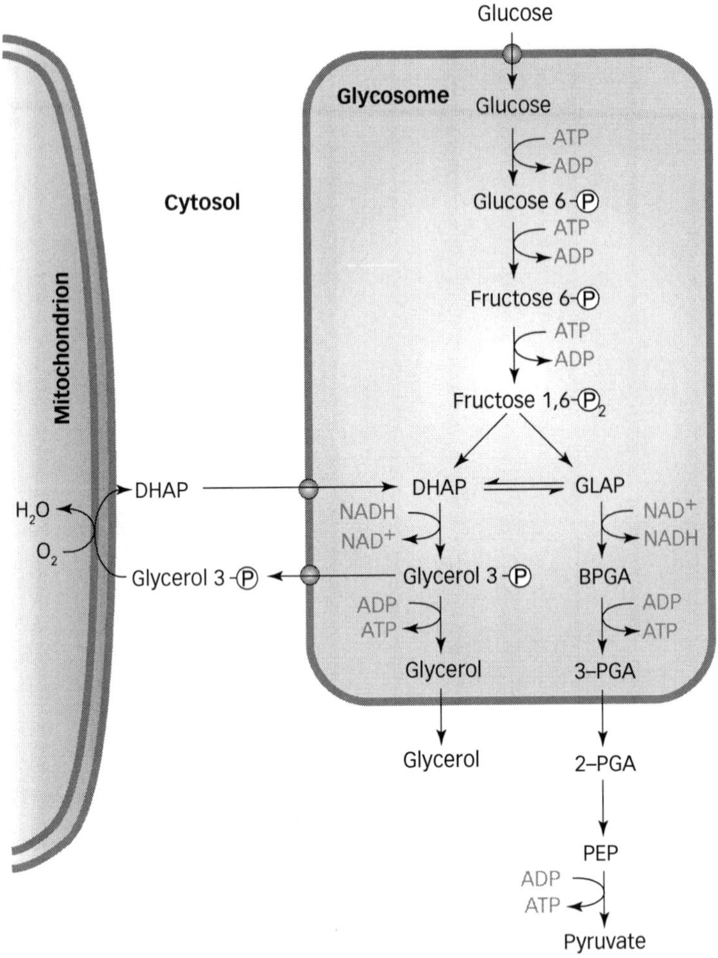

Figure 12.8 Overview of glycolysis in the parasitic trypanosome, *Trypanosoma brucei*. Redrawn from Helfert, S., Estevez, A.M., Bakker, B., Michel, P., and Clayton, C. (2001) Roles of triosephosphate isomerase and aerobic metabolism in *Trypanosoma brucei*. *Biochemical Journal*, 357, 117–25. BPGA, glycerate 1,3-bisphosphate; DHAP, dihydroxyacetone phosphate; GLAP, glyceraldehyde 3-phosphate; PEP, phosphoenolpyruvate; 2-PGA, glycerate 2-phosphate; 3-PGA, glycerate 3-phosphate. Compare with Figure 7.2.

- Under aerobic conditions, 90% of the glucose is metabolized to pyruvate and approximately 10% forms glycerol. Both products are released from the glycosome.

➔ *Biochemical details of these pathways can be found in the companion volume,* Thrive in Biochemistry and Molecular Biology.

Looking for extra marks?

Need excellent photographs and diagrams relating to pathogens and their vectors for an assignment? Try the Public Health Image Library (PHIL) of the Centers for Disease Control and Prevention at: http://phil.cdc.gov/phil/quicksearch.asp Remember to fully acknowledge your source!

Check your understanding

12.4 Which of the following statements is/are false?

a. Glyoxysomes contain enzymes that catalyse the glyoxylate cycle.

b. Glycosomes are restricted to a relatively few species of protozoa, such as human pathogenic trypanosomes.

c. The glyoxylate cycle converts acetyl CoA to succinate.

d. All organisms can perform the glyoxylate cycle.

e. Glucose can be metabolized to glyceraldehyde 3-phosphate in glycosomes.

Revision tip

Do browse through the further reading references given for this chapter online. Go to http://www.oxfordtextbooks.co.uk/orc/thrive/

Check your understanding

12.5 Which of the following statements is/are true or is/are false?

a. The major defining feature of peroxisomes is the presence of catalase, urate oxidase, and D-amino acid oxidase activities.

b. Peroxisomes catalyse the β-oxidation of certain long chain, very long chain, and some branched fatty acids to form acetyl CoA.

c. Catalase shows both catalase and peroxidase activities.

d. The peroxidase activity of catalase degrades hydrogen peroxide to water and dioxygen.

e. Catalase, peroxidases, and superoxide dismutase protect the cell from reactive oxygen species.

f. The *carboxylase* activity of rubisco catalyses the fixation of CO_2 forming 2-phosphoglycerate.

g. The *oxygenase* activity of rubisco catalyses the formation of 3-phosphoglycolate.

h. Photorespiration is promoted by high light intensity on hot, dry days.

i. Peroxisomal proteins are formed on ribosomes of the endoplasmic reticulum.

j. The major peroxisomal targetting signal (PTS1) consists of three amino terminal amino acids residues with the consensus sequence –Ser–Lys–Leu–COO^-.

13 Eukaryotic cell walls

Key features of cell walls

- Eukaryotic cell walls are found in:
  - plants
  - algae
  - fungi
  - a number of unicellular organisms.
- However, plant cell walls are the best studied (Figure 13.1).
- Cell walls are to some extent semipermeable: they are relatively impermeable to large macromolecules but water-soluble micromolecules can freely diffuse through them.
- Cell walls (and animal extracellular matrices (Chapter 14)) have a generalized structure that can be described as 'fibres in a ground substance or matrix'.
- Structural materials based on fibres in a matrix are extremely strong, able to bear extensive stresses, and do not propagate any fractures that might occur.
- These types of structures are not only able to withstand mechanical stresses but can be modified to suit the changing needs of the organism.
- The composition of eukaryotic cell walls varies between different organisms.
- Carbohydrates predominate but glycoproteins are also found in the cell walls of all groups.

- Cellulose is the major carbohydrate in the cell walls of plants and some other groups but alginates, agar, or carrageenans are found in algae, glucans and chitin in fungi, and yeasts have mannose-rich cell walls.
- Mature plant cell walls also contain lignin.

Looking for extra marks?

All cells are surrounded by a cell wall or an extracellular matrix (ECM). The majority of cells are enclosed by a cell wall.

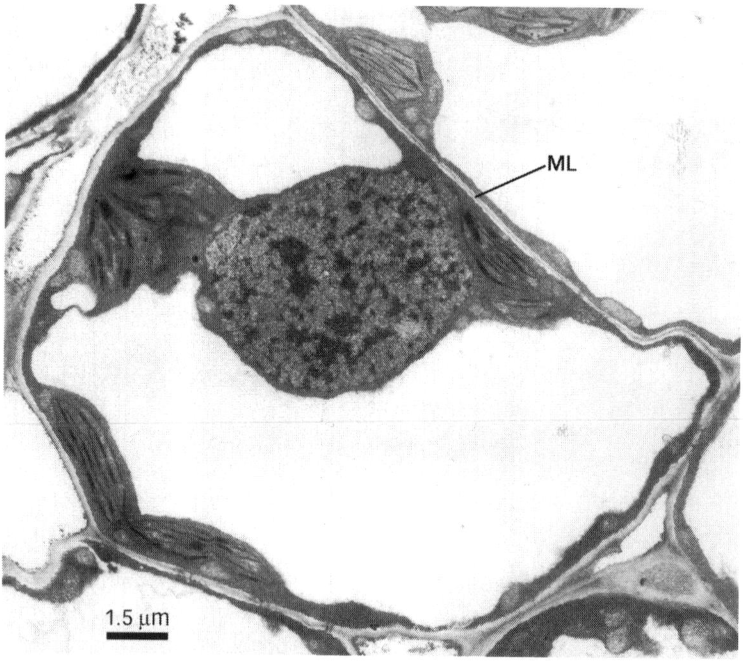

Figure 13.1 Electron micrograph of a plant cell. Note the cell wall enclosing the cell. The middle lamella (ML) forms the boundary between adjacent cells. Courtesy Professor D.G. Robinson, Pflanzenphysiologisches Institut der Universität, Gottingen, Germany.

Key roles of eukaryotic cell walls

- All eukaryotic cell walls perform similar functions:
  - protect the cell
  - support the cell
  - prevent osmotic lysis

continued

- prevent desiccation
- give shape to the cell.

- The two material phases of a cell wall, *fibres* and *matrix*, complement one another; fibres resist tension and the matrix compression produces a strong material able to withstand mechanical stresses.
- Cell walls provide a rigid framework against which the **turgor pressure** arising from osmotic pressure of the vacuole (Chapter 9) provides support for the cell and in plants for non-woody tissues.
- Cell wall pressure counterbalances the osmotic pressure that tends to drive water into the cytoplasm, and so prevents an excessive uptake of water by the cell.
- Cell walls are to some extent semipermeable: large macromolecules diffuse through it with relative difficulty but gases, ions, and the water-soluble micromolecules of sugars, amino acids, and many plant hormones can move freely through it.
- Cell walls are hydrated structures and oppose excessive water losses from the cell in dehydrating conditions.
- Cell walls also impose a shape to the **protoplast**, the portion of cell enclosed within the wall.

13.1 PLANT CELL WALLS

Plant cell walls are the best studied of the eukaryotic types.

- Mature plant cell walls consist of three basic layers:
  1. middle lamella
  2. primary cell wall
  3. secondary cell wall.
- Plant cells are glued together by a **middle lamella**, which separates their walls.
- Within the middle lamella is a **primary cell wall**.
- In some cells, a **secondary cell wall** may be deposited between the plasma membrane and the primary wall (Figure 13.2).

Components of plant cell walls

The principal components of plant cell walls are:

- Cellulose
- Hemicelluloses
- Pectins
- Ca^{2+}
- Lignin
- Proteins.

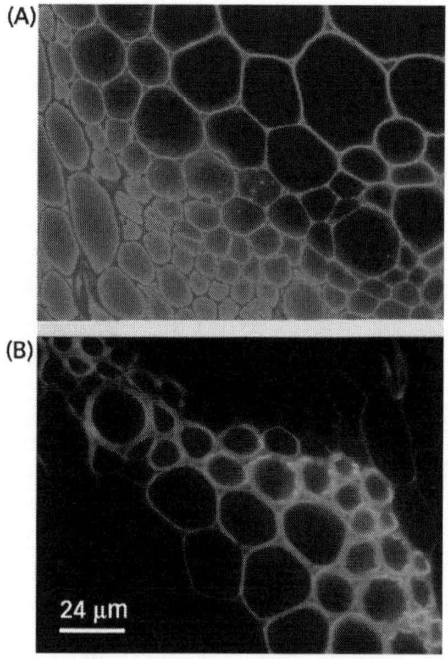

Figure 13.2 Photomicrographs using probes specific for (A) pectin and (B) cellulose to show the positions of the primary and secondary cells walls. Courtesy of Professor J.M. Macurulla, Universidad del Pais Vasco, Bilbao, Spain.

Cellulose

- Cellulose is the major carbohydrate of most plant cell walls, and is the most abundant carbohydrate on earth.
- It is an unbranched polymer of glucose residues linked by β,1→4 glycosidic bonds (Figure 13.3).
- Cellulose molecules can be 8000–15,000 sugar residues long, and have extended conformations that are stabilized by numerous intermolecular hydrogen bonds.
- Within the cell wall, approximately 50–60 cellulose molecules are twisted around one another to form a cellulose **microfibril** of 3–5 nm diameter (Figure 13.4).
- Each layer of cellulose microfibrils in the wall may be laid down as parallel bundles, which are oriented at an angle to the layers above and below to give a strong polylaminate structure (Figure 13.4).

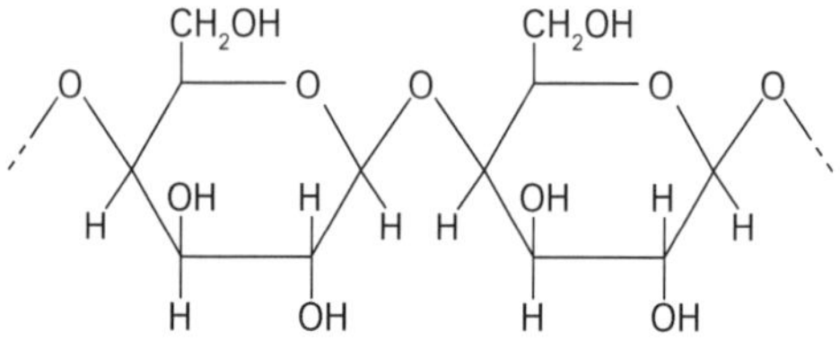

Figure 13.3 Repeating disaccharide unit of cellulose.

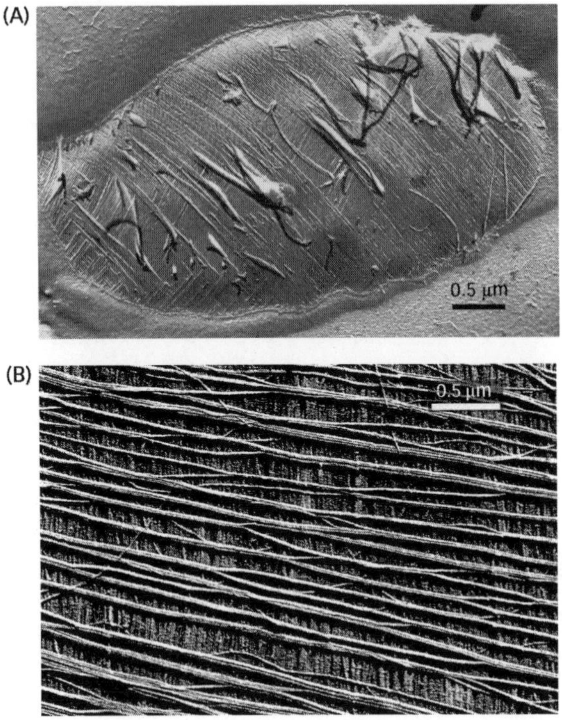

Figure 13.4 (A) Electron micrograph following freeze fracture of the cell wall of the green alga, *Oocystis lacustris* showing its polylaminate cellulose microfibrils. Courtesy of Professor D.G. Robinson, Pflanzenphysiologisches Institut der Universität, Gottingen, Germany. (B) Electron micrograph detail showing cellulose fibres in a cell wall of the coarse green sea hair seaweed, *Chaetomorpha melagonium*. Reprinted from Frei, E. and Preston, R. D. (1961) *Proceedings of the Royal Society* B, 155, 55–77.

- Intermolecular hydrogen bonds produce stable molecules with a high resistance to tensile (stretching) forces.
- Microfibrils are partially crystalline in nature and essentially free of water.
- In turn, microfibrils are usually twisted around one another to give macrofibrils.

Hemicelluloses

- Hemicelluloses consist of extended chains of glucose or xylose sugar residues with short side chains.
- The side chains contain a variety of different sugars: glucose, galactose, and mannose (all hexoses) and xylose and arabinose (both pentoses).

➔ *Carbohydrates are reviewed in the companion volume,* Thrive in Biochemistry and Molecular Biology.

- Hemicelluloses are a diverse group of materials and very different to cellulose despite their similar names.

Pectins

- Pectins are the most complex of all the carbohydrates of the cell wall.
- They also are branched molecules but unlike the neutral hemicelluloses are acidic carbohydrates.
- The backbone consists of chains of α,1→4 galacturonates and rhamnose sugars; the side chains resemble those of hemicellulose.
- The major acidic portion of the backbone consists of α,1→4 galacturonates (see 'Calcium ions'). Other portions are alternating α,1→4 galacturonate and α,1→2 rhamnose residues. These are sometimes referred to as homogalacturonan and rhamnogalacturonan respectively.

> **Check your understanding**
>
> 13.1 Why are side chains composed of sugar residues able to produce a greater diversity of structures than ones formed of amino acids?

Calcium ions

- Calcium ions can cross-link and align acidic regions of homogalacturonan and produce hydrated gels (Figure 13.5).
- The extent of Ca^{2+} cross-linking can be adjusted by chemical modifications of homogalacturonan that stabilize the links or prevent them forming.
- Calcium pectate gels are essential glues for cell-to-cell adhesion and control the porosity of the wall.

> **Check your understanding**
>
> 13.2 How would the methylation carboxylic acids ($R\text{-}COO^- \rightarrow R\text{-}COOCH_3$) in pectin molecules affect the plant cell?

Lignin

- Mature plant cell walls contain lignin, an insoluble phenolic polymer made from aromatic alcohols of high M_r (Figure 13.6).

> **Looking for extra marks?**
>
> Lignin can form up to 30% of plant cell wall material making it the second most abundant biological material on earth.

Proteins

- The major structural proteins of the cell wall are glycoproteins called **extensins** or **hydroxyproline-rich glycoproteins**. (Hydroxyproline is also used in animal extracellular matrices (see Figure 14.1).)

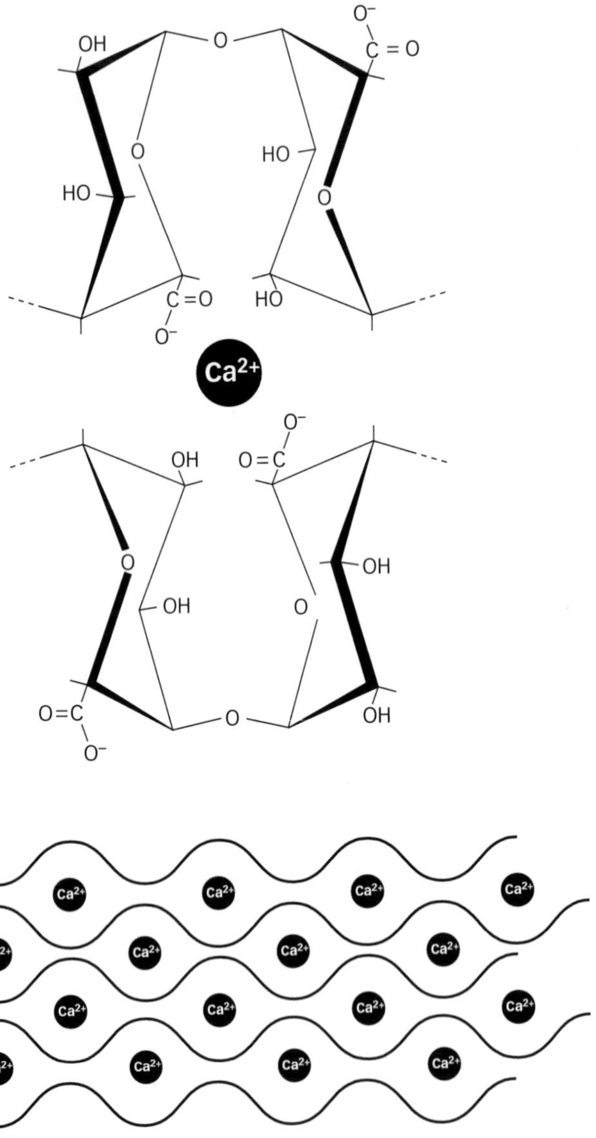

Figure 13.5 Simplified model of Ca^{2+} cross-linked homogalacturonans of pectin molecules.

- Extensins contain up to 50% carbohydrate, much more than 'conventional' glycoproteins that typically have approximately 2–10%. (Again, the proteoglycans of the extracellular matrix are carbohydrate rich; in this case up to 95% polysaccharides.)

➔ *You can revise the structure of proteins and glycoproteins in the companion volume,* Thrive in Biochemistry and Molecular Biology.

- Other groups of structural proteins are rich in proline or glycine or are enzymes concerned with the formation of the wall.

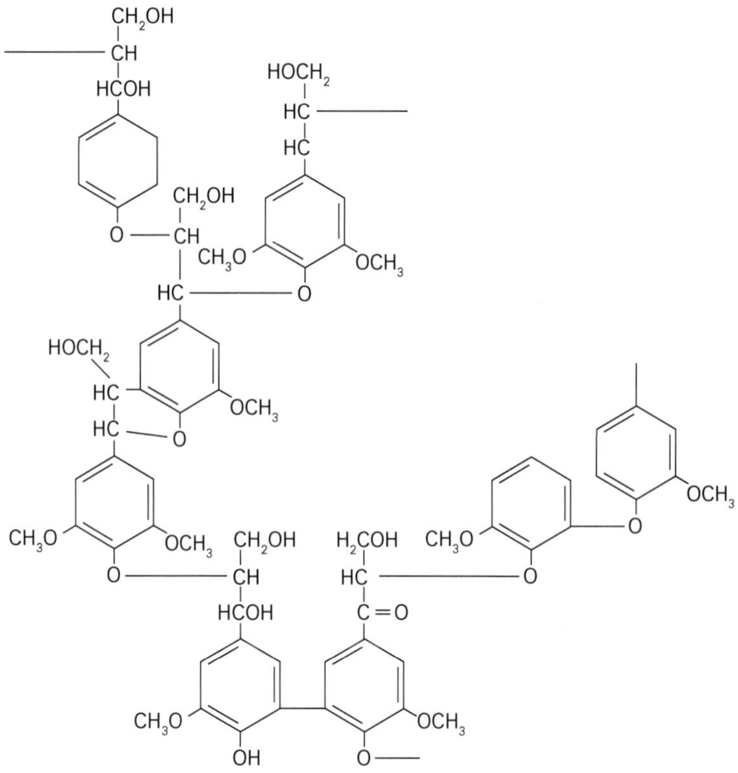

Figure 13.6 Possible partial structure of a lignin molecule.

Structure of plant cell walls

The primary cell wall of plants is a complex three-dimensional network of macromolecules only 100–200 nm thick (Figure 13.7).

- Polysaccharides are the major polymers present, with much less protein and little lignin.
- Cellulose microfibrils form the fibres of the plant cell wall, and are embedded in a complex matrix composed of neutral hemicelluloses, complexes of Ca^{2+} and pectin, and relatively small amounts of structural glycoproteins and enzymes.
- Hydrogen bonds link the cellulose microfibrils and hemicelluloses; Ca^{2+} link together various homogalacturonan regions of pectin molecules.
- As the cell walls develop, increasing amounts of lignin and extensins are deposited in the wall.

Cell wall formation and development

- The synthesis of new cell walls begins during plant cell division (Chapter 16).
- The new cell walls that will form between the daughter cells begin as a **phragmoplast**.

Plant cell walls

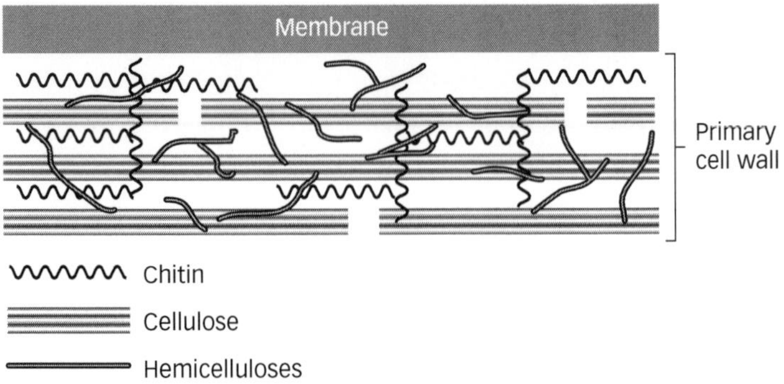

Figure 13.7 Simplified schematic of a plant primary cell wall. See text for explanation.

- Motor proteins on microtubules that originated from the mitotic spindle (Chapter 16) move Golgi-derived vesicles to form a layer across the middle of the dividing cell.
- The layer extends in all directions and fusion of the vesicles converts the phragmoplast into a developing **cell plate**.
- Microtubules guide the growth of the cell plate outwards until it fuses with the cell wall and divides the cell into two daughter cells.
- Both daughter cells deposit wall material on each side of the cell plate converting it to a middle lamella, which is shared between them.
- Middle lamellae are largely composed of pectin and hold adjacent cells together.

Cell wall development

- Walls grow from the outside inwards and therefore the middle lamella forms the outermost layer of each cell wall.
- Vesicles produced by the Golgi apparatus containing cell wall material are deposited immediately outside the membrane and fuse to form a thin wall layer.
- Following the completion of the middle lamellas, the wall is thickened by the addition of a primary, and then possibly a secondary, cell wall layer.

Primary cell walls

- Primary cell walls are relatively flexible and can be extended to allow the new cells to grow.
- Cellulose microfibrils are formed by **cellulose synthase** complexes called **rosettes** as illustrated in Figure 13.8.
- Rosettes are found within the plasma membrane and use *cytoplasmic* UDP-glucose as substrate to form microfibrils directly on the *outer surface* of the membrane.
- Growing microfibrils are anchored to wall components and so they are extended as rosettes move in the plane of the membrane.
- The *direction* of extension is related to the orientation of cortical arrays of microtubules lying just below the plasma membrane.

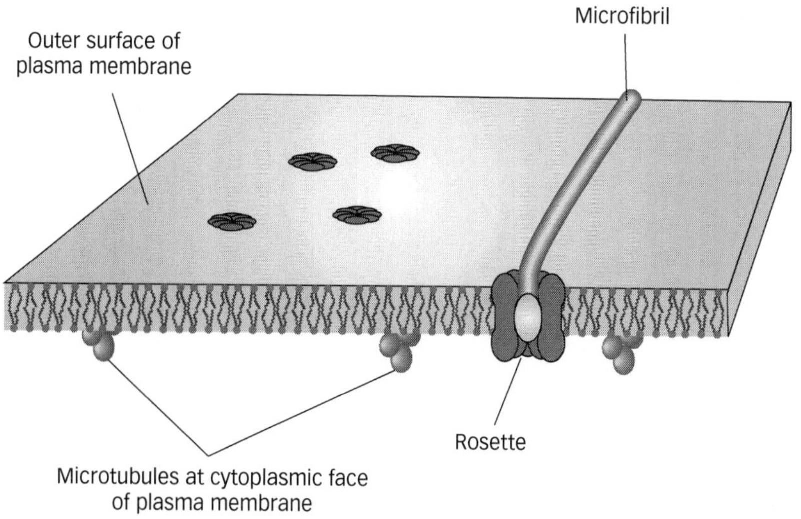

Figure 13.8 Schematic illustrating the roles of rosettes and microtubules in the deposition of cellulose microfibrils at the cell surface. See text for explanation.

- Matrix polysaccharides are synthesized in the Golgi apparatus and exported to the cell wall by exocytosis (Chapters 3 and 9). Here, they intercalate among the cellulose microfibrils that have been directly deposited into the wall by rosette activity.
- In many cell types, development of the cell wall does not proceed beyond the deposition of the primary cell wall; in others a secondary wall is also formed.

Secondary cell walls

- The addition of cell wall material to the plasma membrane side of a primary wall produces a secondary wall.
- Secondary walls lack pectin but contain extra cellulose, up to 60% of the content, and are lignified.
- The fibres of lignin are deposited between those of cellulose in the wall as are additional amounts of structural proteins such as extensins.
- The secondary wall typically consists of three layers called S1, S2, and S3 named in the order of their deposition (Figure 13.9), each of which has a different and characteristic composition.
- In each layer, the cellulose microfibrils are laid down as parallel bundles and are oriented in different directions, making the secondary cell wall extremely resistant to mechanical stresses.
- Secondary walls are not extensible and therefore confer a final size and shape to the cell.
- Plant supporting tissues, such as sclerenchyma, may have cells with extremely thick lignified cell walls.
- Wood is primarily lignified plant tissue. It is composed almost entirely of secondary cell walls.

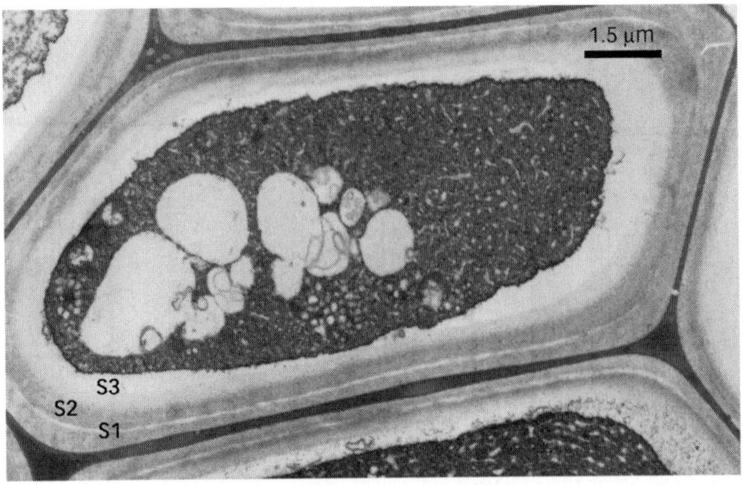

Figure 13.9 Electron micrograph of apple wood with the S1, S2, and S3 layers of the secondary cell wall indicated. Courtesy of Dr B. Nelmes, Department of Biochemistry and Molecular Biology, University of Leeds, UK.

Check your understanding

13.3 Which of the following statements is/are false?

a. Hydrogen bonds bind hemicelluloses to cellulose microfibrils.

b. A secondary cell wall surrounds all mature plant cells.

c. Secondary cell walls are composed mainly of cellulose and lignin.

d. Hemicelluloses form fibrils stabilized by hydrogen bonds.

e. The Golgi apparatus secretes cellulose microfibrils.

Plasmodesmata

Plant cells communicate with one another through channels of cytoplasm called **plasmodesmata.**

- Plasmodesmata pass through adjacent holes in the cell walls of neighbouring cells; they are analogous to the gap junctions of animal cells (Chapter 3).
- Plasmodesmata consist of channels with variable diameters of 30–50 nm through which ions and molecules can be transferred from cell to cell.
- The upper M_r range of molecules able to pass through a plasmodesma is larger than that for gap junctions: it includes RNA molecules, transcription factors, and signalling molecules. They have even been exploited by plant viruses (Chapter 18) and possibly some phytopathogenic fungi.
- Plants can therefore be considered as consisting of two integrated domains:
  1. the **apoplast** composed of cell walls and intercellular spaces
  2. the **symplast** consisting of protoplasts connected by plasmodesmata.

Looking for extra marks?

If plant cells were surrounded by a continuous cell wall then communications between them would be difficult and the plant would be more of a colony of cells than an organism capable of integrated activities.

Formation and structure of plasmodesmata

- During cell plate formation (see 'Cell wall formation and development') partitioning between the daughter cells is incomplete; connections of plasma membrane and cytoplasm between the cells are maintained and eventually form plasmodesmata.
- The structure of a typical plasmodesma is shown in Figure 13.10.
- The hole through the cell walls is lined with plasma membrane, which is continuous with the plasma membranes of both adjacent cells and encloses the channel of cytoplasm.
- Extensions of the rough endoplasmic reticulum (Chapter 8) form a single tubule called the **desmotubule** through the centre of the channel.
- The continuous ring of cytosol between the desmotubule and plasma membrane of a plasmodesma is called the **annulus**.
- The annulus is the route for the free passage of materials between the connected cells.

Check your understanding

13.4 Which of the following statements is/are true?

a. Plasmodesmata are homologous with gap junctions.
b. Plasmodesmata and gap junctions have similar overall roles in organisms.
c. Desmotubules are extensions of the plasma membrane.
d. The symplast is composed of cell walls and intercellular spaces.
e. The apoplast consists of protoplasts connected by plasmodesmata.

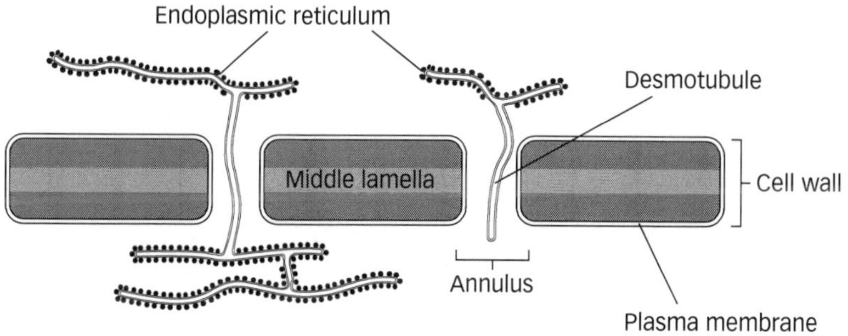

Figure 13.10 Simplified schematic illustration of a plasmodesma.

13.2 ALGAL CELL WALLS

Algal cell walls display an incredibly diverse composition. Many contain cellulose microfibrils but others use mannose or xylan-rich carbohydrates.

- Marine algae have walls that are rich in matrix-forming polysaccharides, presumably as an evolutionary adaptation to the buffering actions of waves and strong tides.
- Brown algal cell walls contain less than 10% cellulose and are composed mainly of polymers of β,1→4 linked mannuronic and guluronic acids called alginic acids.
- Red algae use galactose-based polysaccharides such as agar or carrageenans.
- Diatoms are unicellular phytoplankton with silicate-based (SiO_2, sand) cell walls.
- Their wall consists of two dissimilar sized portions or frustules that interlock like a Petri dish or pill box (Figure 13.11).
- During asexual reproduction, the frustules separate and each acts as a template for the formation of a new wall portion within it, causing the cells to decrease in size with each successive generation until, eventually, a critical size is reached that initiates sexual reproduction.
- The resulting zygote develops to form large cells and restarts the asexual cycles.

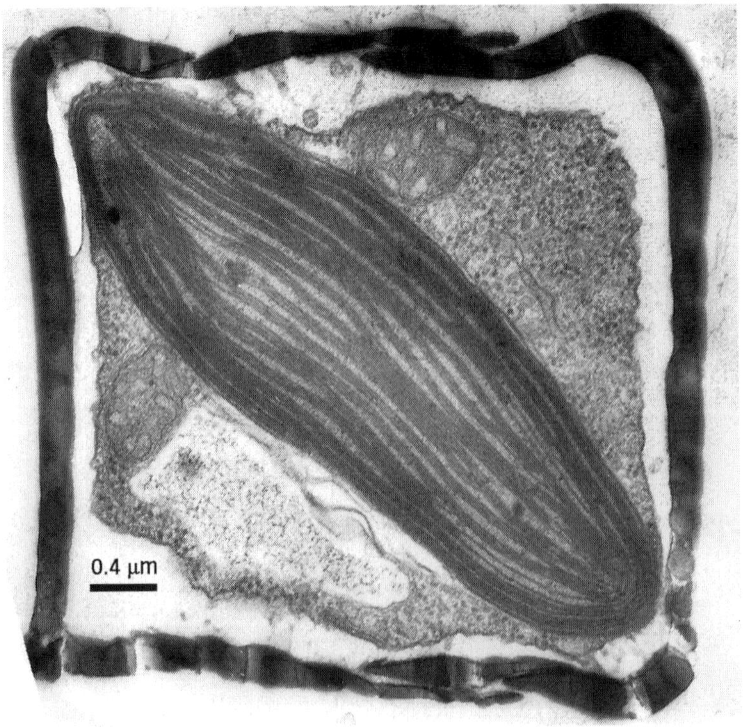

Figure 13.11 Electron micrograph of a bacillariophyceae diatom. Note the cell wall consists of two overlapping structures. Courtesy of Dr G. Beakes, Department of Biology, Newcastle University, UK.

13.3 FUNGAL CELL WALLS

All fungi possess cell walls. Some have walls containing cellulose microfibrils but they differ considerably in composition from plant cell walls.

- Filamentous fungal cell walls are composed mainly of the polysaccharides **glucans** and **chitin**, and glycoproteins.
- Extensive cross-linking of these components integrates the overall structure of the wall.

Components and structure of fungal cell walls

Glucans (polymers of glucose) form 50–60% of the dry weight of fungal cell walls and are its major structural components.

- They are thought to extend throughout the bulk of the fungal cell wall (Figure 13.12).
- Glucans are predominantly β,1→3 linked polysaccharides of approximately 1500 glucose residues.
- The C-6 position at every 50–60 residues forms branch sites to which other β,1→3 glucans can attach.
- Glucan branches can be cross-linked to chitin and mannoproteins (see 'Chitin' section) conferring great mechanical strength to the wall.

Chitin

- Chitin, which forms 10–40% of the wall of filamentous fungi, resembles cellulose but consists of β,1→4 linked *N*-acetylglucosamine (GlcNAc) residues not β glucose (Figure 13.13).
- Like cellulose, chitin's long polymeric molecules are partially crystalline.

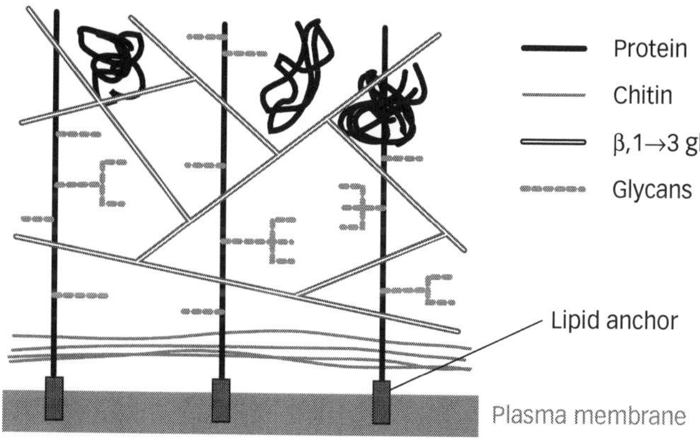

Figure 13.12 Schematic model of a filamentous fungal cell wall.

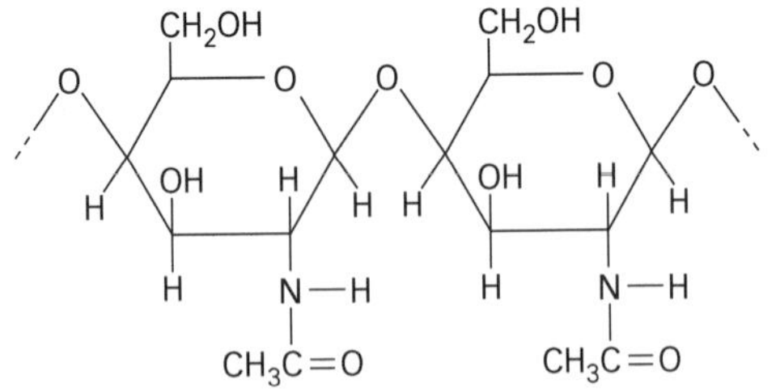

Figure 13.13 The repeating disaccharide of chitin. Compare with Figure 13.3.

- It is mainly located in parts of the wall near to the plasma membrane (Figure 13.13).
- Glycoproteins form approximately 15–30% of the dry weight of filamentous fungal walls, and are all extensively glycosylated, with their carbohydrate portions forming links to other wall constituents.
- The type of glycosylation varies between species. Some are the traditional types of carbohydrates found in 'typical' glycoproteins; others contain both galactose and mannose residues and are called **galactomannans**.

➔ *You can revise glycosidic bonds in the companion volume,* Thrive in Biochemistry and Molecular Biology.

- A number of wall proteins are lipid-anchored proteins of the plasma membrane (see Chapter 3).
- Cell wall proteins have several recognized roles. For example, they:
  - mediate adhesion during cell migration
  - influence the absorption of molecules by the cell
  - transmit signals arising from external stimuli
  - remodel cell wall components.
- Components of the fungal cell wall matrix are variable between species but may include polysaccharides and glycoproteins called **mannoproteins** whose carbohydrate portions are rich in mannose residues and therefore called **mannans**.

13.4 YEAST CELL WALLS

The cell wall of yeasts such as *Candida albicans* and *Saccharomyces cerevisiae* differ from those of filamentous fungi.

- The major polysaccharides present are branched β,1→3 and β,1→6 glucans and contain only 1–2% chitin.
- Approximately 30–50% of the yeast wall is glycoprotein and its outer layer is particularly rich in mannoproteins (Figure 13.14).
- Mannoproteins in the yeast cell wall have been linked with adherence and pathogenesis.

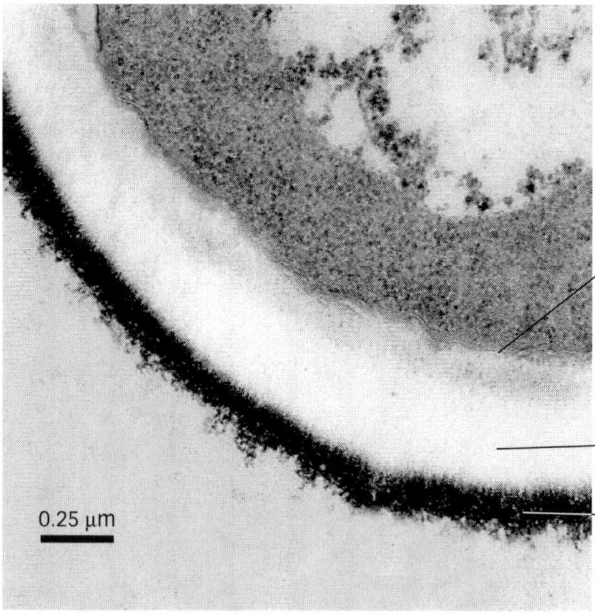

Figure 13.14 Electron micrograph of the cell wall of the pathogenic yeast *Candida albicans*. Courtesy Drs A. Curry and A. Fox, Public Health Laboratory, Manchester, UK.

Check your understanding

13.5 Which of the following statements is/are true?

a. The main components of most filamentous fungal cell walls are glucans.

b. Chitin resembles cellulose but consists of β,1→4 linked glucose residues not *N*-acetylgalatosamine (GlcNAc).

c. The major polysaccharides of brown algal cell walls are agar or carrageenans.

d. Red algae use the galactose-based polysaccharides called alginic acids in their cell walls.

e. Yeast cell walls differ from those of filamentous fungi in being composed mainly of chitin.

Revision tip

Do browse through the further reading references given for this chapter online. Go to http://www.oxfordtextbooks.co.uk/orc/thrive/

Check your understanding

13.6 Which of the following statements is/are true or is/are false?

a. During plant cell expansion, cellulose molecules in the primary cell are hydrolysed.
b. Cells walls are resistant to tension but cannot resist compressive forces.
c. The middle lamella, primary cell wall, and secondary cell wall are features of all plant cell walls.
d. New cellulose microfibrils are formed at the cell surface.
e. The secondary cell wall is external to the primary.
f. The hydrated gels produced by pectins are formed when Ca^{2+} complexes of homogalacturonan occur.
g. Both homogalacturonan and rhamnogalacturonans are acidic regions of pectin molecules.
h. Microfibrils are formed by chitin synthase complexes called rosettes.
i. The major component of wood is lignin.
j. Sclerenchyma cells have extremely thick supported primary cell walls.

14 Extracellular matrix and connective tissues

Key features of the extracellular matrix

- The extracellular matrix (ECM) is the ground substance/connective tissue surrounding animal cells.
- Its thickness varies from being very thin between hepatocytes, which are in close contact, to the much thicker areolar connective tissue, which contains comparatively few cells.
- It is produced largely by fibroblasts but does contain other types of cells.
- The ECM consists of a water-rich gel or matrix containing a complex network of structural and adhesive proteins and glycoproteins (Figure 14.1) and so, like cell walls (Chapter 13), its structural basis is one of 'fibres in a ground substance'.
- The proteins of the ECM are mainly fibrous glycoproteins and proteins with extended conformations, which are usually partially immobilized by being bound to other proteins or to cells.
- Proteoglycans are able to bind enormous amounts of water and form the ground substance of the ECM, which resists *compressive* forces.
- Collagens and elastin are the major structural proteins of the matrix, which give it *mechanical strength* and *elasticity* respectively.

continued

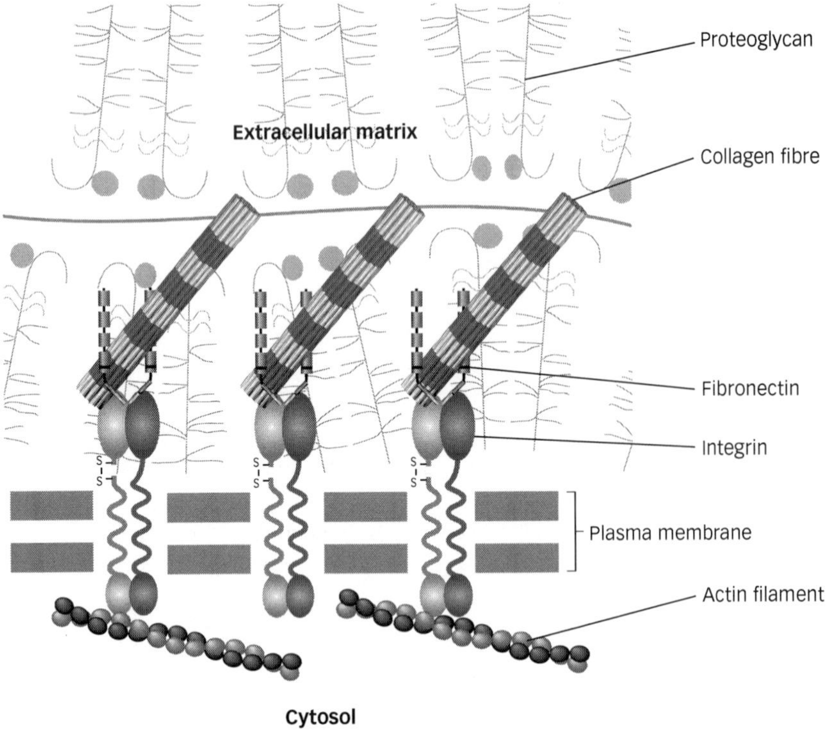

Figure 14.1 Schematic overview of the extracellular matrix and its attachment to the surface of the cell.

- **Fibronectins** and **laminins** are *adhesive* proteins that interconnect the various components of ECMs.
- The ECM is produced by the cells it contains, such as **fibroblasts** and other connective tissues cells.
- The proportions of proteoglycans and the different proteins vary between different types of tissues producing a variety of matrices and connective tissues. Indeed, the ECM is not static but is a dynamic tissue, constantly being remodelled by resynthesis and degradation.
- The ECM is mainly degraded by matrix metalloproteases and serine proteases. These are usually local acting and hydrolyse ECM components only in the region of the cell secreting them.
- The major role of the extracellular matrix is support, but extracellular matrices can form permeability barriers and regulate numerous other cell activities.
- Different types of connective tissues are formed by modifying the proportions of the different constituents of the ECM.
- In some cases, the ECM is highly specialized, such as in the basement lamina of epithelia or fibrous connective tissues or bone.

14.1 KEY FUNCTIONS OF THE EXTRACELLULAR MATRIX

The major function of the ECM is *support*; it is a tough material that can carry the mechanical loads imposed on tissues.

- The variety of types and the various ways in which they are formed or deposited means the ECM of a cell may also influence its activities directly by mechanical signals to the cytoskeleton and indirectly by binding growth factors.
- The ECM of tissues may help regulate cell motility, recognition and adhesion, and division and differentiation during embryonic development.
- The gel-like ground substance of the ECM is sufficiently porous to permit the rapid diffusion of nutrients, metabolites, hormones, and cellular waste material between cells of the tissues and the blood.
- The ECM forms filters that regulate movement of material through the extracellular medium.
- Specialized types of ECMs form a variety of connective tissues, with distinctive roles in organisms.

Revision tip

Construct a simple table to compare and contrast the functions of the extracellular matrices and cell walls.

14.2 MAJOR CELLS OF THE EXTRACELLULAR MATRIX

The *resident* cell population of the ECM are mainly:

- **Fibroblasts**
- **Macrophages** and **mast cells**
- **Adipose cells**
- **Mesenchymal stem cells.**

In addition, various types of leucocytes form a *transient* population. These enter the ECM when needed for specific purposes.

- Fibroblasts are the principal cells of the ECM, and are responsible for synthesizing the key glycoproteins and proteins of the ECM including **proteoglycans**, **collagens**, and **elastins** (see Section 14.3).
- It is thought that any one fibroblast is capable of producing and secreting all the components of the ECM.
- Macrophages and mast cells are cells of the immune system.
- Macrophages are specialized phagocytes that arise when blood monocytes migrate into the ECM and differentiate.

- Mast cells resemble basophils and contain granules whose contents are released when the cell is stimulated, and which often promote inflammation.
- Adipose cells contain extensive smooth endoplasmic reticulum and are able to synthesize and store triacylglycerols (Chapter 8).
- Mesenchymal stem cells are multiple potential cells that can give rise to differentiated cells when new tissue must be synthesized or existing tissue repaired (Chapter 16).

14.3 KEY GLYCOPROTEINS AND PROTEINS OF THE EXTRACELLULAR MATRIX

- The ECM contains a variety of protein components. However, this section will revise only the major glycoproteins and proteins present:
  - proteoglycans
  - collagens
  - elastins
  - **fibronectins**
  - **laminins.**
- The roles of some of these proteins have been clarified using **knock-out mice**, in which the expression of a gene for a particular protein (typically in mouse embryonic stem cells (Chapter 16)) is destroyed and the effects on the subsequent mutant mouse determined.

Proteoglycans

Proteoglycans form the ground substance of the ECM. They are complex macromolecules comprising approximately 95% carbohydrate and only 5% protein.

- The overall structure of a proteoglycan is shown in Figure 14.2.
- Proteoglycan molecules consist of numerous carbohydrate chains called **glycosaminoglycans (GAGs)** that are attached to core proteins.
- Glycosaminoglycans are heteropolysaccharides usually 80–100 sugar residues long, made up of repeating disaccharide residues.
- The first sugar of the repeat is usually an amino sugar: *N*-acetylglucosamine (GlcNAc) or *N*-acetylgalatosamine (GalNAc)
- The second is usually a sugar acid, for example glucuronate (GlcUA), with one or more sulphate groups attached to it.
- Up to 200 GAGs can be attached to a single core protein (Figure 14.2).
- The five major GAGS are:
  1. chondroitin sulphate
  2. dermatan sulphate
  3. heparan sulphate

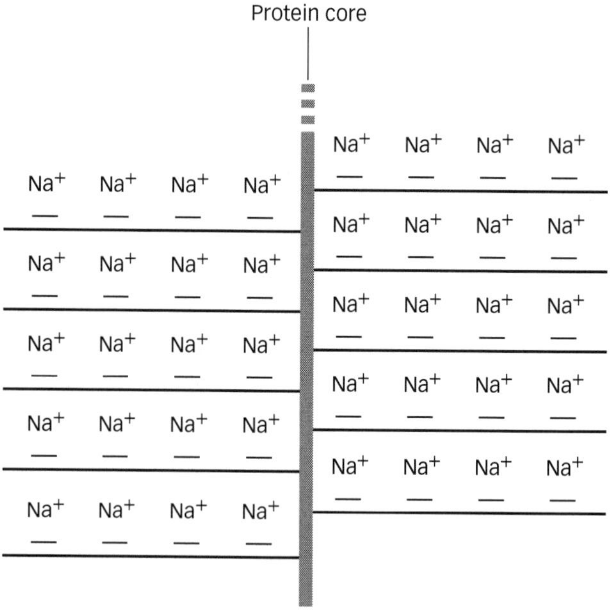

Figure 14.2 Schematic structure of a proteoglycan molecule. See text for explanation.

4. keratan sulphate
5. hyaluronan.

- The sizes of core proteins in different proteoglycans vary from 1000 to 5000 amino acid residues.
- Hyaluronan differs from the other types in not being attached to a core protein and lacking sulphated sugars. Its molecules can be up to 50,000 sugar residues and 25 μm in length.
- Proteoglycans have M_r of 250,000 to more than 3×10^6 and so can be huge molecules, as illustrated by the major proteoglycan of cartilage, which is essentially an aggregate of **aggrecan** molecules (Figure 14.3).
- Aggrecan is a proteoglycan consisting of numerous chondroitin sulphate and keratan sulphate attached to a common protein core.
- In cartilage, large numbers of aggrecan units are non-covalently attached to a hyaluronan molecule by link proteins (Figure 14.3).
- The M_r of this aggregate can exceed 1×10^8, and have a volume approaching that of many prokaryotic cells.
- *Free* hyaluronan is an essential component of joints where it functions in resisting compressive forces and as a lubricant.

Key functions of proteoglycans

- The negative charges on GAG molecules are mutually repulsive and so proteoglycan adopts extended conformations.

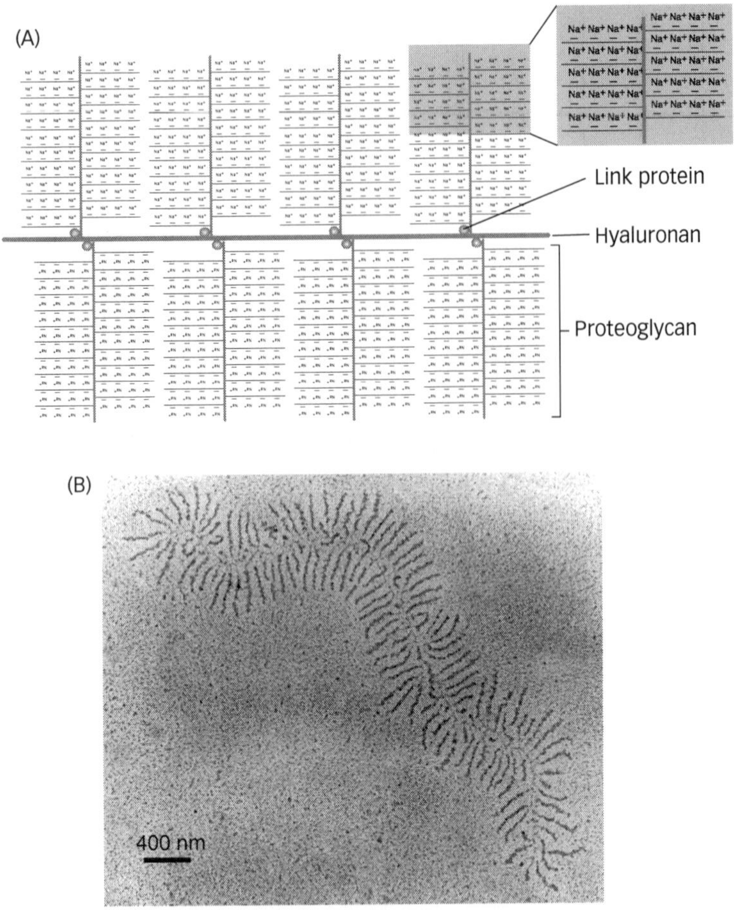

Figure 14.3 (A) Schematic showing how the linking of multiple glycosaminoglycans to a central hyaluronan (itself a GAG) by link proteins forms the major proteoglycan aggregate of cartilage. (B) Electron micrograph of a proteoglycan aggregate from bovine fetal cartilage. The thicker side branches consist of numerous chondroitin sulphate and keratan sulphate glycosaminoglycans collapsed around their core proteins; each of which is attached to the central hyaluronan molecule. Courtesy of Dr J.A. Buckwalter, Department of Orthopedics, Veterans Medical Center and University of Iowa, USA.

- Their many negative charges attract numerous counter ions, mainly Na^+ that attract even larger quantities of water.
- Proteoglycan and the associated huge amount of water form porous hydrated gels whose turgor pressure means they expand and fill most of the extracellular space.
- Proteoglycans therefore form most of the volume of the ECM despite being only approximately 10% of the protein content.
- The major role of proteoglycan is to resist compressive forces because their hydrated gel structure, like all fluids, resists compression and supports the tissue.

- Proteoglycan gels lack mechanical strength but are complemented by structural glycoproteins, such as collagen (reviewed in the 'Collagen' section), which resist tension.
- Gels of proteoglycans have varying pore sizes and charge densities and can form selective molecular sieves, as in the kidney glomerulus.
- Proteoglycans bind a number of signalling molecules and control their diffusion through the ECM, which limits both their range of actions and half-lives.

➔ *More information about signalling molecules is given in Chapter 15.*

- Proteolytic enzymes and their inhibitors are also bound by proteoglycans which help control the remodelling of the ECM.

Membrane-bound proteoglycans

- Some proteoglycans are associated with the plasma membrane.
- A number of these are integral proteins with their core proteins inserted into the lipid bilayer; others are covalently linked to phospholipids of the outer leaflet.
- Some plasma membrane proteoglycans act as coreceptors on the cell surface.

➔ *Membrane structure is reviewed in Chapter 3.*

- The best studied proteoglycan integral proteins are the **syndecans**.
- Syndecans are found on fibroblasts and epithelial and other types of cells.
- Their intracellular domain can interact with actin filaments of the cytoskeleton and with signalling molecules, while the surface portion is able to associate with integrins and fibronectins (see later sections).

Check your understanding

14.1 Intervertebral discs have a central proteoglycan gel region. Humans are tallest in the morning but decrease in height by approximately 30 mm on average during the day. Account for this decrease.

Collagens

Collagen is the most abundant protein of the ECM and in most connective tissues. It provides tensile strength to the ECM: it resists stretching and so complements the compression resistance of proteoglycans.

- About 20 different types of collagens have been described; most are tissue-specific.
- Individual types of collagen are secreted by fibroblasts as precursors called **procollagens**, which are formed from polypeptides about 1000 residues long called **pro-α-chains.**

- Pro-α-chains are synthesized on the rough endoplasmic reticulum (Chapter 8) of fibroblasts, and are subsequently modified by post-translational modifications such as hydroxylations and glycosylations to form **α chains**.

Structure of collagen

- The amino and carboxyl termini of α chains are called **extension peptides**, which flank an extended central region with an unusual amino acid composition:
  - glycine residues are found at every third position
  - proline residues form about 30% of the residues present.
- The primary structure is therefore essentially a repeat of the three residue sequence Gly–X–X, for example:

  … –Gly–Pro–Hyp–Gly–Pro–Met–Gly–Pro–Hyp–Gly–Leu–Ala– …

  where Hyp stands for hydroxyproline.
- The extension polypeptides promote the twisting of the central regions of three collagens around each other to form a right-handed **superhelix**, although each of the three helices has a left-handed twist (Figure 14.4).
- The high proportion of glycine residues, which lack side chains, facilitates the close contact necessary to form the superhelix.
- Hydroxyproline and hydroxylysine residues are formed by hydroxylating proline and lysine residues after they have been incorporated into the protein.
- 4-Hydroxyproline is commonest and is formed by the action of prolyl 4-hydroxylase which requires the presence of vitamin C (ascorbic acid) to maintain an essential Fe^{2+} ion in the enzyme in its reduced form (Figure 14.5a).
- A lack of dietary vitamin C results in scurvy.
- 3-Hydroxypoline and 5-hydroxylysine are also formed (Figure 14.5b) but in smaller amounts.
- Hydroxyproline and hydroxylysine residues on adjacent chains in the triple helix are able to form hydrogen bonds with one another and help stabilize its structure.
- Other hydroxylysine residues are glycosylated by the attachment of galactose or galactose-containing disaccharides.
- The resulting structure (triple helix and extension peptides) is called **procollagen** which is secreted into the ECM by the Golgi apparatus (Chapter 9).
- Within the ECM, procollagen aminopetidase and procollagen carboxypeptidase catalyse the hydrolytic removal of the extension peptides to leave a triple helical **tropocollagen** molecule.
- Tropocollagens are 300 nm long and 1.5 nm in diameter and spontaneously align in the same direction with one another to form **microfibrils** approximately 50 nm in diameter.
- Microfibrils pack loosely together to form mature collagen fibres (Figures 14.1 and 14.4).
- Covalent bonds are formed between adjacent lysine residues and form strong cross links both within and between microfibrils, which contribute mechanical strength to the ECM (Figures 14.6).
- Within the fibre, adjacent microfibrils are displaced by a distance of 64 nm, which gives a distinctive pattern when viewed by electron microscopy (Figure 14.7).

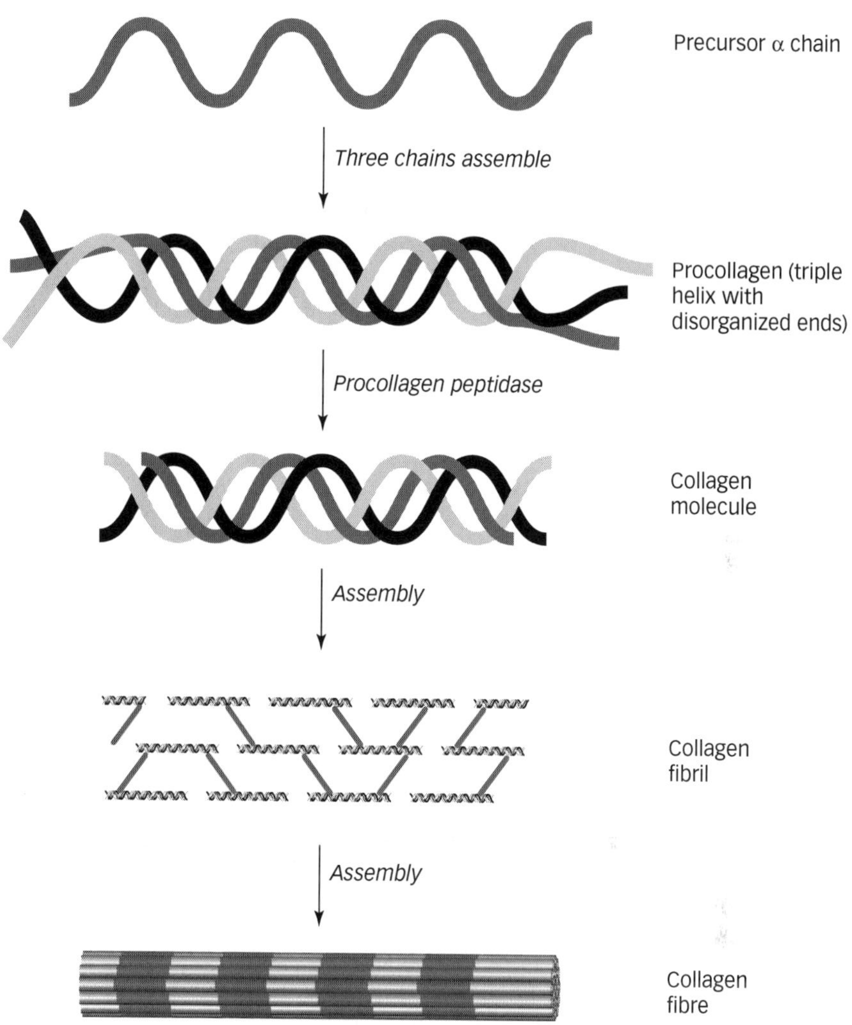

Figure 14.4 Conversion of pro-α-collagen to a collagen fibre. This process is described in the text.

Check your understanding

14.2 Would you expect to recover radioactively-labelled collagen from an experimental rat if it was fed (a) ^{14}C-labelled proline or (b) hydroxyproline with its hydroxyl group labelled with ^{3}H?

Looking for extra marks?

Each α-collagen chain is coded for by a separate gene. It is thought the many genes for α-chains and a number of collagen-like proteins, arose by successive gene duplication mutations of the original ancestral gene.

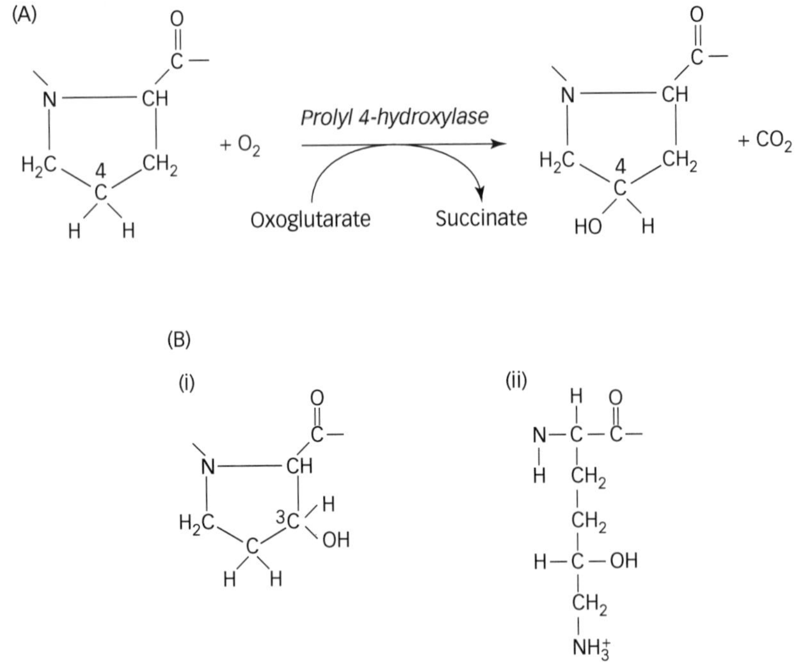

Figure 14.5 (A) Catalysed hydroxylation of proline by prolyl 4-hydroxylase to form 4-hydroxyproline. (B) (i) 3-Hydroxyproline and (ii) 5-hydroxylysine.

Elastin

Elastin proteins allow tissues to be stretched and then relax to their original dimensions without damage.

- It provides elasticity to the ECM and certain connective tissues such as those in arterial walls, skin, ligaments, and lungs. Thus, it complements the role of collagen.
- Elastin lacks the repeated amino acid motifs of collagen but is rich in alanine and lysine residues.
- It is secreted into the ECM by fibroblasts as the soluble protein **proelastin** where it is crossed-linked by the formation of desmosine between lysine residues on four polypeptide chains (Figure 14.8).
- Cross-linking converts proelastin molecules into an insoluble three-dimensional network of elastic fibres that are largely helical or lack secondary structure.
- When fully hydrated, the cross-linked roughly globular elastin molecules can stretch under tension in any direction with about 70% extensibility to form ellipsoid, almost straight chain structures (Figure 14.9).

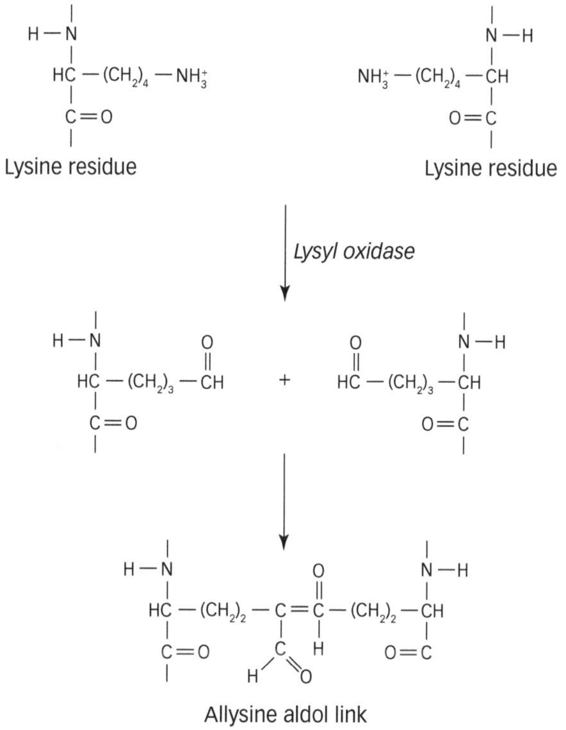

Figure 14.6 Formation of one type of lysine-based cross link associated with collagen microfibrils.

- When the tension is released, the cross links allow the structure to relax to its original conformation.
- Within the ECM and connective tissues, the elastin network is interwoven with long collagen fibres that prevent overstretching and damage to the tissues.

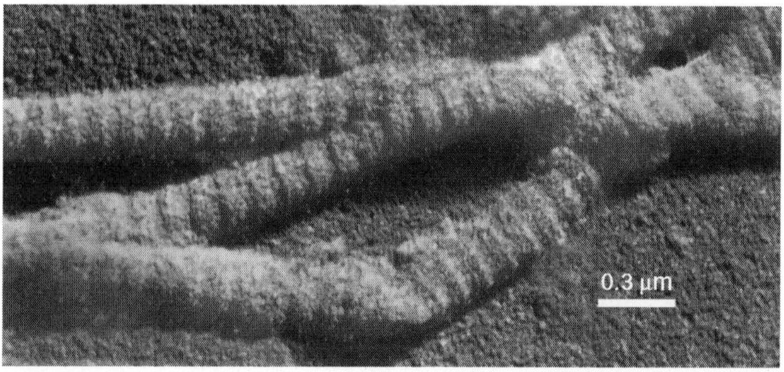

Figure 14.7 Electron micrograph of calf skin collagen fibrils. Note the characteristic banding pattern.

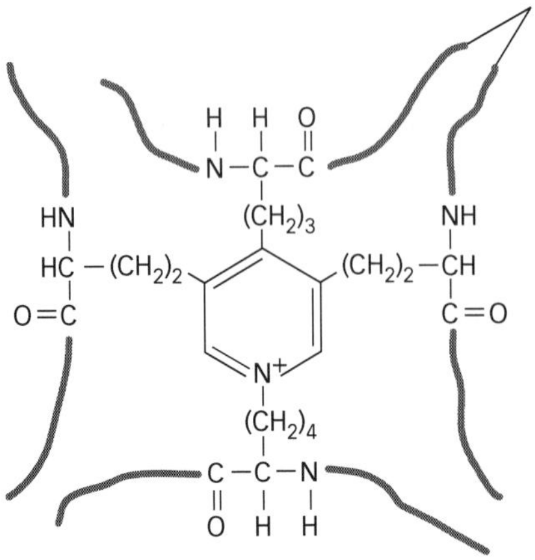

Figure 14.8 Desmosine cross-linking lysine residues on four elastin polypeptides.

Adhesive glycoproteins

Adhesive glycoproteins bind proteins of the ECM together and form direct links between the ECM and the plasma membrane. The best known adhesive glycoproteins are fibronectins and laminins.

Fibronectins

- Fibronectins link cells to the ECM and occur as:
  - insoluble fibrils in the ECM

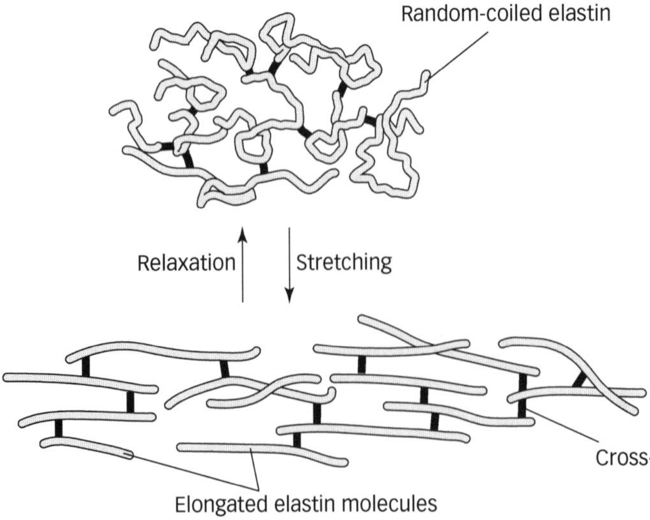

Figure 14.9 Highly cross-linked elastin can be stretched but upon relaxation returns to its original structure.

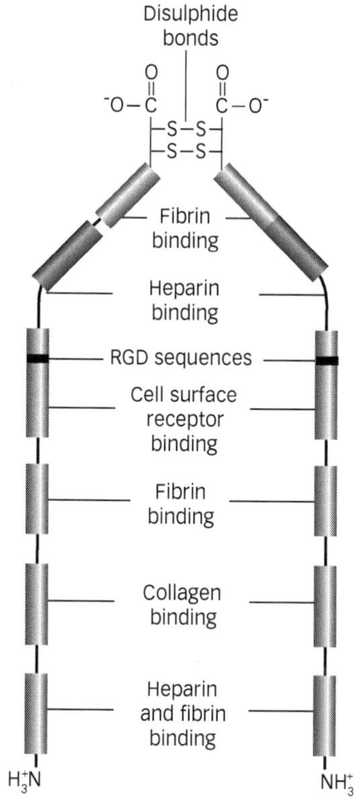

Figure 14.10 Schematic structure of a fibronectin dimer.

- a soluble form in blood that functions in clotting
- a form intermediate between the two that is associated with the surface of the cell.

- Fibronectins are essential for interactions between cells and their ECM.
- Fibronectin molecules are dimers consisting of two elongated subunits, each approximately 2500 amino acid residues long, linked by two disulphide bridges situated near their carboxyl termini.
- Each subunit is composed of a series of binding domains joined by short flexible regions of the polypeptide chain (Figure 14.10).
- The binding domains have sites specific for other fibronectins, for collagens, and for receptors on the cell surfaces.
- The latter contains an RGD (–Arg–Gly–Asp–) sequence that can be recognized by a major group of fibronectin receptors on the cell surface called **integrins** (see 'Integrins' section).

Check your understanding

14.3 Which of the following statements is/are true?

a. The ECM is mainly degraded by matrix metalloproteases and serine proteases released from lysosomes.

b. The collagen superhelix is formed within fibroblasts.
c. Elastins contain significant quantities of alanine residues.
d. Weight-bearing tissues lack blood vessels and obtain nutrients through their extracellular matrix.
e. Fibronectin is able to bind specifically to laminins.

Laminins

- Laminins are adhesive glycoproteins of the **basal lamina** (sometimes called the **basement membrane**), a sheet of specialized ECM which underlays epithelial cells and also surrounds muscle, fat, and Schwann cells.
- The basal lamina:
  - supports and maintains the integrity of the epithelium
  - forms permeability barriers.
- Tissue integrity is maintained because the basal laminae of tissues and cells to which they are attached is a layer of ECM approximately 50 nm thick (range 40–120 nm) that separates them from surrounding tissues.
- Basal laminae may show selective permeability:
  - Cells generally cannot move through it into epithelia, although phagocyte cells of the immune system can migrate through it when they are needed to fight infection.
  - The Bowman's capsules of the kidney have an extremely specialized basal lamina that filters the blood removing ions and micromolecules to the provisional urine while retaining macromolecules within the blood plasma.
- Laminin molecules (M_r approximately 850,000) are trimers of α, β, and γ subunits that are linked together by disulphide bonds in a roughly cross-shaped conformation as shown in Figure 14.11.
- Laminins form molecular bridges between the cell surface and the ECM.
- Laminins are able to bind epithelial cells to the basal lamina because:
  - their longest 'arm' is recognized and bound by integrins (see 'Integrins' section) on the cell surface
  - the cross arms have binding sites for other laminins, enabling them to associate and form aggregates
  - the cross arms are also able to recognize and bind components of the ECM such as collagen molecules.

Revision tip

Do not confuse *laminins* with the lamin proteins of the nuclear lamina.

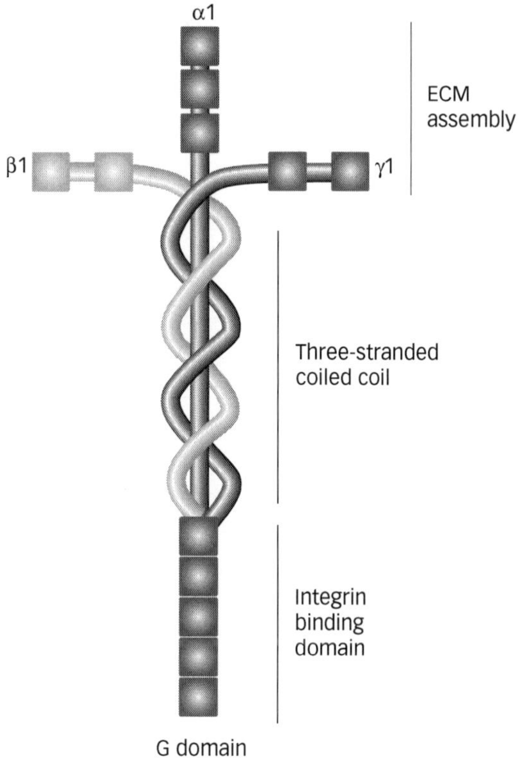

Figure 14.11 Schematic structure of a laminin molecule.

Check your understanding

14.4 Which of the following statements is/are true?

a. Laminins are adhesive glycoproteins of the basal lamina.

b. Basal laminae are able to form permeable barriers.

c. The cytoplasmic part of the integrin molecule is able to interact directly with fibres of the cytoskeleton.

d. The basal lamina is a sheet of specialized ECM that is only found underlying epithelial cells.

e. Laminin molecules are trimers of α, β, and γ subunits; each about 2500 amino acid residues long.

Integrins

Integrins are the major cell surface receptors for fibronectins, laminins, or indeed other cells. They are the main mediators of cell–ECM interactions and are major signalling proteins.

Key glycoproteins and proteins of the extracellular matrix

- Integrins are a large heterogeneous family of transmembrane proteins, which differ in their binding specificities and sizes. All are dimers, consisting of α (M_r 110,000–140,000) and β (M_r 85,000–91,000) subunits.
- Different types of cells express different integrins: more than 20 different ones have been observed from varying combinations of 18 α and eight β gene products.
- Extracellular portions of integrins differ in their specificity for different ligands in the ECM or for binding sites on the surfaces of other cells.
- The cytoplasmic part of the integrin molecule is able to interact with cytosolic proteins that can link the integrin indirectly with fibres of the cytoskeleton.
- There are two main connections with the cytoskeleton.
  1. Interactions between the ECM and actin filaments of non-epithelial and migratory cells are mediated by clusters of integrins in **focal adhesion** sites and by the cytosolic proteins, talin, vinculin, and α actinin.
  2. Epithelial cells interactions between laminin and an integrin occur through proteins of a **hemidesmosome** (this structure resembles half of a desmosome, described in Chapter 3).
- The linker proteins of the hemidesmosome form a plaque in the cytoplasm that attaches to keratin intermediate filaments (Chapter 7).
- Given they bind ligands of the ECM or to other cells and can also interact with the cytoskeleton, integrins can directly transmit (signal) events in the ECM to the interior of the cell and from the interior of the cell to the ECM.

Key roles of integrins

- Integrins influence directly or indirectly numerous aspects of cell behaviour such as cell adhesion or migration.
- Integrins function in several signalling mechanisms. This includes 'inside out' and 'outside in' effects, the latter being where a change in the interior of the cell results in a change on the cell surface.
- For example, the binding of growth factors is signalled to the interior of the cell, leading to contractions of actin filaments.
- The resulting tension on integrin in the plasma membrane exposes their integrin binding sites in the ECM and results in integrins clustering at the cell surface.
- 'Outside in' effects result from the function of integrins as cell surface receptors; for example, **anchorage dependent growth**, when cells can *only* grow and divide if they are attached to a substratum (Chapter 2). This type of growth depends on the integrin clustering as just described.
- Following integrin clustering at focal adhesions, a number of kinases are activated that are necessary for anchorage dependent growth.

  ➔ *The roles of kinases in cell signalling are revised in Chapter 15.*

14.4 CONNECTIVE TISSUES

The proportion of the different components of the ECM varies enormously between different types of connective tissues, which connect other tissues and organs together.

- Almost all connective tissues arise from the **mesoderm** of the embryo, but they can be divided into two major groups:
  1. general connective tissues
  2. specialized connective tissues.

➔ *Development is briefly revised in Chapter 16.*

General connective tissues

There are two groups of general connective tissues:

1. areolar or loose connective tissues
2. dense connective tissues.

Areolar connective tissues

- Areolar or loose connective tissue forms the basic packing material of organisms (Figure 14.12).
- It consists largely of a proteoglycan matrix forming a gel-like material, which contains a few seemingly randomly orientated dispersed collagen and elastic fibres.
- Interspersed between the fibres are a small number of cells such as fibroblasts.

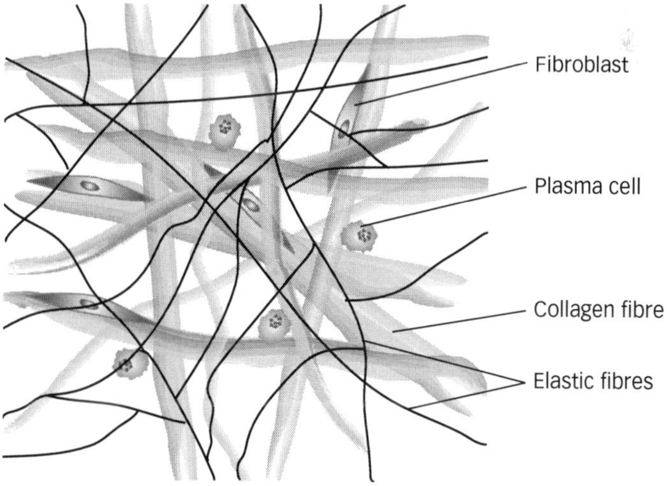

Figure 14.12 Schematic structure of areolar or loose connective tissue.

Connective tissues

Dense connective tissues

- Dense connective tissues may be **irregular** or **regular**.
- Irregular connective tissue consists mainly of bundles of collagen fibres that are oriented in all directions and has a sparse number of fibroblasts and little matrix.
- Irregular connective tissue is a significant strengthening material in organs subject to tension such as the dermis of the skin or the submucosa of the gastrointestinal tract.
- Regular dense connective tissues are characterized by regimented, densely packed arrays of fibres and fibroblasts with little ECM.
- Two major types occur, which are found in **tendons** and **ligaments** respectively.
- Tendons are composed of collagen fibres, and attach muscles to bone and transmit the contraction of muscles (Chapter 7) to bones, which produces movements of the jointed skeleton.
- Ligaments are largely elastin fibres, and join bones to bones and allow movements of the joints.

Specialized connective tissues

In specialized connective tissues the basic ECM structure is heavily modified to perform specific roles in organisms.

- Three examples of specialized connective tissues are:
  1. cartilage
  2. bone
  3. adipose tissue.

Cartilage

- Cartilage is a pliable but firm skeletal material in which the cells producing it (chondrocytes (Figure 14.13)) are embedded within a matrix of proteoglycan chondroitin and collagen fibres.
- Several types of cartilage occur, each adapted to specific functions.

Bone

- Bone has a mineralized, rigid ECM with relatively few interspersed cells (Figure 14.14).
- It is produced by specialized cells called **osteoblasts** and degraded by **osteoclasts.**
- In bone, the ECM is heavily calcified with calcium phosphate as crystal of **hydroxyapatite** [$Ca_{10}(PO_4)_6(OH)_2$], which produces a hard, supportive skeletal structure.

Adipose tissue

- **White adipose tissue** contains large numbers of adipose cells or **adipocytes** (Figure 14.15).
- Each cell contains large amounts of triacylglycerols stored in the cytoplasm and so constitutes an energy-rich reserve.

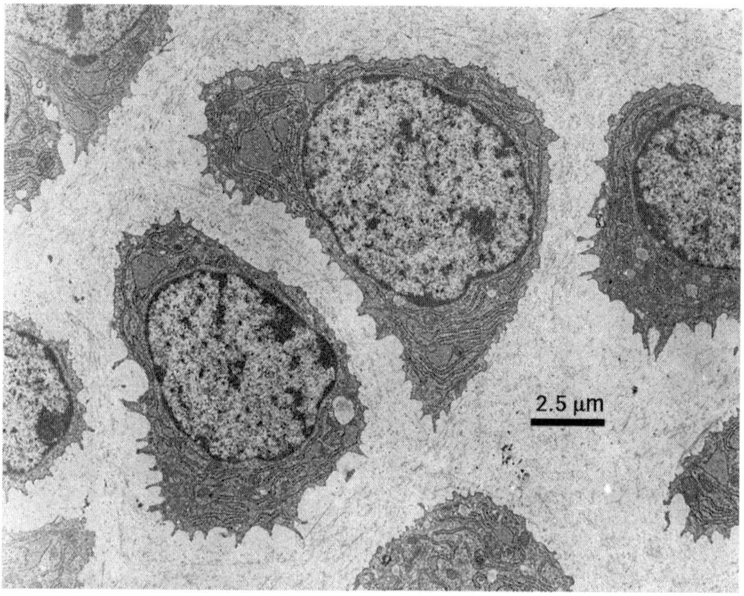

Figure 14.13 Electron micrograph of chondrocytes embedded in cartilage matrix. Reproduced from Seegmillar, R., Ferguson, C.C., and Sheldon, H. (1972) Studies on cartilage VI. A genetically determined defect in tracheal cartilage. *Journal of Ultrastructural Research*, 38, 288–301.

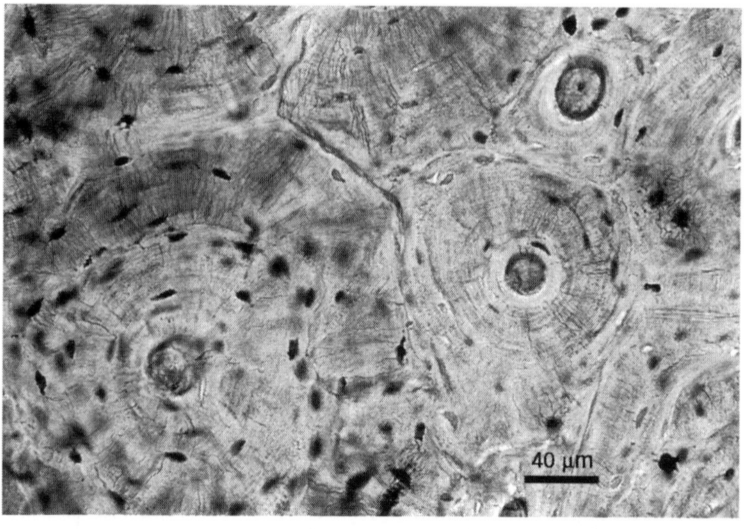

Figure 14.14 Photomicrograph of a section of hard bone. The osteocytes appear as concentric circles of black-stained cells contained in lacunae (spaces) surrounding central Haversian canals.

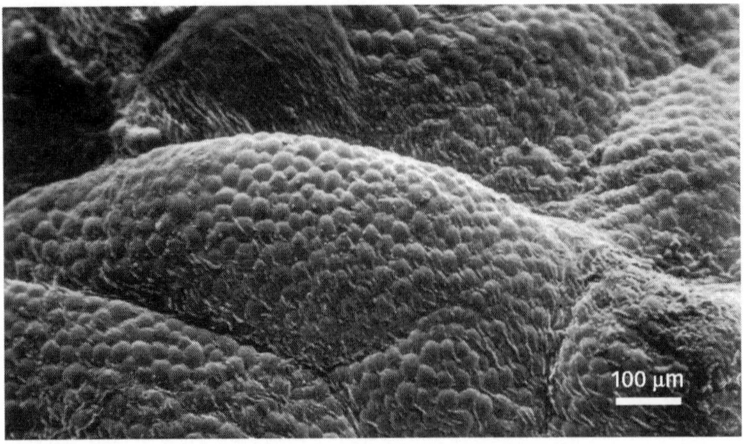

Figure 14.15 Scanning electron micrograph of white adipose cells from rat epididymis. Courtesy of Dr N.O. Nilsson, Department of Physiological Chemistry, Lund University, Sweden.

Check your understanding

14.5 Which of the following statements is/are false?

a. Adipocytes produce and store large quantities of phospholipids for use as an energy reserve.
b. Almost all connective tissues arise from the mesoderm of the embryo.
c. Areolar connective tissues form the basic packing material of organisms.
d. Specialized connective tissues include cartilage, bone, and adipose tissue.
e. Fibronectins occur in soluble and insoluble forms.

Revision tip

Do browse through the further reading references given for this chapter online. Go to http://www.oxfordtextbooks.co.uk/orc/thrive/

Check your understanding

14.6 Which of the following statements is/are true or is/are false?

a. The ECM is a dynamic tissue that is constantly remodelled by resynthesis and degradation.
b. Some forms of ECMs are highly specialized, such as the basement lamina or fibrous connective tissues.
c. Collagens and elastins are the major structural proteins of the matrix, which give it elasticity and mechanical strength respectively.

d. The major function of the ECM is to support cells and tissues.
e. Fibronectins and laminins are connective tissues and interconnect the various components of ECMs.
f. Fibroblasts are the principal cells of the ECM and can synthesize all the major components of the extracellular matrix.
g. Proteoglycans form the ground substance of the ECM and are comprised of approximately 5% carbohydrate and 95% protein.
h. Four major GAGS are: chondroitin sulphate, dermatan sulphate, heparan sulphate, and keratan sulphate.
i. Proteoglycans form most of the volume of the ECM despite being only approximately 10% of the protein content.
j. Elastin lacks the repeated amino acid motifs of collagen but is rich in arginine, glycine, and aspartate residues.

15 Cell signalling

Key features of cell signalling

- Cells communicate with one another using chemical or electrical signals that activate complex signalling pathways.
- For cells to respond to these signals, they must be able to perceive them, communicate that signal to the interior of the cell, and, if the signal received is weak, amplify it.
- The binding of the signal messenger to its receptor initiates a cascade of reactions called a signal transduction pathway that produces the physiological effects associated with the signal.
- The perception of stimuli by sensory cells is also relayed by signal transduction pathways.
- The basic events in cell signalling do not differ greatly between different organisms. Indeed, cell signalling molecules and pathways are seemingly similar in many species.
- Signalling pathways are best documented in multicellular animals, on which this chapter will focus.

15.1 LOCAL AND LONG-DISTANCE SIGNALLING

The cells of animals and plants communicate using **extracellular messenger molecules**, which are also called **extracellular messengers** or **ligands**.

- *Local* communication may, however, be by direct contact or by secreting and receiving local-acting extracellular messengers such as **growth factors**, **cytokines**, or **neurotransmitters**.
- *Long-distance* signalling between cells uses **hormones**; animal cells also communicate using **electrical** signals.
- Extracellular chemical messengers may travel only a short distance from the secretory cell and only stimulate cells in close proximity, or may travel throughout the organism and stimulate cells that are relatively far away.

Local signalling between cells

Animal and plant cells communicate at the local level using:

- Direct cell–cell connections
- Local acting extracellular messenger molecules.

Direct cell–cell connections

- Animals and plants use gap junctions and plasmodesmata (see Chapter 3, Figure 3.28 and Chapter 13, Figure 13.10) respectively to directly connect the cytoplasm of adjacent cells.
- Gap junctions and plasmodesmata allow signal molecules or ions of low M_r to pass freely between adjacent cells.
- Animal cells can communicate by cell–cell recognition, whereby membrane-bound cell-surface molecules on one cell specifically recognize and bind to receptors of another cell (see Chapter 3, Figure 3.24).
- These types of local signalling are essential in stimulating, for example, appropriate immune responses as described in the companion volume, *Thrive in Immunology*.

Local acting extracellular messenger molecules

- Local acting extracellular messenger molecules are involved in **autocrine**, **paracrine**, and **synaptic** signalling (Figure 15.1).
- Autocrine signalling occurs when the cell that produces the chemical messenger expresses receptors for it on its surface such that the activities of the secretory cell will be stimulated or inhibited by its own signal.
- Paracrine signalling involves the use of chemical messengers that usually have a limited ability to travel around the body because they are unstable or are degraded by enzymes, or they bind to the extracellular matrix (Chapter 14).

Local and long-distance signalling

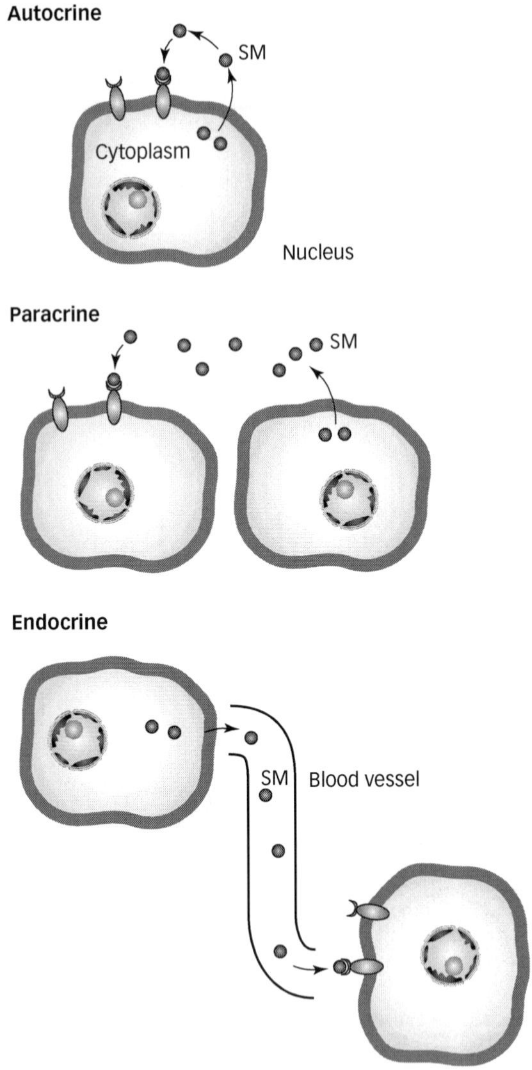

Figure 15.1 Autocrine, paracrine, and endocrine types of intercellular signalling. SM, signalling molecule.

- Given this limitation of distances, paracrine signalling stimulates only neighbouring cells.
- Synaptic signalling is a specialized type of local signalling that occurs in the nervous system across synaptic gaps (see later in this chapter).

Long-distance signalling

Long-distance signalling between cells in multicellular organisms involves communications using **hormones** or the nervous system.

- Specialized cells in the **endocrine glands** of animals release hormones, which travel in the circulatory system to act on distant target cells in the body.
- Plant hormones, which are often called **plant growth regulators**, sometimes travel in plant vessels but more often reach their targets by moving from cell to cell, or by gaseous diffusion through the air spaces in the plant.
- For example, the plant hormone ethylene (C_2H_4) is a gas that promotes fruit ripening and also regulates shoot and root differentiation and growth.
- Animal nerve cells or **neurons** can send messages over long distances due to their **axons**, which can be extensive in length; some elephant axons extending from the spinal cord to the foot can be several metres long.

15.2 SIGNALLING MOLECULES

A variety of biological molecules function as extracellular messengers or ligands. These include:

- Hormones
- Derivatives of vitamins A and D
- Growth factors and cytokines
- Eicosanoids
- Neurotransmitters.

Hormones

Hormones, from the Greek *to arouse* or *excite*, are chemical messages produced and secreted by endocrine glands that are transported to their target tissues by circulating blood.

They can be classified into several broad groups:

- Small water-soluble types, for example, **adrenalin** and **noradrenalin** (Figure 15.2a)
- Peptide or protein hormones such as **human growth hormone**, **insulin**, and **glucagon** (Figure 15.2b)
- Hydrophobic (lipophilic) hormones (Figure 15.2c) including:
    - steroid hormones such as **glucocorticosteroid**, **mineralocorticoids**, and sex hormones
    - thyroid hormones
    - the derivatives of vitamin A and D, for example, **retinoic acid**, and **1,25-dihydroxycholecalcipherol** respectively.
- Derivatives of vitamin A and D—including retinoic acid and retinol, and 1,25-dihydroxycholecalcipherol (Figure 15.3)—are also signalling molecules.
- Although they are not produced by endocrine glands, they are often regarded as 'honorary' hormones because they produce their stimulatory effects by similar molecular mechanisms.

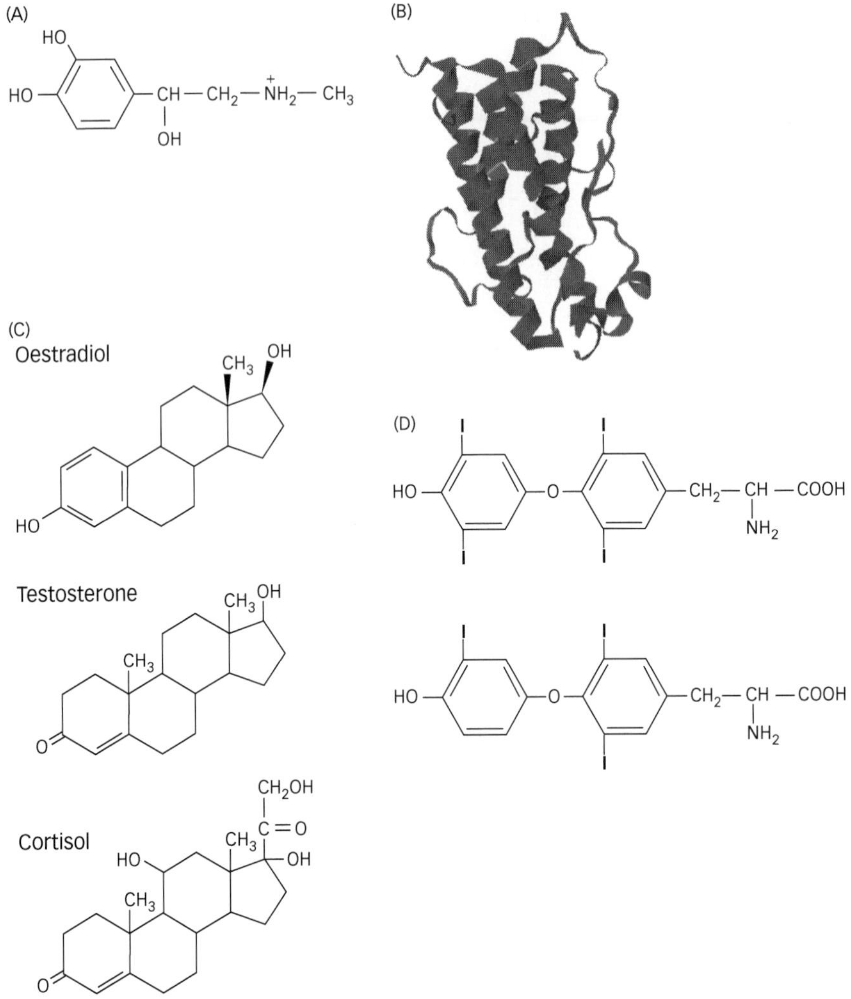

Figure 15.2 Structure of representative hormones. (A) Adrenalin. (B) Molecular model of human growth hormone. Model constructed using PDB file 1HGU. (C) The steroid hormones, oestradiol, testosterone, and cortisol. (D) Thyroid hormone (T4).

Hormones have wide-ranging effects

- Hormones are involved in regulating or controlling a broad range of metabolic and developmental processes.
- Adrenalin has been referred to as the 'fright, fight, and flight' hormone and is produced by the adrenal medulla.
- Adrenalin causes increases in blood pressure and pulse rate, contraction of smooth muscles, an increase in glycogen breakdown, and an increase in lipid hydrolysis in adipose tissue; effects that ready the body for a situation where greater energy is required.

- Glucagon and insulin are responsible for the regulation of concentration of glucose in blood plasma. However, insulin also stimulates a wide range of responses in target cells including altering their gene expression.
- Steroid hormones are all derived from cholesterol (see Chapter 3, Figure 3.1).
- **Cortisol** or **hydrocortisone** is a glucocorticosteroid produced by the cortex of the adrenal gland. It has significant effects upon carbohydrate metabolism and also acts as an immunosuppressant.
- **Aldosterone** is a mineralocorticoid also produced by the cortex of the adrenal gland. Mineralocorticoids act on the kidneys to regulate salt and water balance.
- **Progesterone** is synthesized in the ovaries and placenta. It is involved in the development of the uterus in preparation for implantation of the embryo, as well as stabilizing the early stages of pregnancy, and the development of the mammary glands.
- **Testosterone** produced by the testes, is responsible for the development and functioning of the male sex organs.
- **Thyroxine** (T_4) is produced by the thyroid gland from the amino acid tyrosine. It stimulates metabolism in many cells and increases heat production.

Looking for extra marks?

Many steroid hormones are used to treat clinical skin problems. They are applied topically and their hydrophobicity means they can be taken up by skin cells where needed.

Derivatives of vitamins A and D

- Derivatives of vitamin A are collectively called **retinoids**.
- Two retinoids, retinoic acid (Figure 15.3a) and retinal have well documented roles in embryonic development (see Chapter 17) and vision (described later) respectively.
- The active form of vitamin D, 1,25-dihydroxycholecalcipherol (Figure 15.3b) has an essential role in the absorption of Ca^{2+} from the gastrointestinal tract (GIT).

Transport of lipophilic hormones in the blood

- The transport of lipophilic hormones, retinoids, and 1,25-dihydroxycholecalcipherol between cells is not as simple as that of water-soluble ones.
- In the blood, these compounds are stabilized by their association with specific carrier proteins (Figure 15.4).
- Dissociation of the hormone from its carrier must then occur when they reach their target tissues before they cross the plasma membrane and enter the cells.

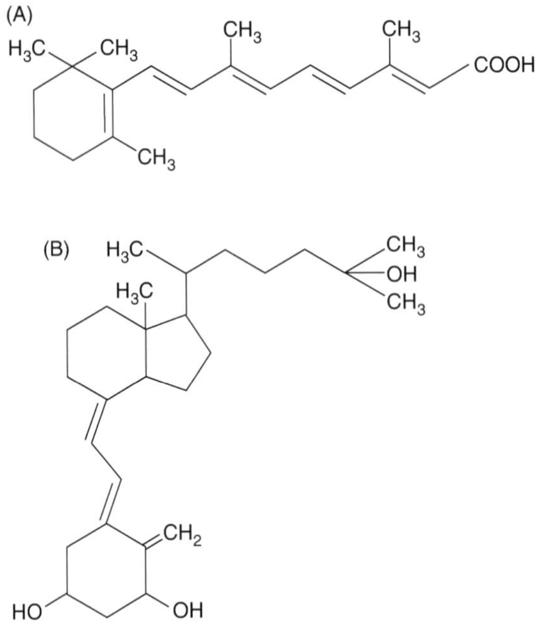

Figure 15.3 (A) Retinoic acid and (B) 1,25-dihydroxycholecalcipherol.

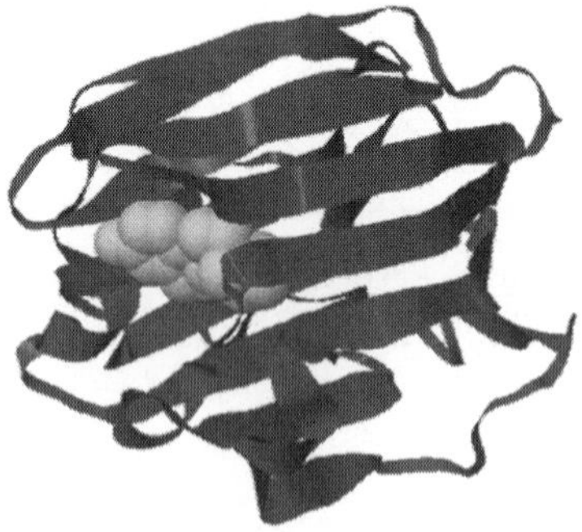

Figure 15.4 Molecular model of a steroid hormone bound to the blood plasma sex hormone transport globulin (SHBG). Model constructed using PDB file 1LHU.

Growth factors and cytokines

Both growth factors and cytokines are regulatory chemical messengers; however, they are not produced by specialized tissues like hormones but are secreted by many types of cells.

- Growth factors are a group of polypeptide signalling molecules that induce or inhibit cell division; they also help regulate cell proliferation, differentiation, and cell mobility (see Chapter 16).

- The names of growth factors usually reflect their individual roles—for example, epidermal growth factor (EGF), fibroblast growth factor (FGF), nerve growth factor (NGF), platelet-derived growth factor (PDGF), and vascular endothelial growth factor (VEGF).
- Cytokines are also protein or peptide molecules produced by a number of cell types, including macrophages, T cells, fibroblasts, and epithelial cells.
- Cytokines control the development of blood cells and are intimately involved in immune responses. For example, cytokines such as interferons, interleukins, and tumour necrosis factor are principally responsible for the coordination of the immune response and the development of the cells used in animal host defence.

➔ *The cell cycle is revised in Chapter 16. The immune system is revised in the companion volume,* Thrive in Immunology.

- Most growth factors and cytokines are local acting and affect the growth or differentiation of the types of cells producing them or other cells within a short distance.

Eicosanoids

All eicosanoids are derived from the $C_{20:4}$ fatty acid, arachidonic acid (Figure 15.5).

- They include:
  - prostaglandins
  - thromboxanes
  - leukotrienes.
- All eicosanoids act as local chemical signalling agents that affect a wide range of physiological activities.
- Prostaglandins influence inflammation, blood pressure, and smooth muscle contraction.
- Thromboxanes are also involved in mediating inflammation. They also stimulate the aggregation of blood platelets and hence blood clotting.
- Leukotrienes influence the contraction of smooth muscle tissues and are able to stimulate the migration of polymorphonuclear leukocytes and eosinophils to sites of injury or infection.

Neurotransmitters

Neurotransmitters are small signalling molecules, for example, acetylcholine; dopamine; the amino acids γ aminobutyric acid (GABA), glutamate, and glycine; neuropeptides, and the gas nitrous oxide (Figure 15.6).

- They are synthesized by the presynaptic neurons.
- Following synthesis, and prior to their use, neurotransmitters are packaged into multiple membrane-bounded compartments called **synaptic vesicles**.

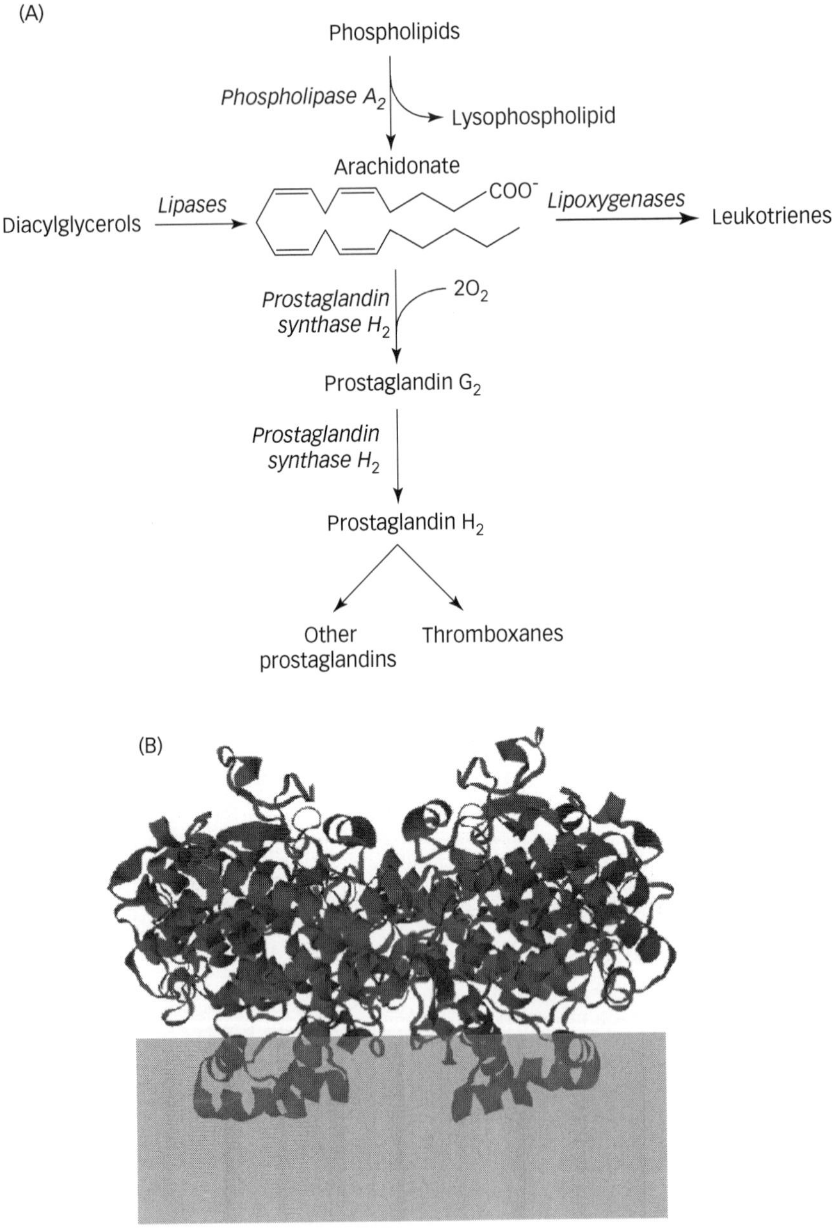

Figure 15.5 (A) Outline of the biosynthesis of prostaglandins, thromboxanes, and leukotrienes. (B) Molecular model of prostaglandin synthase H_2 of the endoplasmic reticulum membrane, a key enzyme of prostaglandin synthesis. The model was constructed using PDB file 1CQE.

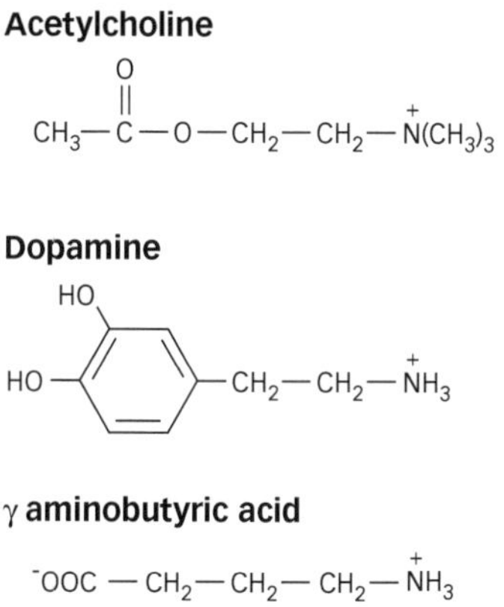

Figure 15.6 Examples of neurotransmitters: the structures of acetylcholine, dopamine, and γ aminobutyric acid (GABA).

Looking for extra marks?

In addition to its role in energy transfer reactions, adenosine triphosphate (ATP) also functions as an extracellular signalling molecule. It can be released from cells involved in the secretion of neurotransmitters, in storage granules from cells of the adrenal medulla, or by lymphocytes. Cell death and lysis can also lead to the release of ATP into the extracellular medium. Platelets, neutrophils, and fibroblasts have cell surface **purinoceptors** that detect and bind ATP; this is thought to be involved in apoptosis (Chapter 17).

Check your understanding

15.1 Which of the following statements is/are false?

a. The derivatives of vitamin A, 1,25-dihydroxycalcipherol and vitamin D, retinoic acid are signalling molecules.

b. Glucagon and insulin are largely responsible for regulating the concentration of glucose in blood plasma.

c. The mineralocorticoid cortisol is produced by the cortex of the adrenal gland from cholesterol.

d. Growth factors and cytokines require blood transport proteins because they are lipophilic.

e. Neurotransmitters are synthesized by postsynaptic neurons.

15.3 RECEPTORS AND SECOND MESSENGERS

Signalling molecules bind to specific protein receptors, which transmit the signal to the rest of the cell.

Receptors and second messengers

- Some hormones, growth factors, cytokines, prostaglandins, and neurotransmitters bind to receptors that are integral transplasma membrane proteins with extracellular binding sites.
- Receptors for steroid hormones, and vitamin A (retinoids) and D derivatives are located in the cytosol or nucleus and the ligands must therefore enter the cell to bind to them.

Common features of receptors

The receptors of all types of signal molecule share a number of common features.

- The binding of a signalling molecule and receptor is specific: a receptor detects only one signalling molecule or a range of closely related molecules.
- The sensitivity of a cell to the concentration of the signalling molecule can be varied by altering the density of receptors.
- Binding of the signal to a receptor results in a specific conformational change in a receptor, which is often the initial transduction of the signal.
- The binding of a hormone to its receptor is extremely strong: dissociation constants (K_a) for the equilibrium binding of hormone-receptor are 10^7–10^{11} mol dm^{-3}. This is the same range of concentrations as that of hormones in plasma.
- This initial transduction event allows the receptor to transmit the receipt of the message to the rest of the cell, usually using components of a **signalling cascade** or **signal transduction pathway**.
- The binding of the signalling molecule in most signal transduction pathways stimulates the production of small M_r, water-soluble molecules or the release of ions.
- The initial signalling molecule is often referred to as the **first messenger** and these small molecules or ions as **second messengers**.
- Once the message has been received and induced its physiological responses, the receptor needs to be turned off.
- The appropriate activity of receptors is crucial to proper cell functioning, and abnormal activities of receptors are associated with many clinical conditions.

Cell-surface receptors

Three major types of cell surface transmembrane receptors are involved in signal detection and transduction:

1. **G protein-coupled receptors** (GPCRs)
2. **Receptor tyrosine kinases** (RTKs)
3. **Ligand-gated ion channels**.

G protein-coupled receptors

- G protein-coupled receptors or **G protein-linked receptors** belong to a group of integral membrane proteins containing seven transmembrane α helices (Figure 15.7).

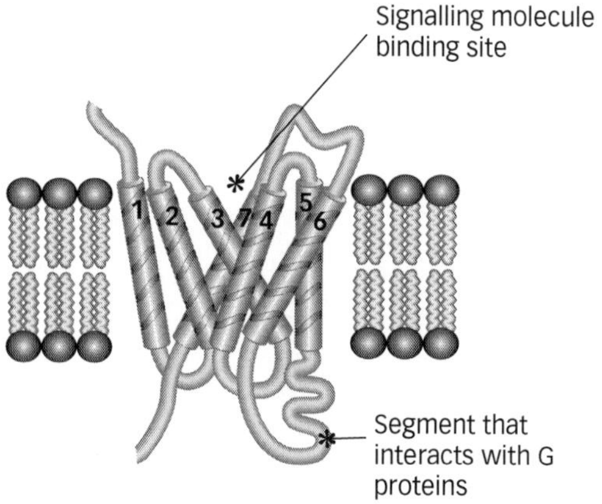

Figure 15.7 Schematic representation of a G protein-coupled receptor.

- Specific loops between the helices form a binding site for the ligand on the extracellular surface and one for a G protein on the cytosolic side of the membrane.

  ➔ *G proteins are encountered in many chapters, especially Chapters 5–8.*

- Many different signalling molecules, including adrenalin, glucagon, thyroid stimulating hormone (TSH), and vasopressin, most eicosanoids, and the neurotransmitter acetylcholine, stimulate their effects through GPCRs.
- GPCRs vary in the binding sites for their ligands and also for different types of G proteins inside the cell.
- The binding of the hormone or neurotransmitter activates the GPCR, which then activates a specific G protein.
- The G protein, in turn, activates yet another protein, thus propagating the signal along a signal transduction pathway or cascade.
- Information processing systems using GPCRs are extremely widespread in different organisms and diverse in their functions—for example, they help regulate embryonic development and are essential in the sensory reception of vision, sound, smell, taste, and touch.

Receptor tyrosine kinases

- RTKs possess a cytoplasmic domain capable of protein kinase activity and catalysing the transfer of a phosphate group from ATP to tyrosine residues of the receptor itself.
- Signalling molecules that bind to RTKs include the hormones insulin and growth hormone (Figure 15.2b and 15.8), many growth factors such as epidermal growth factor, and cytokines (for example, **erythropoietin** (EPO)).
- The binding of a specific extracellular signal to RTKs usually results in two receptor-signal molecules associating to form a dimer (Figure 15.9), which leads to conformational changes and activation of the protein kinase domain.
- The insulin receptor is an exception to the dimerization rule; its structure is described later.

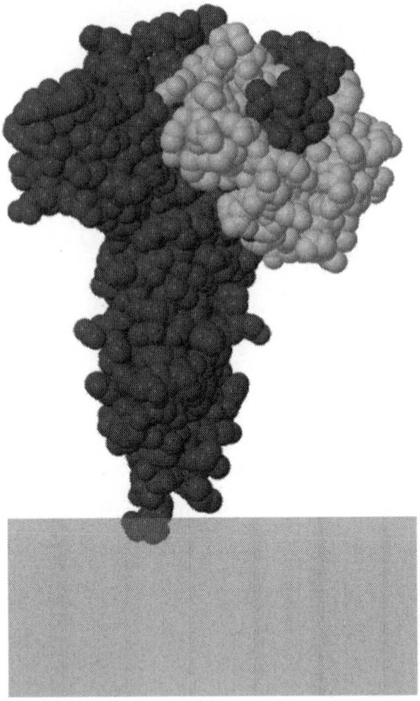

Figure 15.8 Molecular model of a human growth hormone (lighter tone) bound to the extracellular domain of its receptor. Model constructed using PDB file 3HHR.

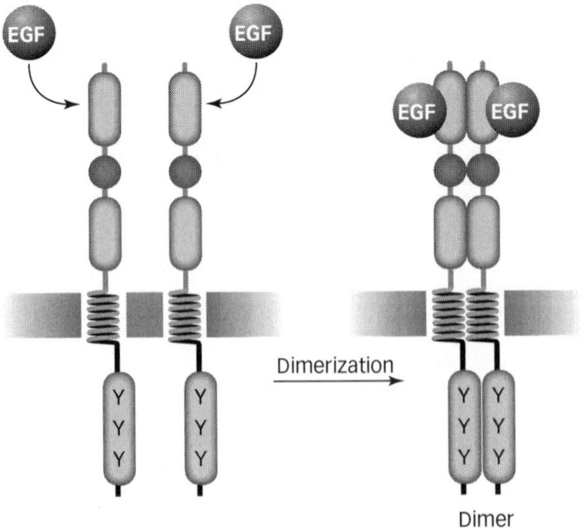

Figure 15.9 Schematic representation of the dimerization of the epidermal growth factor, a tyrosine kinase receptor (TKR), following binding of epidermal growth factor (EGF). Y, tyrosine residue phosphorylation sites.

- Relay proteins in the cell are then activated by binding to different phosphorylated tyrosine residues, which allows one receptor dimer to trigger several different intracellular signalling pathways simultaneously.

- The ability of a single ligand-binding event to trigger many pathways is a major difference between RTKs and GPCRs.
- Abnormal RTKs that function even in the absence of signalling molecules (such as growth factor) are associated with cancers.

Ligand-gated ion channels

- Ligand-gated ion channels are transmembrane proteins that open or close in response to binding of a specific ligand, allowing or blocking the flow of specific ions across the membrane.

➔ *Ligand-gated ion channels are introduced in Chapter 3.*

- Ligand-gated ion channels are often involved in the response to neurotransmitters, such as acetylcholine, and so are sometimes referred to as **transmitter-gated ion channels** or **ionotropic receptors**.
- They are composed of several polypeptide chains that pass through the membrane and form a pore (Figure 15.10).
- A flow of ions across the membrane can result in a temporary change in the membrane potential (Chapter 3), which can alter the activity of specific cytoplasmic enzymes.
- The flow is also essential in the transmission of nervous impulses.

Intracellular receptors

Intracellular receptors recognize and bind signalling molecules that are hydrophobic or small enough to cross the plasma membrane.

- They include the receptors for steroid and thyroid hormones, retinoic acids and 1,25-dihydroxycholecalcipherol.
- Binding of the signal to its receptor leads to a conformation change, which activates the receptor and eventually leads to changes in the transcription of specific genes.

➔ *Transcription in eukaryotes is revised in Chapter 6.*

Second messengers

Second messengers are produced in response to stimulation of the target cell by the initial signalling molecule.

- They include cyclic adenosine monophosphate (cAMP), cyclic guanosine monophosphate (cGMP), inositol trisphosphate (IP_3), Ca^{2+}, and diacylglycerol (DAG).
- Second messengers are able to diffuse rapidly in the cytosol and quickly relay the information that a signalling molecule has bound by an extracellular receptor.

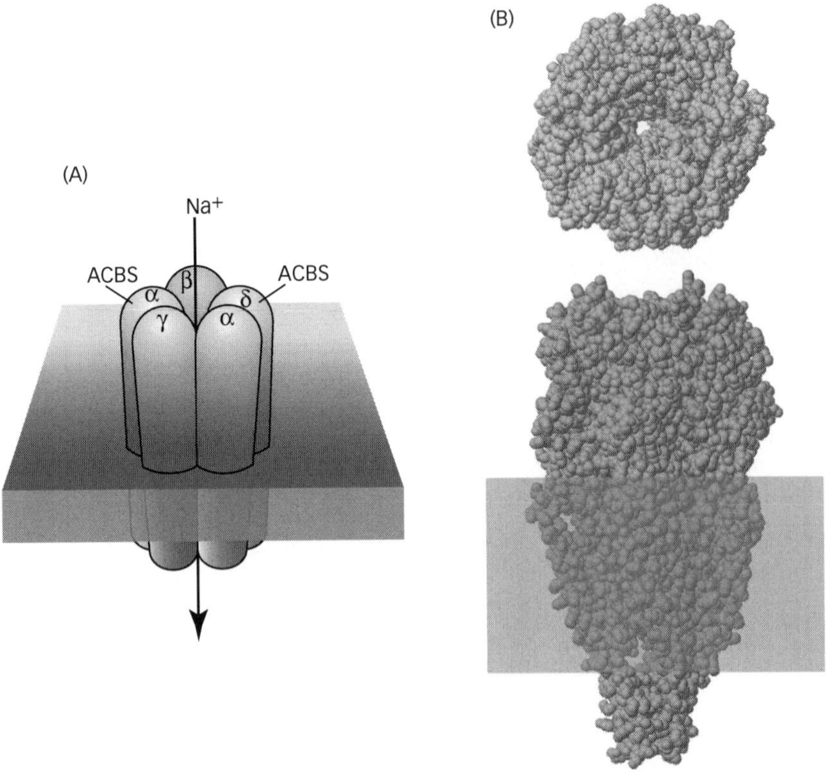

Figure 15.10 (A) Schematic structure of an acetylcholine receptor. This ligand-gated ion channel is composed of five polypeptides ($\alpha_2\beta\gamma\delta$). The two identical α polypeptides contain the acetylcholine binding sites (ACBS). (B) Top and side views of a molecular model of the acetylcholine receptor. Note the channel passing through the molecule. Model constructed using PDB file 2BG9.

Check your understanding

15.2 Which of the following statements is/are true?

a. Hormones, growth factors, cytokines, prostaglandins, neurotransmitters, and retinoic acid bind to receptors that are integral transplasma membrane proteins with extracellular binding sites.

b. All ligand-gated ion channels are integral membrane proteins containing seven transmembrane α helices.

c. Most growth factors and cytokines bind to tyrosine kinase receptors.

d. Hydrophobic chemical messengers are generally recognized by intracellular receptors.

e. Cyclic adenosine monophosphate (cAMP), cyclic guanosine monophosphate (cGMP), and cyclic inositol monophosphate are all second messengers in signal transduction pathways.

15.4 EXAMPLES OF CELL SIGNAL TRANSDUCTION PATHWAYS

Multicellular organisms are composed of billions of cells: the human body has an estimated 10^{13} cells. Coordinating the activities of such large populations requires extensive information processing abilities.

- Signal transduction pathways are one of the many ways organisms coordinate their activities.
- This section will concentrate on only the following to illustrate their similarities and differences:
  - adrenalin and glucagon signalling pathways
  - insulin and growth factor signalling pathways
  - erythropoietin signalling pathway
  - steroid hormone signalling
  - transmission of nervous impulses
  - vision perception.

Adrenalin and noradrenalin signalling pathways

The hormones adrenalin and noradrenalin referred to earlier as the 'fright, fight and flight hormones' allow organisms to respond to stress. They act through specific GPCRs, using cAMP as a second messenger.

- Cells can have two types of adrenalin receptors, α and β **adrenoreceptors** (or α and β **adrenergic** receptors).
- A simplified response of adrenal binding to β adrenoreceptors of muscle cells is illustrated in Figure 15.11.
- Muscle cells have β adrenoreceptors on their surfaces.
- Binding of adrenalin to a β adrenoreceptor promotes a conformational change in the receptor that stimulates a trimeric **guanosine nucleotide binding (G) protein.**
- Membrane-bound G proteins generally consist of three subunits: α, β, and γ subunits (Figure 15.11).
- They bind guanosine nucleotides: they are *active* when GTP is bound, *inactive* if the nucleotide is GDP.
- The adrenalin-receptor complex opens the GDP/GTP binding site of the G_α subunit allowing it to exchange its bound GDP for a GTP (Figure 15.11).
- The binding of GTP, in turn, alters the conformation of the G_α subunit, which reduces its affinity for the $G_{\beta\gamma}$ subunits.
- The dissociation of the G_α subunit from the $G_{\alpha\beta\gamma}$ complex is the transmembrane signal that the hormone has bound to the receptor.

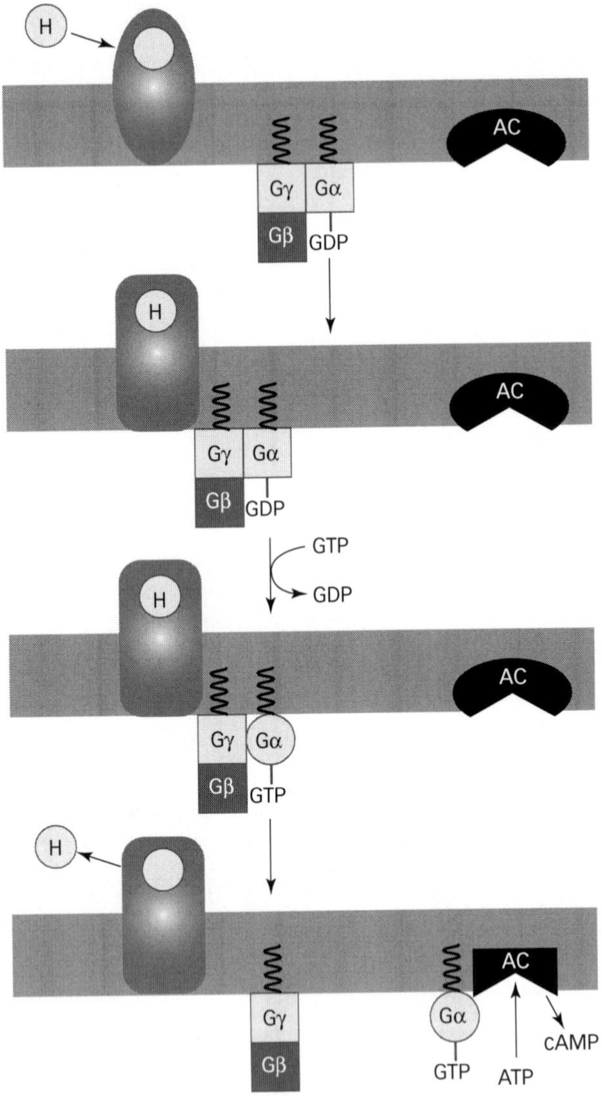

Figure 15.11 Simplified schematic response of adrenalin (H) binding to β adrenoreceptors of muscle cells (R). AC, adenylate kinase. See text for explanations of each step.

- The activated G_{α} subunit is able to move laterally in the membrane and bind to the membrane-bound **adenylate cyclase**, inducing a conformational change in the enzyme.
- This allows it to catalyse the conversion of cytosolic ATP to cyclic adenosine monophosphate (cyclic AMP or cAMP) as illustrated in Figures 15.11 and 15.12.
- This activity increases the concentration of cAMP in the cytosol.
- Cyclic AMP is the second messenger for adrenalin and, indeed, for a number of other hormones.

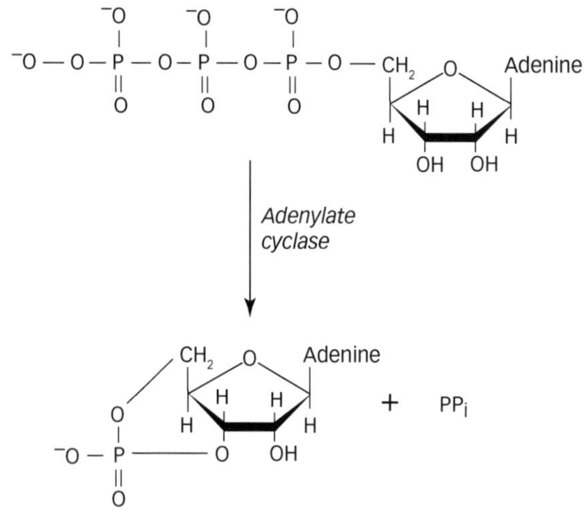

Figure 15.12 Conversion of ATP to cyclic adenosine triphosphate (cAMP).

- The increased concentration of cAMP activates or deactivates many effectors within the cell, producing the physiological effects associated with stimulation by adrenalin.

Looking for extra marks?

Cyclic AMP is an ancient signalling molecule. Its release by single cells of the slime mould, *Dictyostelium* allows hundreds of cells to move towards each other and form a reproductive multicellular aggregate called a fruiting body.

Signal cascade

- The strength of the initial signal (first messenger) is increased by using **cascade mechanisms** during transduction; each step amplifies the effects of the previous as shown in Figure 15.13.
- A single adrenalin molecule bound to a receptor is able to activate many hundreds of G proteins.
- In turn, each activated adenylate cyclase increases the intracellular concentration of cAMP, which activates **protein kinase A (PKA)**.
- In its inactive form, PKA is a tetrameric holoenzyme consisting of two regulatory subunits and two (inactive) catalytic ones.
- The binding of two cAMP molecules to each regulatory subunit alters their conformations, allowing them to dissociate from the holoenzyme and releasing the now active catalytic apoenzyme subunits (Figure 15.14).
- Phosphorylase kinase A catalyses the phosphorylation of phosphorylase, converting it from its inactive form, phosphorylase *b*, to its active form phosphorylase *a* (see Chapter 7, Figure 7.8).

Examples of cell signal transduction pathways

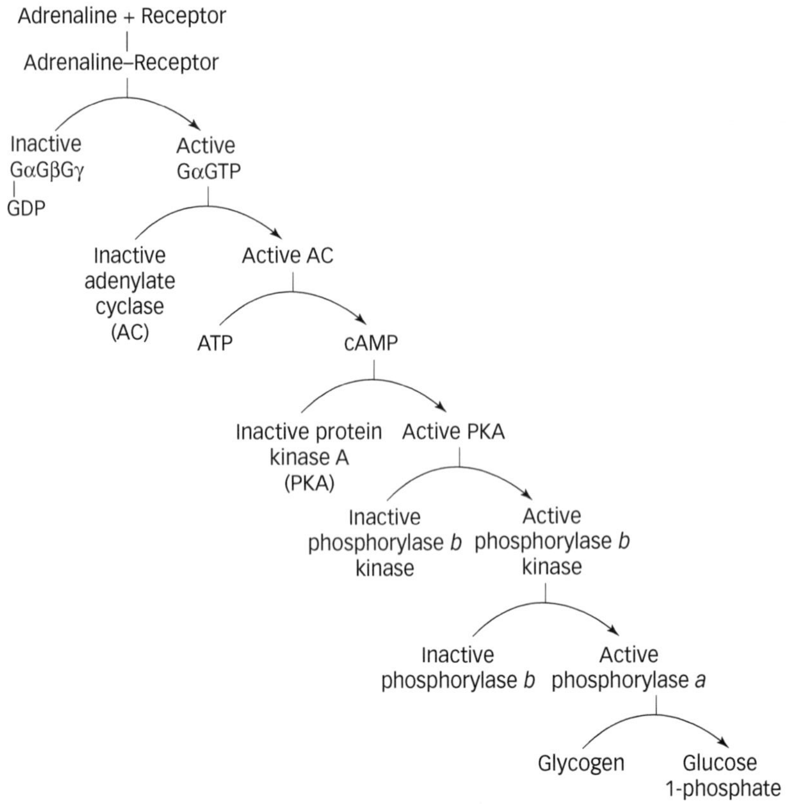

Figure 15.13 Adrenalin stimulated degradation of glycogen to glucose 1-phosphate. Each step amplifies the response to adrenalin as explained in the main text.

- Each activated PKA can stimulate the activity of many phosphorylase molecules.
- Finally, each active phosphorylase *a* can catalyse the release of glucose 1-phosphate from glycogen stores.
- Glucose 1-phosphate can enter glycolysis (see Chapter 7, Figure 7.2), generating the ATP that the cell needs to respond to the stress that led to the secretion of adrenalin into the blood.
- The same system simultaneously switches off the synthesis of glycogen.
- In this case, the PKA-catalysed phosphorylation of **glycogen synthase** inactivates the enzyme.

Check your understanding

15.3 If the binding of a single adrenalin molecule to a receptor leads to the production of 10^4 molecules of cAMP, how many active PKA molecules will be produced?

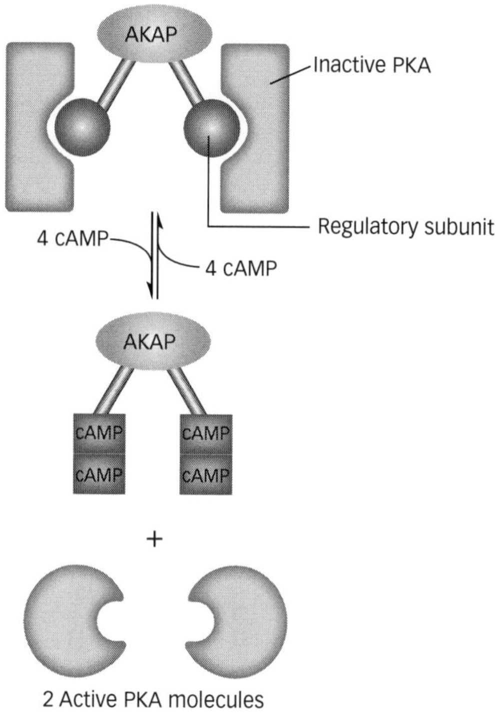

Figure 15.14 Role of cyclic adenosine monophosphate (cAMP) in activating protein kinase A (PKA). AKAP, A kinase anchoring protein, cAMP, cyclic adenosine monophosphate.

Signal deactivation

- The system is deactivated because G_α is a slow acting GTPase: within seconds to minutes its intrinsic GTPase activity hydrolyses the bound GTP to GDP.
- This allows it to dissociate from the adenylate cyclase, reassociate with the $G_{\beta\gamma}$, and reform the inactive $G_{\alpha\beta\gamma}$.
- Also, phosphorylation of the adrenalin receptor by a specific G protein-coupled receptor kinase (GCRK) allows the protein **arrestin** to bind to it. Binding of arrestin prevents further activation of the G protein.
- Cyclic AMP is hydrolysed to AMP by a specific cAMP phosphodiesterase:

$$\text{cAMP} + H_2O \xrightarrow{\text{cAMP phosphodiesterase}} \text{AMP}$$

- The reduced cytosolic concentration of cAMP allows the regulatory subunits of PKA to bind to and inactivate its catalytic ones.
- Active phosphorylase *a* is deactivated by **protein phosphatase 1**, which catalyses the removal of its phosphate groups.
- These events switch the system off and prevent the cellular responses to adrenalin being permanent.

Revision tip

Help unravel the complex cell signalling pathways by downloading the free slides provided by Sigma-Aldrich Life Science from the website: http://www.sigmaaldrich.com/life-science/cell-biology/learning-center/pathway-slides-and.html

Adrenalin and glucagon stimulation of hepatocytes

Adrenalin binds to both α and β adrenoreceptors located on the surface of hepatocytes.

- Binding to the β adrenoreceptor increases the intracellular concentration of cAMP, leading to the activation of PKA and glycogen breakdown, although hepatocytes are less responsive to adrenalin than muscle cells.
- The binding of adrenalin to an α adrenoreceptor stimulates **phospholipase C**, which leads to an increase in the second messengers, IP_3, DAG, and Ca^{2+}; all of these promote the effects of cAMP on glycogen breakdown.
- The glucose 1-phosphate produced need not necessarily enter glycolysis but may be dephosphorylated to free glucose:

$$\text{Glucose 1-phosphate} \xrightarrow{\text{Phosphoglucomutase}} \text{glucose 6-phosphate}$$

$$\text{Glucose 6-phosphate} + H_2O \xrightarrow{\text{Glucose-6-phosphatase}} \text{glucose} + P_i$$

- The glucose produced can then be transported out of the hepatocyte to increase the concentration in blood.
- Muscle cells lack glucose-6-phosphatase and so cannot contribute to replenishing blood glucose.
- The peptide hormone, glucagon (M_r 3500) acts in an identical manner to adrenalin following binding to its GPCR on the surface of hepatocytes (Figure 15.15).
- Glucagon is released by α cells of the islets of Langerhans in the pancreas when blood glucose concentrations are low.

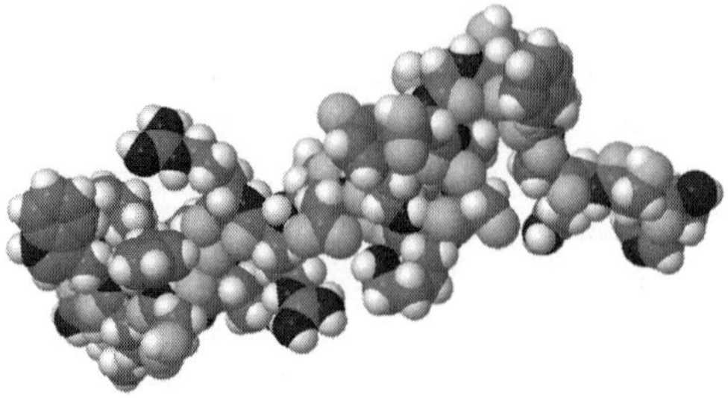

Figure 15.15 Molecular model of glucagon. Model constructed using PDB file 1KX6.

- Hepatocytes are more responsive to glucagon than adrenalin and so their stimulation increases the concentration of glucose in blood plasma (**hyperglycaemia** effect).
- Glucagon, together with insulin, therefore regulates the concentration of glucose in the blood.

Looking for extra marks?

Cholera is an acute diarrhoeal disease caused by drinking water contaminated with the bacterium, *Vibrio cholerae*. The cholera toxin (choleragen) is an enzyme that chemically modifies a G protein involved in regulating salt and water loss from the GIT. The modified G protein is unable to hydrolyse GTP to GDP and remains active and continuously stimulates adenylate cyclase. The resulting high concentration of cAMP continuously activates PKA causing the loss of large amounts of salts and water from the GIT. An infected person quickly develops profuse diarrhoea and, if left untreated, can soon die from the loss of water and salts.

Check your understanding

15.4 Which of the following statements is/are false?

a. The binding of GTP reduces the affinity between G_α and $G_{\beta\gamma}$ subunits of trimeric G proteins.

b. The G_αGTP stimulates a membrane-bound adenylate cyclase.

c. Cyclic AMP actives the tetrameric holoenzyme form of protein kinase A (PKA).

d. The inactive form of protein kinase A is a tetrameric apoenzyme.

e. All G proteins are relatively slow-acting GTPases.

Insulin signalling pathways

Insulin (Figure 15.16) is produced by the β cells of the islets of Langerhans in the pancreas. Its effects on metabolism are numerous and complex; only two will be discussed here.

- Perhaps the most immediate effect of insulin stimulation is to *increase* the rate of glucose uptake by cells and so *reduce* its concentration in blood (**hypoglycaemia** effect).
- The *opposing* actions of insulin and glucagon (see 'Adrenalin and glucagon stimulation of hepatocytes' section) are responsible for regulating the concentration of glucose in the blood plasma.

Figure 15.16 Molecular model of pig (*Porcus domesticus*) insulin. The lighter coloured regions are the cysteine residues forming disulphide bonds that stabilize the structure. Model constructed using PDB file 4INS.

- However, while glucagon acts by binding to its GPCR and mobilizing glycogen reserves through the cAMP activation of PKA, insulin produces its effect by binding to a specific RTK.
- Longer-term effects associated with insulin stimulation are affected by changing gene expression in target cells.

Insulin receptor protein

- The RTK that specifically recognizes and binds insulin, the **insulin receptor protein** (**IRP**), is a glycoprotein consisting of two α subunits of M_r 135,000 and two β subunits of M_r 95,000.
- The α subunits are on the outside of the cell and are responsible for insulin binding, whereas the β subunits contain transmembrane domains along with tyrosine kinase domains (Figure 15.17).

Looking for extra marks?

Insulin-like growth factor (IGF-1) is produced by the liver; not surprisingly given its name, its structure resembles that of insulin. Its function is helping to control growth. The receptor for IGF-1 also resembles the IRP. Again unsurprisingly, each can bind to the other's receptor. However, in both cases, binding is approximately 100 times less strong than to its cognate receptor.

Check your understanding

15.5 Use the Jmol program at the Protein Database (http://www.rcsb.org/pdb/home/home.do) to examine the structure of insulin-like growth factor (PDB file 1GZR) and compare its structure with that of insulin (Figure 15.16).

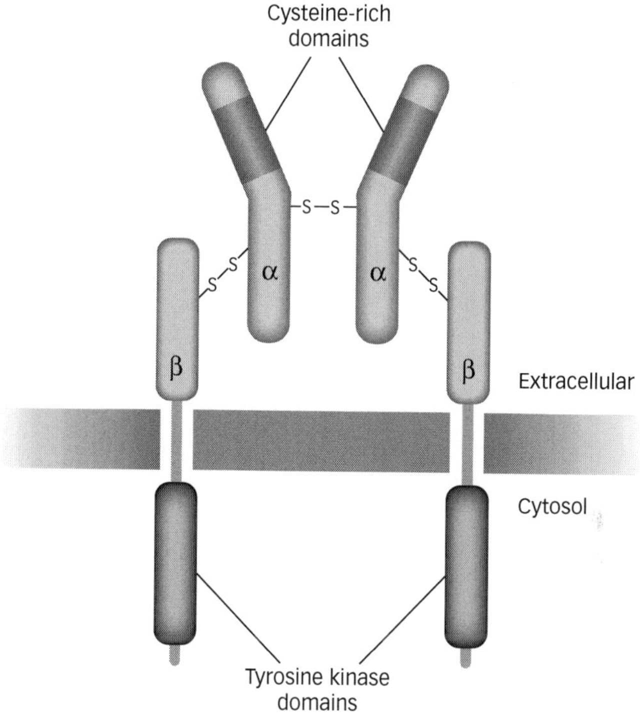

Figure 15.17 Schematic structure of the insulin receptor protein (IRP). The receptor site is formed by the two α subunits, which are situated on the outside of the plasma membrane. These are linked to the two β subunits, which have transmembrane and tyrosine kinase domains, by disulphide bonds.

Insulin transduction pathways

- The binding of insulin to its receptor produces a conformational change that activates the tyrosine kinase domains of the β subunits (Figure 15.18).
- Once activated, each β subunit autophosphorylates three tyrosine residues located near the carboxyl terminus of the other subunit.
- Autophosphorylation opens the active site allowing the enzyme to catalyse the phosphorylation of other cellular proteins involved in the insulin signal transduction pathways.
- **Insulin receptor substrate 1** (**IRS1**) is a key protein in insulin signal pathways because it acts as a relay protein upon phosphorylation and can stimulate two major insulin-dependent pathways:
  - one pathway activates glycogen synthase and increases the facilitated diffusion of glucose into target cells
  - the other pathway regulates gene expression (transcription).

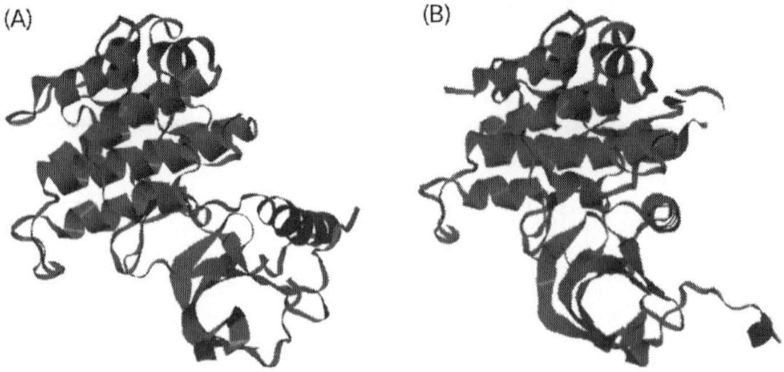

Figure 15.18 Molecular models of (A) the inactive and (B) active conformations of the insulin receptor tyrosine kinase domain. Note the difference in conformations. Models constructed using PDB files 1IRK and 1IR3 respectively.

Insulin-dependent hypoglycaemia

- Activated IRS1 can bind to the Src (pronounced sark) homology 2 (SH2) domain of the regulatory subunit of phosphoinositide 3-kinase (PI-3K) as shown in Figure 15.19.
- (The SH2 protein domain is named because of its similarity to a domain in the tyrosine kinase, Src. The domain consists of a small number of amino acid residues containing a phosphorylated tyrosine residue and is found in many intracellular signalling proteins.)
- Phosphoinositide 3-kinase, in turn, catalyses the conversion of the membrane lipid, phosphatidylinositol 4,5-bisphosphate (PIP_2) to phosphatidylinositol 3,4,5-trisphosphate (PIP_3).
- Phosphatidylinositol 3,4,5-trisphosphate remains part of the lipid bilayer and is able to bind the next protein (enzyme) of the pathway, protein kinase B (PKB) onto the cytosolic side of the plasma membrane (Figure 15.19).
- Here, another kinase, PDK1 phosphorylates and activates PKB.
- Activated PKB is now able to add phosphate groups from ATP to specific serine and threonine residues on target proteins such as glycogen synthase kinase 3 (GSK3).
- *Non*-phosphorylated glycogen synthase kinase is active; thus, its phosphorylation leads to its *in*activation.
- Inactive GSK3 is unable to inactivate glycogen synthase (GS) by phosphorylation; hence the effect of insulin stimulation is to increase the synthesis of glycogen.
- Insulin-dependent signalling also leads to translocation of the glucose transporter 4, GLUT4.
- In non-stimulated cells, GLUT4 are found predominantly in intracellular membrane vesicles.
- Activate PKB signalling leads to the transport of the vesicles to the plasma membrane, where the membranes fuse and allow GLUT4 transporters to become integral proteins of the plasma membrane.
- Here, GLUT4 facilitates the uptake of glucose into the cell.

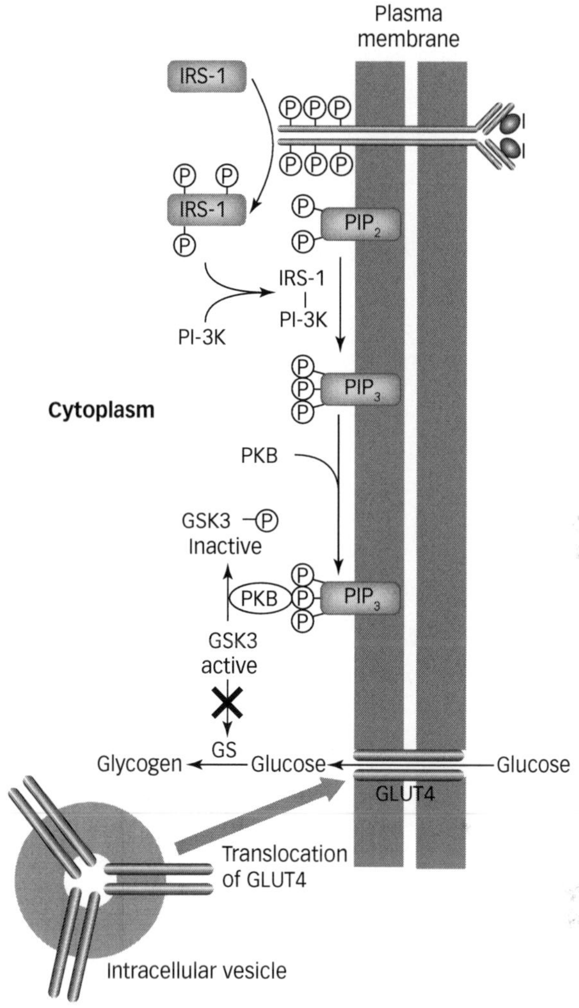

Figure 15.19 Schematic outline of the insulin-dependent hypoglycaemia signalling pathway. See text for explanation. GLUT4, glucose transporter 4; GS, glycogen synthase; GSK3, glycogen synthase kinase 3; I, insulin; PIP_2, phosphatidylinositol 4,5-bisphosphate; PIP_3, phosphatidylinositol 3,4,5-trisphosphate; PI-3K, phosphoinositide 3-kinase; PKB, protein kinase B.

Deactivation of insulin-dependent hypoglycaemia

- Removal of the insulin signal deactivates the system because:
  - GSK3 is activated by dephosphorylation and so is able to inactivate GS by catalysing its phosphorylation reducing glycogen synthesis
  - the supply of glucose into the cell is reduced because GLUT4 is endocytosed (Chapter 9) into intracellular vesicles decreasing the facilitated uptake of glucose.

Looking for extra marks?

Defects in the production of insulin or in insulin signalling can lead to the clinical condition, diabetes mellitus.

Insulin-dependent regulation of gene expression

- Despite remaining extracellular, insulin is able to stimulate a signalling pathway that alters the genetic expression of target cells through the phosphorylation and activation of IRS-1.
- This pathway is dependent on the actions of a rather large number of proteins that aggregate to form the complex shown in Figure 15.20.
- Phosphorylated IRS-1 binds to the SH2 domain of **growth factor receptor-bound protein 2** (**GRB2**), which then recruits the protein, **son of sevenless** or **SOS**.
- This trimeric complex is an active guanosine nucleotide-exchange factor that catalyses the exchange of a GDP for a GTP on the small (M_r approximately 21,000) membrane-bound G protein, **Ras** (Figure 15.20).

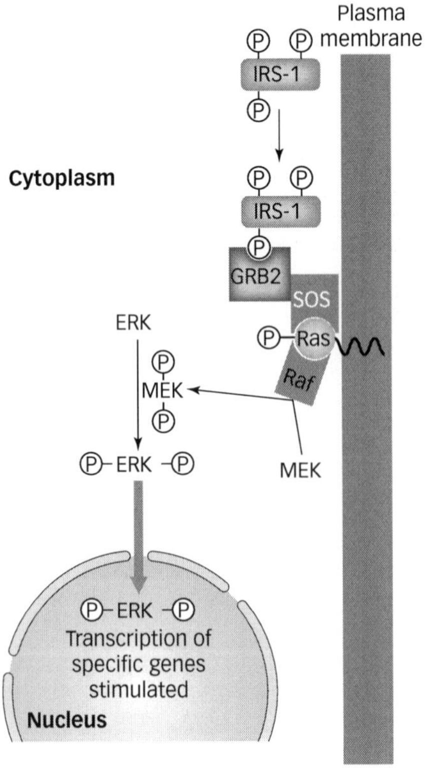

Figure 15.20 Insulin-dependent regulation of gene expression. ERK, type of MAP kinase; GRB2, growth factor receptor-bound protein 2; IRS-1, insulin receptor substrate 1; MEK, type of MAP-kinase-kinase and SOS, son of sevenless. See text for explanation.

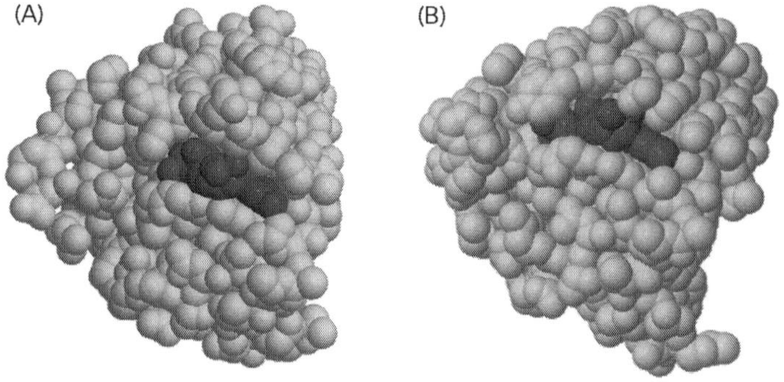

Figure 15.21 Molecular models showing the different conformations of (A) RasGDP and (B) RasGTP. In each model, the darker-toned structures are the guanosine nucleotides (GDP and GTP respectively). Models constructed using PDB files 4Q21 and 5P21 respectively.

- The GDP-GTP exchange alters the conformation of Ras from an inactive one to an active state (Figure 15.21).
- Kinase activity of RasGTP leads to the subsequent cascade of phosphorylations and activations of three kinases, collectively called **mitogen-activated protein (MAP) kinases: Raf-1**, **MEK**, and **ERK** (see Box 15.1).
- Once phosphorylated by RasGTP, active Raf-1 phosphorylates and activates the kinase MEK, which, in turn, activates the ERK kinase through phosphorylation (Figure 15.20).
- Phosphorylated (active) ERK can be translocated through nuclear pores into the nucleus where it catalyses the phosphorylation of transcription factors that mediate the transcription of approximately 100 insulin-dependent genes.

➔ *Nucleocytoplasmic transport is revised in Chapter 6.*

Box 15.1 Nomenclature of Ras pathway kinases

- Mitogen-activated protein (MAP) kinases stimulate cell division, hence the term *mitogen*. The substrate of the kinase, MEK is the **extracellular signal-regulated protein kinase** or ERK. Thus, the abbreviation, MEK means MAP kinase/ERK kinase.
- The three groups of MAP kinases, Raf, MEK, and ERK, found in different signalling pathways also have collective terms. It is easiest to understand these rather clumsy abbreviations by describing them in the reverse order.
- Thus:
  - ERK is a MAP kinase or **MAPK**
  - MEK is also a protein kinase; since it phosphorylates ERK (a MAPK) it is a MAP-kinase-kinase (**MAPKK**)
  - Ras protein kinase activity phosphorylates MEK (a MAPKK) therefore Ras is a MAP kinase-kinase-kinase or **MAPKKK**.

Check your understanding

15.6 Which of the following statements is/are true?

a. Insulin is produced by the β cells of the pancreas.

b. Insulin is the only peptide hormone that regulates the concentration of glucose in the blood.

c. The insulin receptor, a glycoprotein that spans the plasma membrane, undergoes a conformational change upon insulin binding.

d. The insulin receptor substrate 1 is subject to autophosphorylation, which is a key event in the insulin signal transduction pathway.

e. Phosphoinositide 3-kinase catalyses the conversion of phosphatidylinositol 4,5-bisphosphate to phosphatidylinositol 3,4,5-trisphosphate.

Growth factor receptors and Ras signalling

A number of growth factors, including EGF, FGF, NGF, PDGF, and VEGF, bind to receptors that are integral membrane proteins with extracellular binding sites.

- The receptors for these growth factors are monomeric proteins, and dimerize following binding of their respective factor (for example, see Figure 15.9).

Cytokine signalling pathways

The receptor for the cytokine, erythropoietin (M_r 30,400) is a transmembrane protein with an extracellular binding domain. Binding of EPO stimulates two receptors to dimerize (Figure 15.22).

- Unlike the receptors for growth factors, the EPO receptor does not have any intrinsic tyrosine kinase activity, although binding of EPO does stimulate a signal transduction pathway (Figure 15.23).
- The dimerized receptor is able to recruit a soluble tyrosine kinase from the cytosol called **Janus protein kinase** or **JAK** (Figure 15.23).
- Binding of JAK to the EPO receptor activates the enzyme and allows it to phosphorylate:
  - the EPO receptor
  - a family of transcription factors called **signal transducers and activators of transcription** (**STATs**).
- The phosphorylated EPO receptor binds GRB2 and so initiates the MAPK cascade signalling pathway, which alters gene expression in the target cell.
- Phosphorylated STAT molecules are able to dimerize (Figure 15.23) because they possess an SH2 domain that recognizes and binds tyrosine-phosphate groups on the other molecule.

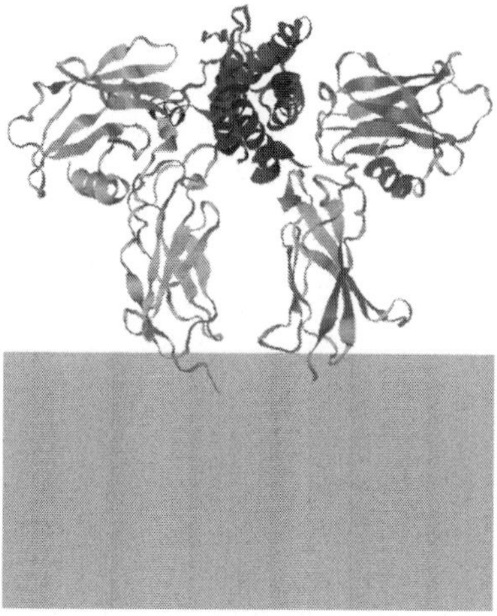

Figure 15.22 Molecular model of erythropoietin (darker tone) bound to the receptor site of the extracellular domains of its dimerized receptor. Model constructed using PDB file 1CN4.

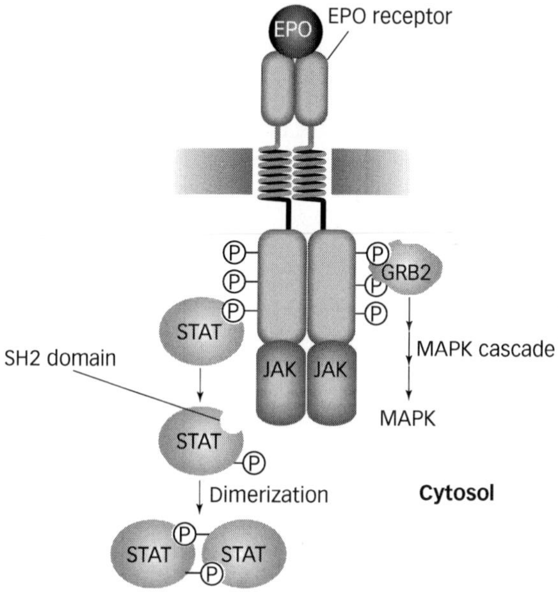

Figure 15.23 The erythropoietin signal transduction pathway. See text for details. EPO, erythropoietin; GRB2, growth factor receptor-bound protein 2; JAK, Janus protein kinase; MAPK, mitogen-activated protein kinase; STAT, signal transducers and activators of transcription.

- This altered conformation makes them a target for nucleocytoplasmic transport (Chapter 6).
- Within the nucleus, STAT dimers stimulate the transcription of specific genes essential for the maturation of erythrocytes.

Deactivation of the JAK pathway

- Cytokine signalling pathways are deactivated by several methods.
- The cytokine bound to the receptor may be internalized and hydrolysed to small peptides and amino acids (Chapters 3 and 9).
- Phosphatases deactivate cascade proteins by hydrolytically removing their phosphate groups.
- The cytokine-receptor interactions switch on positive and negative controls simultaneously.
- Active STAT dimers not only lead to the production of proteins that are required to give the physiological response to cytokine-receptor binding, but also lead to the formation of **suppressors of cytokine signalling** (SOCS) **proteins**.
- These proteins bind to JAKs attached to receptors and inhibit their kinase activity by blocking the active sites.
- In addition, **protein inhibitors of activated STATs** (**PIAS**) can bind to active STAT dimers and prevent them stimulating transcription.

Steroid and thyroid hormone, and retinoid signalling pathways

Steroid and thyroid hormones, and retinoids all bind to intracytosolic or nuclear receptors and mediate the transcription of specific genes.

- Responses to these hormones may take hours or several days to become apparent, reflecting the time needed for transcription and translation to produce the necessary changes in protein expression in the target cell.

Steroid hormone signalling

- Steroid hormones are extremely hydrophobic and are transported to target cells in the blood bound to specific carrier proteins.
- At their targets, they are able to diffuse directly into the cytoplasm upon release.
- Within the cytoplasm, their fate depends upon the nature of the hormone.
- Glucocorticoid receptors are normally found in the cytosol bound to a complex of heat shock proteins (Hsps).

➔ *Heat shock proteins are revised in Chapters 5, 10, and 11.*

- On entering the cytoplasm, a glucocorticoid, such as cortisol, binds to its receptor and induces a conformational change that releases the Hsps (Figure 15.24).
- This exposes a nuclear locating signal (Chapter 7) and the hormone-receptor complex is transported into the nucleus, where the hormone-receptor binds to specific transcription regulatory elements in the DNA called **glucocorticoid response elements** (**GREs**), which facilitates the binding of a second hormone-receptor to the same GRE.
- The resulting dimer is able to activate the transcription of the adjacent genes as described in Chapter 6.

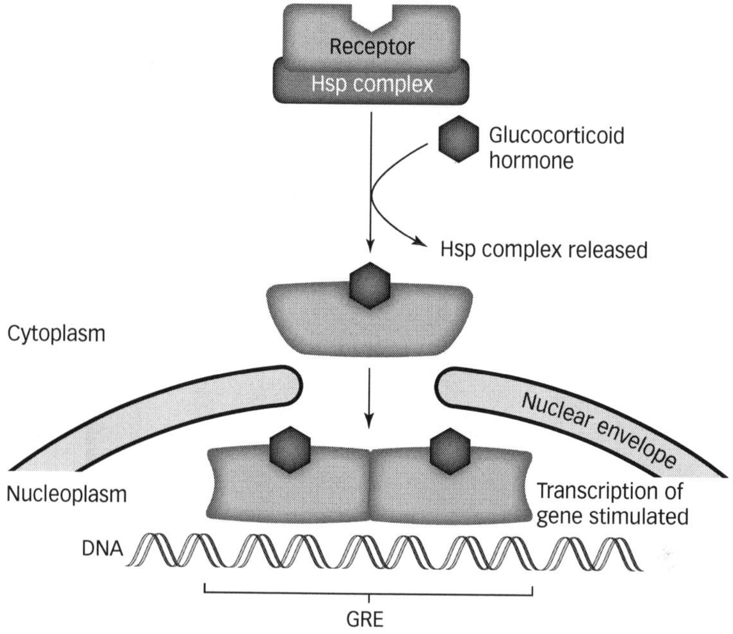

Figure 15.24 Glucocorticoid signalling pathway. GRE, glucocorticoid response element and R, receptor. The events are described in the main text.

- The target cell response to steroid hormones is usually a stimulated increase in transcription. However, in some cases they inhibit transcription.
- For example, the dimer described previously binds to one type of element and activates transcription, but the binding of hormone-receptor to an inhibitory type of GRE does *not* lead to dimerization.
- Instead it recruits **histone deacetylase**, whose activity promotes chromatin condensation (Chapter 6) and inhibits transcription.

Thyroid hormones and retinoic acid

- Thyroid hormones and retinoic acid are lipophilic and directly cross the plasma membrane, enter the cell and migrate into the nucleus.
- Their receptors have accessible nuclear-located signals and so mainly reside in the nucleus.
- Binding of a hormone or retinoic acid to its receptor displaces the Hsps allowing the hormone-receptor to bind to its response element and mediate transcription.

Check your understanding

15.7 Which of the following statements is/are false?

a. Receptors for cytokines and growth factors are generally monomers that dimerize on binding of the cytokine/growth factor.

b. Janus protein kinase or JAK is a membrane-bound enzyme.
c. Steroid hormones are extremely hydrophobic and are transported to target cells in the blood bound to specific carrier proteins.
d. All steroid hormone-receptor complexes bind to specific transcription regulatory elements called glucocorticoid response elements.
e. In the absence of ligand, receptors for steroid and thyroid hormones, and for retinoic acid occur as complexes bound to heat shock proteins.

15.5 NEURONAL SIGNALLING

Communication by neurons (nerve cells) consists of both long- and short-distance electrical signals and short-distance chemical messengers.

- A **membrane potential** consisting of a potential difference is found across the plasma membranes of all cells due to an uneven distribution of ions on either side of the membrane.
- The membrane potential is due largely to the unequal distribution of Na^+ and K^+ generated by the Na^+/K^+-ATPase.

➔ *The role of the Na^+/K^+-ATPase in generating a membrane potential is revised in Chapter 3.*

- The value of the membrane potential is usually −70 to −90 mV, with the interior of the cell being positive with respect to the external environment.
- The membrane potential of most cells does not vary with time, but the value of the membrane potential in excitable cells such as neurons, muscle, and sensory cells can be temporarily altered in response to a given stimulus by the sequential opening and closing of Na^+ and K^+ channels.
- This opening and closing generates a *local* **action potential** of approximately +30 mV.
- The transmission of this localized action potential along the membrane is the basis of the electrical signals called **nerve impulses**.

Neurons and nerve impulses

Neurons consist of a cell body from which processes called **dendrites** and an **axon** extend (Figure 15.25).

- Dendrites receive information in the form of nerve impulses from *pre*synaptic cells.
- Axons pass the receipt of this information on to *post*synaptic cell(s). These may be other neurons or other types of targets such as muscle cells.

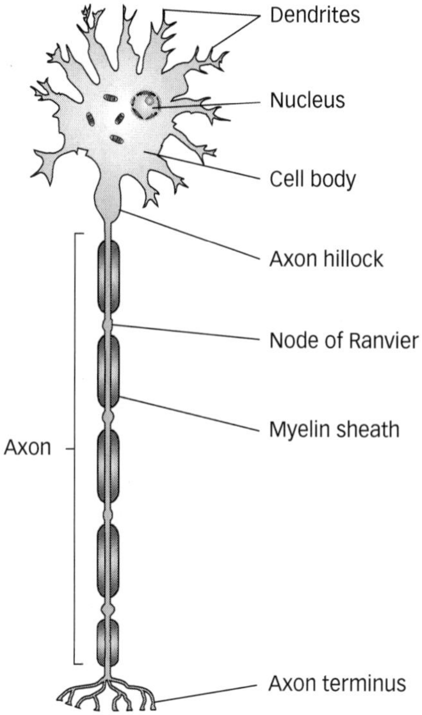

Figure 15.25 Schematic illustration of a typical neuron.

Action potentials

- An action potential is triggered by a stimulus sufficiently large to depolarize the plasma membrane of a neuron beyond a threshold voltage (Figure 15.26).
- Action potentials are triggered at the axon hillock (Figure 15.25) because this region of the neuron has the lowest threshold.
- In resting neurons, the plasma membrane has many open voltage-gated K^+ channels but most voltage-gated Na^+ channels are closed.
- When stimulated, the membrane potential changes from its resting state to the threshold level of approximately −55 to −40 mV, which results in a conformational change in the Na^+ channels of the membrane.
- This change opens the Na^+ channel and allows Na^+ to flow into the cell; this influx further depolarizes the membrane causing more Na^+ channels to open and produces a local excess of positive charges in the cytoplasm.
- The result is a change (depolarization) in the membrane potential from its resting state of −70 mV to one of +30 mV (inside positive).
- Action potentials are transient because the voltage-gated K^+ channels open, allowing the K^+ ions to flow out of the cell and restore the negative resting potential.

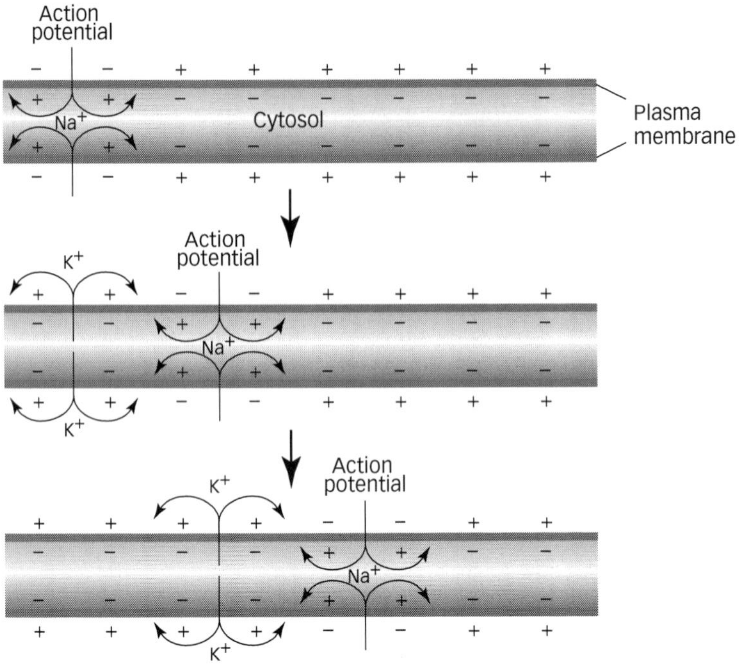

Figure 15.26 Propagation of the action potential along an axon. The action potential is travelling in a left to right direction as Na^+ cross the membrane causing a local depolarization, which is corrected by an outflow of K^+. Once a nerve impulse has passed along the axon, another cannot be transmitted until after the refractory period.

- The movements of Na^+ and K^+ that effect the changes in membrane potential are extremely small and overall concentrations of ions are hardly affected.
- The action potential is produced at a single point on the membrane and is rapidly transmitted along the axon because depolarization causes neighbouring Na^+ channels to open.
- This propagation of the action potential is the nerve impulse.
- In principle, the action potential could be transmitted along the axon in both directions, but channels that have just closed cannot open again for a time termed the **refractory period** (Figure 15.26) such that nerve impulses only travel towards the synapse.

Speeds of transmission

- In non-myelinated axons nerve impulses travel at 0.5–2 ms^{-1} although the speed does increase with the diameter of the axon.
- In many vertebrates, axons are insulated by a myelin sheath and action potentials jump between the nodes of Ranvier (Figure 15.25) in a process called **saltatory conduction**. This can increase the speed of transmission to over 100 ms^{-1}.

Looking for extra marks?

Guanidinium group-containing toxins, such as **saxitoxin** produced by marine dinoflagellates can be isolated from shellfish that feed on them. The toxin selectively blocks the voltage-gated Na^+ channel of neurons without affecting the K^+ channels. It thus specifically inhibits Na^+ channels and prevents the generation of nerve impulses. Saxitoxin is used as a neurotoxin in experimental neurophysiology. In contrast, the alkaloid poison, **veratridine**, produces its effects by binding to a different site on the voltage-gated Na^+ channel to saxitoxin, and fixes the channel in an open conformation, which leads to a permanent depolarization of the axon.

Communications between neurons and other cells at synapses

The electrical information in action potentials is transmitted to other cells at synapses—points of contact between one neuron and a second or between it and a target cell.

- The action potential must cross the synapse if it is to be propagated in the next neuron or stimulate effects in other types of target cells.
- The two main types of synapses are:

   1. chemical synapses
   2. electrical synapses.

Chemical synapses

- The majority of synapses are chemical synapses.
- Information transfer across them involves the release of a neurotransmitter by the presynaptic neuron into the synaptic gap or cleft.
- The axon ends at a nerve terminal where neurotransmitters are stored in synaptic vesicles for release into the synaptic cleft (Figure 15.27).
- The arrival of an action potential at a synaptic terminal depolarizes the plasma membrane, opening voltage-gated channels that allow Ca^{2+} to diffuse into the presynaptic terminal.
- The resulting increase in the concentration of Ca^{2+} in the terminal promotes the fusion of the synaptic vesicles with the terminal plasma membrane releasing the neurotransmitter into the synaptic gap by exocytosis (Chapter 9).
- The neurotransmitter then diffuses across the synaptic gap and can initiate an action potential in the post-synaptic neuron or triggers a response in other types of target cells by binding to a receptor.

Check your understanding

15.8 Are the effects of neurotransmitters autocrine, paracrine, or endocrine?

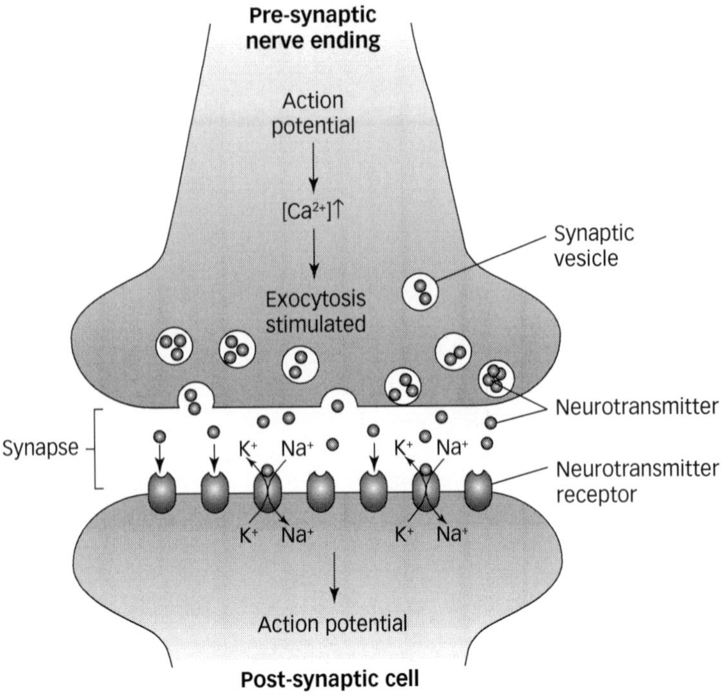

Figure 15.27 Synaptic chemical signalling, which facilitates the transmission of a nerve impulse from a neuron to its target cell.

- Receptors can be:
  - ligand-gated channels
  - GPCRs.
- In the former, the receptor is a transmembrane protein capable of forming a channel through the membrane.
- G protein-coupled receptors do not form channels but can communicate through a G protein with a channel-forming protein.
- The transmission of nerve impulses by ligand-gated ion channels is faster than that by G protein-coupled receptors, but the responses to G protein-coupled receptors are more versatile: the same neurotransmitter can bind to different types of receptors, activate different types of G proteins, and so open different channels or produce cell-specific effects.

Nicotinic acetylcholine receptor

- The neurotransmitter acetylcholine binds to nicotinic acetylcholine receptors. These transmembrane proteins in the post-synaptic cell plasma membrane form acetylcholine-gated channels (Figures 15.10 and 15.28).
- The nicotinic channel comprises five subunits, $\alpha_2\beta\gamma\delta$; the two α subunits have specific acetylcholine binding sites, as described earlier.

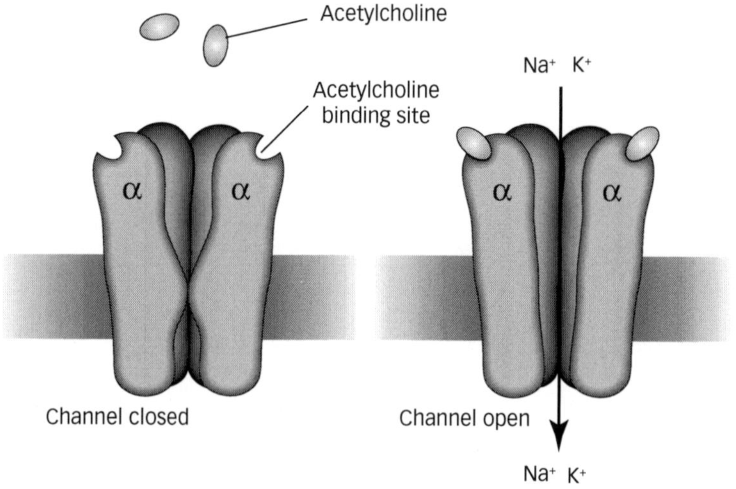

Figure 15.28 Schematic representation of the effects of nicotinic acetylcholine binding to the acetylcholine-gated channels.

- Binding of acetylcholine to the receptor results in a conformational change that opens a pore through the membrane.
- The resting membrane potential favours an influx of Na^+, resulting in a local depolarization and propagation of the nerve impulse.
- The system is desensitized by the closing of the channel within about 1 ms irrespective of whether acetylcholine is bound, and the catalysed hydrolysis of acetylcholine in the synapse by acetylcholinesterase.

$$(CH_3)_3^+N(CH_2)_2OCOCH_3 + H_2O \xrightarrow{\text{Acetylcholinesterase}} (CH_3)_3^+N(CH_2)_2OH + CH_3COOH$$

Looking for extra marks?

The plant alkaloid drugs, **neostigmine (prostigmine)** and **physostigmine (eserine)** are inhibitors of acetylcholinesterase activity and so enhance cholinergic nervous transmission. They are used to treat the autoimmune disease, myasthenia gravis, a condition associated with a low expression of functional receptors.

Muscarinic acetylcholine receptor

- Muscarinic receptors for acetylcholine are G protein-coupled receptors: binding of acetylcholine to a muscarinic receptor initiates a G protein-based signal transduction pathway.
- This signal transduction results in the opening of a separate channel and propagation of the action potential.

- Different muscarinic receptors for acetylcholine produce different effects in different cells.
- Some neurotransmitter receptors activate signal transduction pathways, which can produce long-lasting changes in postsynaptic cells. For example, binding of acetylcholine to cardiac muscarinic receptors results in the opening of specific K^+ channels; the influx of K^+ slows the frequency of heart muscle contractions.
- Like all G protein-coupled receptor-based systems, stimulation of the channel is stopped by degrading the neurotransmitter and hydrolysis of the G_αGTP to G_αGDP.

Electrical synapses

- Electrical synapses allow the passage of the electrical signal by the direct movement of ions from the cytoplasm of one neuron to that of another through gap junctions (see Chapter 3, Figure 3.28).
- Key features of electrical transmission are that:
  - it is extremely rapid compared to chemical synapses
  - signals are transmitted with no distortion
  - information may be transmitted in both directions.
- Electrical synapses are found in both invertebrates and vertebrates where they synchronize the activity of neurons responsible for some rapid and unvarying behaviours—for example, electrical synapses associated with the giant axons of squids and lobsters facilitate the swift execution of escape responses.

Check your understanding

15.9 Which of the following statements is/are true?

a. An action potential is a brief change in membrane potential involving an initial depolarization followed by rapid return to the normal resting potential.

b. The speed of conduction of an action potential increases with axon diameter.

c. An increase in the cytosolic concentration of Ca^{2+} in the presynaptic neuron stimulates the exocytosis of synaptic vesicles.

d. An action potential travelling along a myelinated axon jumps between the nodes of Ranvier.

e. Chemical propagation of an action potential at a synapse always involves a transmembrane ion channel-forming protein.

Sensory cells and action potentials

Sensory cells, like neurons, are excitable: they can generate action potentials in response to changes in the environment.

- Action potentials are transmitted to the central nervous system where they are interpreted as, for example, visual, auditory, olfactory, gustatory, and tactile sensations.
- The detection of sensory stimuli is accomplished by specialized cells using signal transduction pathways that are essentially similar to those used in responding to chemical signals such as hormones.
- For example:
    - the initial signal, in this case, a change in the environment, is detected because it causes a change in conformation of a molecule
    - this initial signal is then amplified using second messenger and ion-gated channel systems.
- The major differences are that:
    - the responses of several receptors is combined before a message is sent to the central nervous system
    - the sensitivity of the system can be modified in response to continual stimulation.
- It is not possible to revise all sensory stimuli; here only the perception of light, that is vision by retinal rods, will be described.

Rods and the perception of vision

- Rods and cones are cells of the retina specialized for the transduction of light stimuli into nerve impulses:
    - rods respond to dim light but cannot detect colour
    - cones detect colour and bright light.
- The location of rods and cones within the retina of the vertebrate eye is shown in Figure 15.29.
- Retinal rods consist of **outer** and **inner segments**, a **nuclear region**, and a **synaptic body** (Figure 15.30).
- The outer segment contains up to 1000 membranous discs that are rich in **rhodopsin** (Figure 15.31), which is a member of the family of GPCRs (see earlier in this chapter).
- Rhodopsin contains the prosthetic group, **11-*cis*-retinal**, which is the photodetector in vision.
- The discs also contain cyclic guanosine monophosphate (cGMP). In the dark, cGMP binds to **cGMP-gated channels** in the plasma membrane, which ensures the channels remain open, allowing the movement of Na^+ or Ca^{2+} into the cell.
- Thus, the resting membrane potential is determined by the difference between the influx of Na^+ through these channels and the quantities of Na^+ and K^+ transported by the Na^+/K^+-ATPase.
- The inner segment contains numerous mitochondria that supply the ATP necessary for phototransduction. Its plasma membrane contains the Na^+/K^+-ATPase whose activity contributes to the membrane potential.
- The synaptic body is in contact with a sensory neuron of the optic nerve.

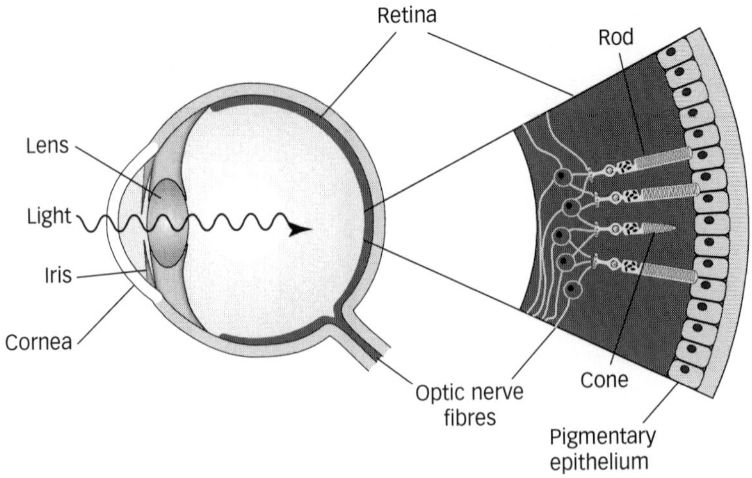

Figure 15.29 Schematic indicating the positions of rods and cones in the retina of the vertebrate eye.

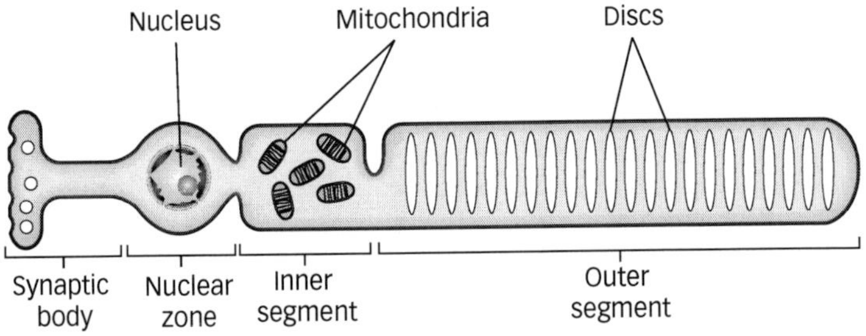

Figure 15.30 Schematic structure of a retinal rod. See text for explanation.

Light perception by rods

- The absorption of a single photon of visible light by a molecule of 11-*cis*-retinal changes its configuration to 11-*trans*-retinal within a few picoseconds (Figure 15.32).
- This change forces a conformational change on the protein portion (an opsin) of rhodopsin.
- This activation allows it to interact with a trimeric G protein on the cytoplasmic face of the disc membrane called **transducin** ($T_\alpha T_\beta T_\gamma$) as shown in Figure 15.33.
- The T_α transducin subunit then exchanges its bound GDP for a GTP and dissociates from the $T_\beta T_\gamma$ portion of the protein.
- The T_αGTP conveys the signal to the next component of the signalling pathway, which is a **cGMP phosphodiesterase** (PDE) specific to rod membranes of retinal cells (Figure 15.33).

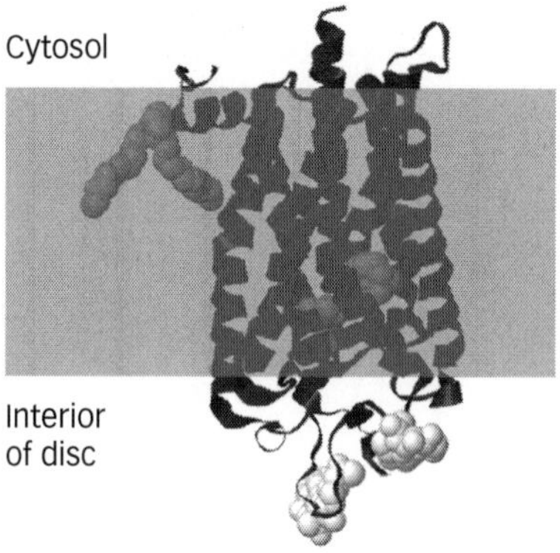

Figure 15.31 Molecular model of rhodopsin. The seven transmembrane helices typical of a G protein-coupled receptor (GPCR) are readily apparent. The retinal prosthetic group is shown in a lighter tone in the centre of the structure. The two attached lipids and two glycans can be seen at the top left and bottom of the structure in the lightest tone. (Rhodopsin is therefore an integral and a lipid-anchored glycoprotein.) Model constructed using PDB file 1HZX.

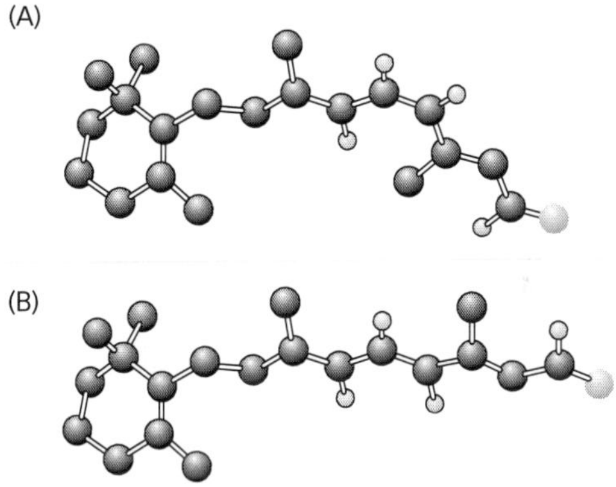

Figure 15.32 Structures of (A) 11-*cis*-retinal and (B) 11-*trans*-retinal.

- In the dark, the PDE is inactive because of a bound inhibitory subunit but T_αGTP is able to displace and bind the inhibitory subunit and so stimulate PDE activity (Figure 15.33).
- The catalytic activity of PDE greatly reduces the cytosolic concentration of cGMP:

$$\text{cGMP} + \text{H}_2\text{O} \xrightarrow{\text{cGMP phosphodiesterase}} \text{GMP}$$

- The decreased concentration of cytosolic cGMP leads to a closure of the cGMP-gated ion channels in the plasma membrane of the outer segment, blocking the entry of Na^+ and Ca^{2+} into the cell such that the plasma membrane becomes *hyper*polarized.

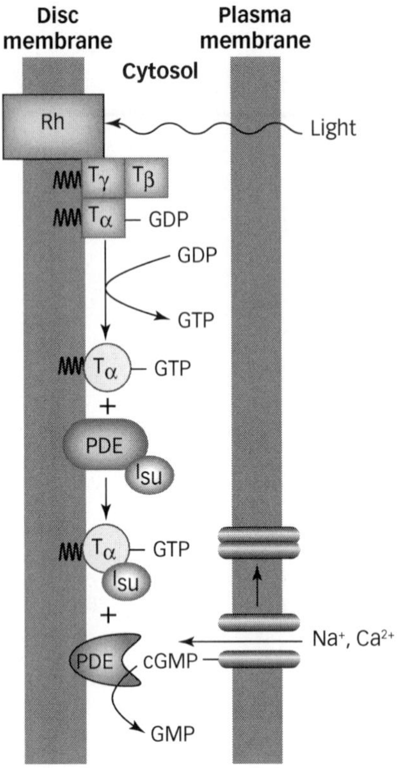

Figure 15.33 Schematic outline of the signalling pathway associated with light perception in retinal rods. I_{SU}, inhibitory subunit; PDE, phosphodiesterase; Rh, rhodopsin. See text for details.

- The change in membrane potential by the light stimulated-hyperpolarization of the rod plasma membrane is about 1 mV.
- This change can be relayed by the synaptic body to sensory neuron where it is propagated as a nerve impulse.
- The initial signal of a single photon is amplified at several stages:
  - one rhodopsin may activate 500 $T_{\alpha}T_{\beta}T_{\gamma}$ molecules
  - the PDE has a turnover number of 4.2×10^2 cGMP s^{-1}
  - binding of cGMP to cGMP-gated channels is cooperative, therefore comparatively small changes in concentration produce relatively large differences in conductance.

Deactivation of the visual signal

- Several mechanisms are involved in deactivating the system and occur extremely rapidly to allow the rhodopsin to respond to another photon.
- The T_{α}GTP hydrolyses its bound nucleotide to GDP, ensuring it releases the inhibitory subunit of PDE.

- **Guanylate cyclase** synthesizes cGMP from GTP and its concentration in the cytosol returns to previous (dark) values.
- Closure of the Na^+ or Ca^{2+} channel during the photon-stimulated response means cytosolic concentrations of Ca^{2+} decrease, leading to its dissociation from the protein **recoverin**.
- Calcium-free recoverin, stimulates guanylate cyclase, and so promotes cGMP formation and the re-opening of the channels.
- Over a refractory period lasting seconds to several minutes, the 11-*trans*-retinal in the activated rhodopsin is replaced by 11-*cis*-retinal.

Changes in sensitivity of light detection

- The sensitivity of the system can adjust to prolonged bright illumination.
- Serine and threonine residues in the carboxyl terminus of rhodopsin are substrates for **rhodopsin kinase**.
- Calcium-recoverin inhibits phosphorylation; however, Ca^{2+}-free recoverin does not.
- Thus, when concentrations of Ca^{2+} decrease following illumination, rhodopsin is phosphorylated forming a binding site for **arrestin 1**, which binds and prevents any interaction between rhodopsin and transducin.
- The brighter the light stimulus the greater the extent of rhodopsin phosphorylation and the lower its affinity for transducin.

Revision tip

Do browse through the further reading references given for this chapter online. Go to http://www.oxfordtextbooks.co.uk/orc/thrive/

Check your understanding

15.10 Which of the following statements is/are true or is/are false?

a. The cytoplasms of adjacent animal and plant cells may be directly connected using plasmodesmata and gap junctions respectively.

b. The effects of most growth factors and cytokines are autocrine or paracrine in action.

c. All eicosanoids are derived from the $C_{18:4}$ fatty acid, arachidonic acid.

d. Many different signalling molecules, including adrenalin, glucagon, and insulin, stimulate their effects through G protein-coupled receptors.

e. Hepatocytes are more responsive to adrenalin than to glucagon.

f. Physiologically, action potentials are triggered at the axon hillock, and can travel in any direction.

g. The muscarinic acetylcholine receptor is an acetylcholine-gated ion channel.
h. The retinal rod cell is rich in rhodopsin and cCMP.
i. The absorption of a single photon of visible light by a molecule of 11-*trans*-retinal changes its configuration to 11-*cis*-retinal within a few picoseconds.
j. Transducin, like all G proteins, is a trimer.

16 Eukaryotic cell cycle, mitosis, and meiosis

Key features of the eukaryotic cell cycle

- The ability to grow and reproduce is a fundamental property of living organisms. In multicellular organisms, cell division either increases the number of cells leading to growth of the organism or replaces cells that have died in tissue renewal.
- The cell cycle is the time between the formation of a cell as the product of a cell division and the subsequent division of that cell itself.
- The cell cycle consists of interphase and mitosis in somatic cells or meiosis in germline cells of gonads, both of which are followed by cytokinesis.
- Not all mitotic divisions are symmetrical; those of a stem cell allow it to self-replicate indefinitely while producing progenitor cells that can differentiate to form specialized somatic cells.
- In mitosis or nuclear division (or meiosis if gamete-producing cells are concerned) the chromosomes are separated.

➔ *Chromosomes are introduced in Chapter 6.*

- Eukaryotic chromosomes consist of chromatin, a complex of DNA and proteins that condenses during cell division (Figure 16.1).

continued

(A)

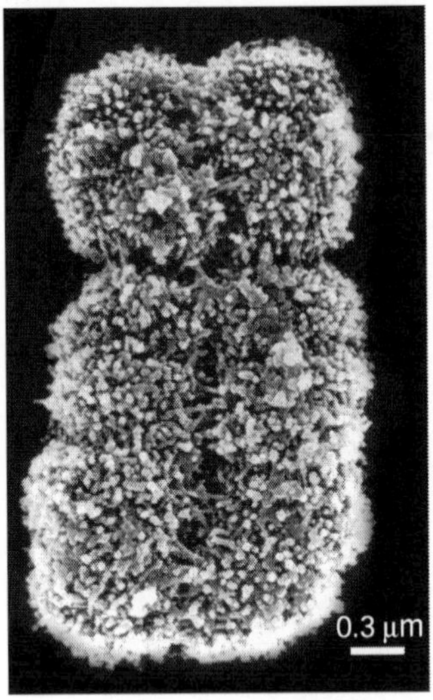

(B)

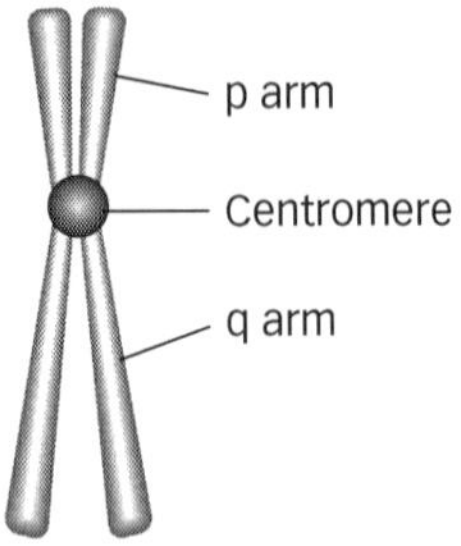

Figure 16.1 (A) Scanning electron micrograph of a eukaryotic chromosome. Courtesy of Dr C.J. Harrison, Christie Hospital, Manchester, UK. (B) Schematic of a typical eukaryotic chromosome showing the relative positions of the long (q) and short (p) arms, and centrosome.

- Eukaryotic somatic (non-reproductive) cells are **diploid** and have two sets of chromosomes. Meiosis is a specialized type of cell division that occurs in the germinative cells of gonads (ovaries or testes) and produces gametes (spermatozoa and ova) that have a haploid chromosome complement.
- Mitosis is followed by cytokinesis, the division of the cytoplasm, when the cell divides into two daughter cells.
- Mitosis/meiosis and cytokinesis form the M phase of the cycle.
- Interphase can be divided into three periods:
  1. gap 1 (G_1)
  2. synthesis (S)
  3. gap 2 (G_2).
- Interphase prepares the cell for division by building up of large energy stores, the synthesis of new organelles, and the growth of the cell. Thus, a high metabolic rate is typical of cells about to undergo division.
- All the nuclear DNA of the parent cell is duplicated prior to mitosis and then equally distributed or segregated to the two new daughter cells.
- The cell cycle is under strict physiological control to ensure its constitutive events take place in the correct sequence and at the appropriate times.
- The key molecules that regulate it are cyclin-dependent kinases, cyclins, and inhibitors of cell cycle.
- In animal cells, interphase lasts about 24 hours but varies considerably between different types of cells. The period of interphase that shows the greatest variation in duration is G_1.

Looking for extra marks?

For single-celled or unicellular organisms, such as bacteria and yeasts, cell division produces new organisms and increases the total number of individuals in a population.

16.1 THE EUKARYOTIC CELL CYCLE

The four phases of the **eukaryotic cell cycle**—G_1, S, G_2, and M—occur generally as a continuous sequence and so are often represented as a circle (Figure 16.2).

- Many cells stop dividing after **mitosis** and enter an extended quiescent period called G_0.
- Durations of cell cycles vary greatly from one cell type to another (Table 16.1).
- In many mammalian cells it lasts 10–30 hours, with a typical time of about 24 hours.
- Typically mammalian G_1, S, and G_2 phases last about 8–10, 10–12, and 4–6 hours respectively.
- The M phase usually lasts approximately an hour in mammalian cells.

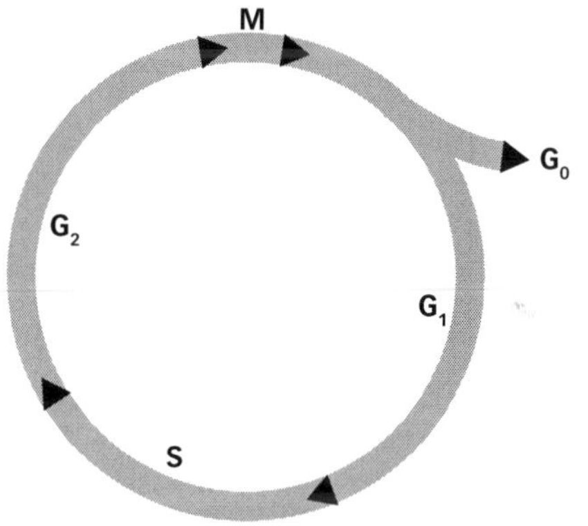

Figure 16.2 An overview of the eukaryotic cell cycle. G_1, first gap; S, synthesis; G_2, second gap; M, mitosis. The first gap, S and G_2 phases constitute interphase. Following mitosis, many cells stop dividing and enter an extended quiescent period called G_0.

Cell type	*Duration of cell cycle (hours)*
[a]Frog embryo (cleavage stage)	0.5
Yeast	1.5–3
[b]Mammalian intestinal epithelial	~12
Cultured mammalian fibroblasts	~20
Human hepatocytes	~1 year
Suspended tobacco (*Nicotiana tabacum*) cells	~12
Primary root tip meristem cells of the common wall cress (*Arabidopsis thaliana*)	~19

Table 16.1 Examples of eukaryotic cell-cycle times

[a] These cells lack G_1 and G_2 phases.
[b] Rapid replacement (proliferation) essential

Interphase

Phases G_1, S, G_2 are called **interphase**, because it is the period between successive mitotic divisions.

- Interphase often accounts for over 90% of the length of the cell cycle.
- In any given somatic tissue, most cells are in interphase; the proportion of cells undergoing mitosis is often reported as its **mitotic index**, the number of cells per 1000 engaged in mitosis at the time of observation.
- Interphase can extend for days, weeks, or longer, depending on the cell type and conditions because many eukaryotes do not divide continuously; instead, they temporarily or permanently enter an arrested quiescent G_0 stage.
- A cell in G_0 must receive a growth-promoting signal if it is to re-enter the cell cycle and proceed from G_0 into G_1 (see later in chapter).

G_1 phase

- In the G_1 phase the cell makes a commitment to divide by the binding of origin recognition complex (ORC) and licensing proteins such as helicase loader proteins and minichromosome maintenance (MCM) helicases to origins of replication (Chapter 7).
- It is also characterized by the synthesis of RNA and protein.

➔ *DNA replication and protein synthesis in eukaryotes are revised in Chapters 6 and 7 respectively.*

- The end of G_1 is characterized by the production of proteins, such as **geminin**, that turn off the DNA replication licensing system by inhibiting the functions of ORC, helicase loader proteins and the MCM helicases.
- Cells in the first cell cycle phase (G_1) do not always continue through the cycle. Instead they may exit from the cell cycle and enter G_0.

S phase

- Nuclear DNA is replicated during the S phase (S = synthesis).
- Additional histones needed by the cell to double the number of nucleosomes in its chromosomes are also synthesized during this period.
- Hence, the chromosome complement is also duplicated during the S phase.
- The centrosome, a microtubule organizing centre, is usually also duplicated at S phase.

➔ *The centrosome and microtubules are revised in Chapter 7.*

G_2 phase

- The G_2 phase is a relatively quiescent period during which organelles are replicated.
- Towards the end of G_2, the extended chromatin fibres of the interphase nucleus begin to condense.

➔ *Condensation of chromatin into compact mitotic chromosomes is revised in Chapter 7.*

- It is followed by the M phase (mitosis and **cytokinesis**) of the cell cycle.

Check your understanding

16.1 Microscopic examination of sections of a somatic tissue gives the results shown in Table 16.2.

Stage of cell cycle	*Number of cells*
Prophase	13
Metaphase	6
Anaphase	3
Telophase	5
Interphase	217

Table 16.2 Numbers of cells in different stages of the cell cycle observed in tissue sections

a. Calculate the mitotic index of a tissue.
b. If the nuclei of this tissue normally contained 30 pg of DNA, how much would be expected in them at the end of S phase?
c. How much DNA per nucleus would be expected in the gametes of the organism from which the tissue was collected?

Mitosis

Mitosis is a major feature of the cell cycle during which chromosomes condense and move apart.

- Mitosis is conventionally divided into four phases (Figure 16.3) called:
  1. prophase
  2. metaphase
  3. anaphase
  4. telophase.
- Some authorities recognize a fifth stage, which they refer to as **prometaphase** but traditionally this has been regarded as *late* prophase.

Revision tip

See the spectacular events of mitosis using contrast microscopy at: http://www.youtube.com/watch?v=DD3IQknCEdc&feature=related

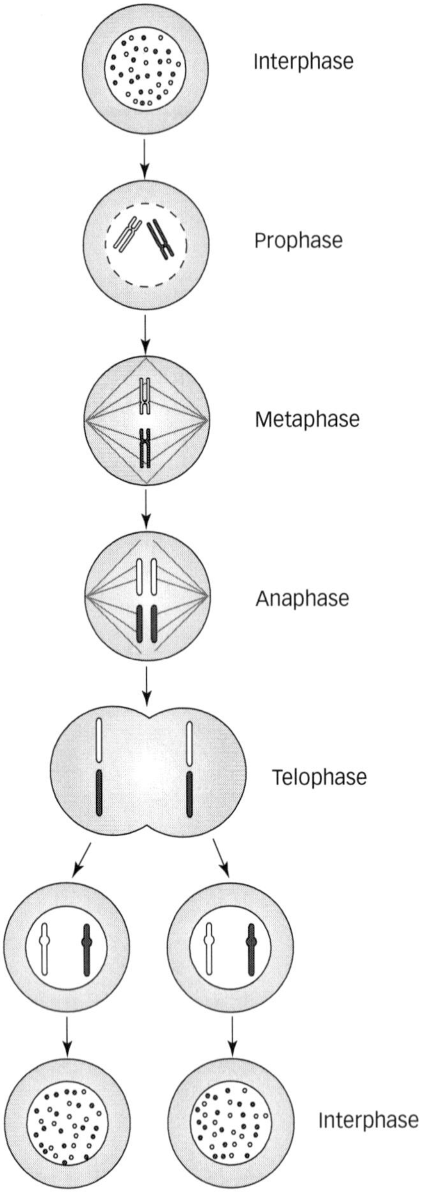

Figure 16.3 Schematic showing the stages of mitosis as outlined in the main text. For simplicity only two chromosomes are shown; each consists of two chromatids.

Prophase

- Prophase is considered to have occurred when the condensation of chromatin fibres that began in G_2 has advanced to a point where individual chromosomes can be observed as discrete structures using light microscopy.

- Each duplicated chromosome appears as two identical sister chromatids joined at their centromeres and, in some species, all along their arms by **cohesion** proteins.
- Meanwhile, the two centrosomes move away from each other towards the poles of the cell on opposite sides of the nucleus.
- In part, their movements are propelled by lengthening microtubules (MTs), which also begin to occupy the space between them to form the **mitotic spindle**.
- The radial arrays of shorter MTs that extend from each centrosome are called **asters** (stars).

Late prophase (prometaphase)

- During late prophase, the nuclear envelope fragments, the nucleoli disappear, and the chromosomes become fully condensed.
- The MTs extending from each centrosome can now invade what was previously the nuclear area and form the mitotic spindle.

Mitotic spindle

- The **mitotic spindle** is composed of centrosomes (now asters) and MTs (Figure 16.4), and controls the movements of chromosomes during mitosis.

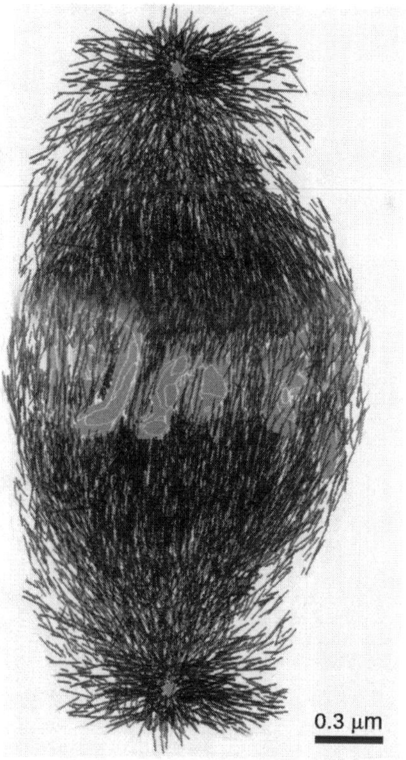

Figure 16.4 A mitotic spindle. Constructed from electron micrographs originally supplied by Dr B. Nicholas, Department of Botany and Zoology, Duke University, North Carolina, USA.

- In animal cells, assembly of spindle microtubules begins in the centrosome, although many cell types, including those of higher plants and many fungi, lack centrosomes.
- In these cells, *chromosomes* induce the formation of the spindle in a Ran-dependent mechanism.

➔ *Ran is a G protein, whose involvement in nucleocytoplasmic transport is revised in Chapter 6.*

- A chromosomal protein stimulates the formation of Ran-GTP.
- In turn, Ran-GTP interacts with importin and promotes the release of importin-bound proteins that stimulate the formation of the spindle.
- Hence in centrosome-deficient cells, the formation of the spindle is not initiated at the cell poles but in the vicinity of chromosomes.
- During late prophase, some spindle MTs attach to parts of the centromeres called **kinetochores.**
- Kinetochores are trilaminar complexes formed of approximately 50 different proteins.
- Each paired set of chromatids has two kinetochores, each associated with one of the chromatids and facing the opposite pole of the cell (Figure 16.5).
- The number of MTs attached to a kinetochore vary, from one in dividing yeast cells to 40–50 in mammalian cells.
- These **kinetochore microtubules** are able to exert forces that agitate the chromosomes in a back and forth manner that gradually draws them towards the centre of the cell.

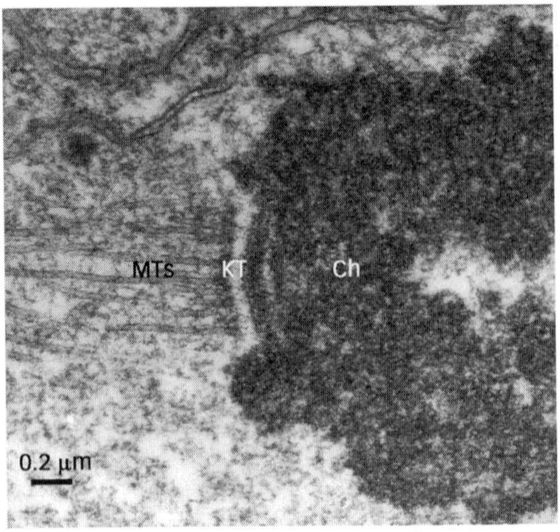

Figure 16.5 Electron micrograph showing the attachment of the spindle to a kinetochore of a chromosome. Ch, chromosome; KT, kinetochore; MTs, spindle microtubules. Courtesy of Dr J.D. Pickett-Heaps, School of Botany, University of Melbourne, Australia.

- Additionally, **polar MTs** extend from the asters towards the centre of the cell where they interact with one another.
- Still other **astral MTs**, attach the asters to the plasma membrane of the cell in the cortex regions.
- Given that the (–) ends of MTs begin in MTOCs, it is the (+) ends of kinetochore, polar, and aster MTs that extend from the asters.

Metaphase

- Metaphase is achieved when the chromosomes are aligned at the **metaphase plate** or **equator**. The plate extends across the centre of the cell and is equidistant from the two poles of the spindle (Figures 16.3 and 16.4).

Anaphase

- Anaphase is the shortest period of mitosis, often lasting only a few minutes.
- It begins when cohesin proteins holding the two sister chromatids of each chromosome pair together are cleaved (see 'M phase checkpoints' section), allowing the chromatids to be pulled apart as their kinetochore microtubules shorten.
- This moves the chromosomes, centromeres first, toward the opposite poles of the cell (see Figure 16.5). This is referred to as anaphase *A* movements.
- Each chromatid is now regarded as a chromosome.
- The lengthening of polar MTs forces the asters apart and causes the cell to elongate. This is referred to as anaphase *B* movements.
- Anaphase A and B movements may occur sequentially or simultaneously, depending upon the type of cell.
- Anaphase is complete when the two poles of the cell have received a complete set of chromosomes.

Mechanisms of anaphase A and B movements

- Figure 16.6 illustrates the mechanisms involved in chromatid separation during anaphase.
- Binding of the (+) ends of kinetochore MTs hampers their growth.
- The lengths of kinetochore MTs is dramatically shortened by their depolymerization at both (+) and (–) ends, which pulls the chromatids apart at the centromeres.
- The depolymerization is stimulated by kinesin-like motor proteins that bind to each end of the kinetochore MT.
- Kinesin motor proteins also bind to pairs of polar MTs in the regions where they overlap (Figure 16.6): the tails of the kinesins are bound to one MT, their heads to the other.
- The ATPase activity of the kinesins forces the pair of MTs to slide along one another, decreasing the length of the overlap and forcing the centromeres apart.
- The action of dynein motors attached to the (+) ends of astral MTs pulls the asters towards the cell cortex and appears to contribute towards their separation (Figure 16.6).

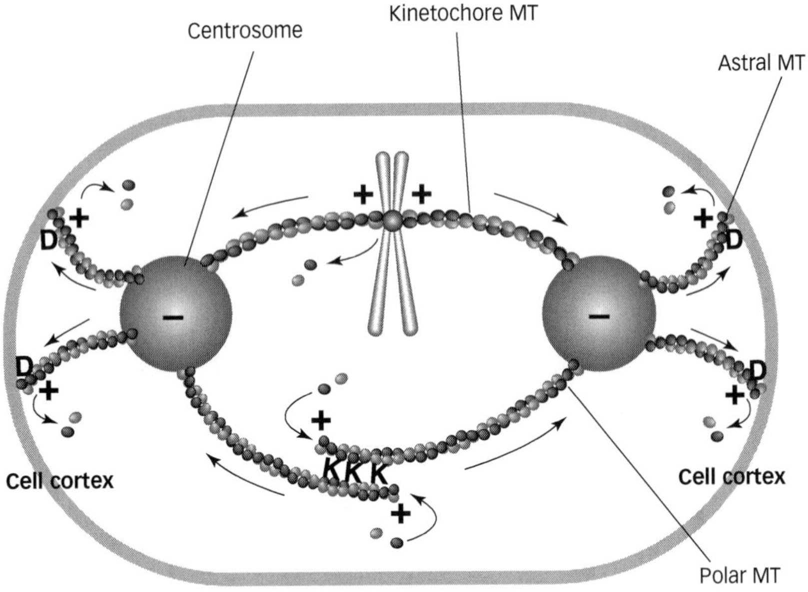

Figure 16.6 Roles of spindle microtubules in anaphase movements. See text for explanations. D, dynein motor proteins; K, kinesin motor proteins; MT, microtubules.

Telophase

- During telophase the same events occur in each of the daughter cells.
- The chromosomes gradually become less condensed and eventually form the euchromatin and heterochromatin typical of the interphase nucleus.
- Nucleoli reappear on nucleoli-organizing regions of nucleolar organizing chromosome.
- In human cells, these initial nucleoli rapidly fuse to give a single larger nucleolus.
- The spindle disappears as its MTs are depolymerized and redistributed in the cell.
- Nuclear envelopes form in each daughter cell.
- The division of one nucleus into two genetically identical nuclei is complete and followed by the remaining M event: cytokinesis.

Cytokinesis

- Cytokinesis, or the process of cytoplasm separation, overlaps with the latter stages of mitosis, beginning during late anaphase or telophase when the spindle disassembles.
- During cytokinesis, the cytoplasm of the parent cell divides.
- In animal cells, cytokinesis occurs by a process known as **cleavage**.
- Initially a **cleavage furrow** forms around the cell (Figure 16.7).
- The cleavage furrow gradually deepens as a ring of actin microfilaments (MFs) called the **contractile ring** begins to contract.

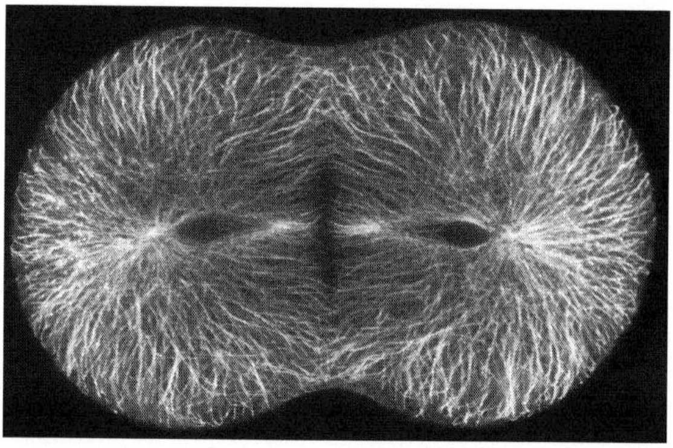

Figure 16.7 Formation of a cleavage furrow in the egg of the green sea urchin, *Strongylocentrotus droebachiensis*. Courtesy of Research Professor V. Foe, Center for Cell Dynamics, University of Washington, USA.

- The ring lies immediately beneath the plasma membrane and parallel to the furrow. As contraction proceeds, it cuts through the equator of the cell and eventually pinches the cell in two.
- Contraction of the microfilaments (MFs) depends upon the actions of myosin motor proteins in catalysing the hydrolysis of ATP.

➔ *The roles of actin microfilaments and motor proteins are revised in Chapter 7.*

- In plant cells, a **phragmoplast** forms across the equator and develops into a **cell plate** that initiates cytokinesis (see Chapter 13).
- Given that cytokinesis occurs at the position previously occupied by the chromosomes at metaphase, it is thought that signals from the central region of the spindle determine where cytokinesis occurs.
- Cytokinesis is usually well established by late telophase, so two discrete daughter cells are formed rapidly following mitosis.
- The completion of telophase produces two genetically identical daughter cells and completes the current cell cycle.

Check your understanding

16.2 Which of the following statements about mitosis is/are true?

a. In prophase, duplicated chromosomes appear as identical pairs of sister chromatids joined together by cohesion proteins.

b. A centrosome is essential to form a mitotic spindle.

c. Mitosis and cytokinesis constitute the nuclear division stage of the cell cycle.

d. During late prophase, astral MTs attach to parts of the centromeres called kinetochores.

e. Cytokinesis is the final event of a round of the cell cycle.

16.2 VARIATIONS ON MITOTIC CELL DIVISION

Variations on mitotic cell division include:

- The variable abilities of cells to grow and divide *in vivo*
- Asymmetric cell division
- Differentiation
- Meiosis.

Abilities of cells to grow and divide *in vivo*

The frequency of cell division varies with the type of cell. The abilities of different types of cells of a multicellular organism to grow and divide fall into three broad categories:

1. Highly specialized cells that lack the ability to divide.
2. Cells that do not normally divide but can be stimulated to begin synthesizing DNA and divide.
3. Cells that essentially divide continuously.

- Examples of the first type include muscle cells and some neurons. Once differentiated, these cells remain in a permanent G_0 until their death.
- Cells that do not normally divide but can be induced to do so include hepatocytes and lymphocytes. Livers cells, for example, can be induced to proliferate by the surgical removal of part of the liver.
- Cells that continuously divide include plant **meristems**.
- Plant *apical* meristems located near the tips of plant roots and stems exhibit rapid and continual cell division and are responsible for elongating roots and shoots. Lateral meristems in cambium are responsible for lateral growth.
- Stem cells of various animal adult tissues, such as haematopoietic stem cells and those at the base of epithelia lining body cavities and covering surfaces, have relatively high levels of mitotic activity.

➔ *The roles of stem cells are revised later in the chapter.*

Asymmetric cell division

Symmetric cytoplasmic division follows the formation of a metaphase plate in the middle of the parental cell and ensures the daughter cells receive approximately equal amounts of cytoplasm. Division can also be **asymmetric** or unequal, and can result in unequal-sized daughter cells (Figure 16.8).

- Asymmetric division occurs in egg formation from oocytes.
- **Meiosis** of the precursor oocyte produces four daughter cells but just one receives most of the parental cytoplasm and forms a mature egg. The other three daughters form polar bodies.
- Asymmetric division is often associated with **differentiation** since the daughter cells can have very different fates even if they are the same size and appear similar.

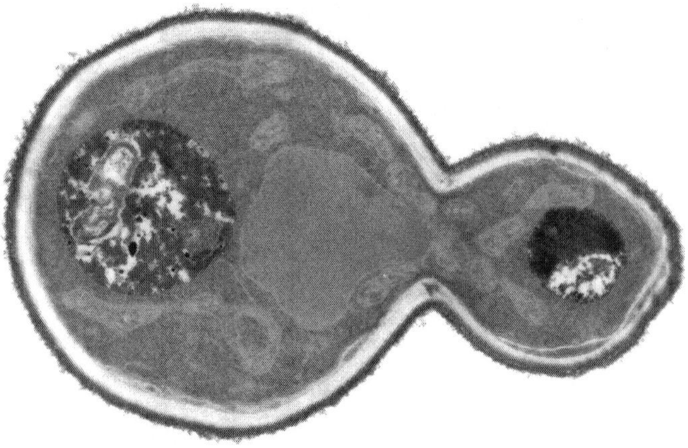

Figure 16.8 Electron micrograph showing asymmetric division in the yeast, *Candida albicans*. Courtesy Drs A. Curry and A. Fox, Public Health Laboratory, Manchester, UK.

Cell differentiation

Cell differentiation is the process by which individual cells or groups of cells acquire specialized ultrastructural features and metabolic functions that distinguish them from other cells in the developing organism.

- Differentiation occurs during **development** and involves mitosis, movements of cells to new locations, and pattern formations, where groups of several types of cells establish and maintain specific spatial relationships to one another.
- In nearly all cases, development begins with a single diploid **zygote** (fertilized egg), which is **totipotent:** it is capable of forming all of the approximately 200 different types of differentiated cells of the adult human body during embryogenesis.
- As development proceeds, the individual cells formed are more and more limited in their ability to differentiate into different cell types.
- With the exception of **germline cells**, which give rise to **gametes** (eggs or ova, and spermatozoa), all the cells of the body are known as **somatic** cells.
- Most somatic cells are differentiated cells and have arisen from unspecialized precursor cells.
- In general, differentiated somatic cells, particularly endothelial cells like hepatocytes, divide and move to a limited extent, a process called proliferative mitosis (Figure 16.9).
- Normally this only occurs when replacing dead cells to maintain the organization of the tissue, or if wounds require repair.
- Some groups of cells are, by contrast, constantly renewed throughout the life of an organism. These include the epidermis of the skin, lining of the gastrointestinal tract (GIT), and blood cells.
- The cells of these tissues are renewed by **stem cells** that remain undifferentiated in the adult.

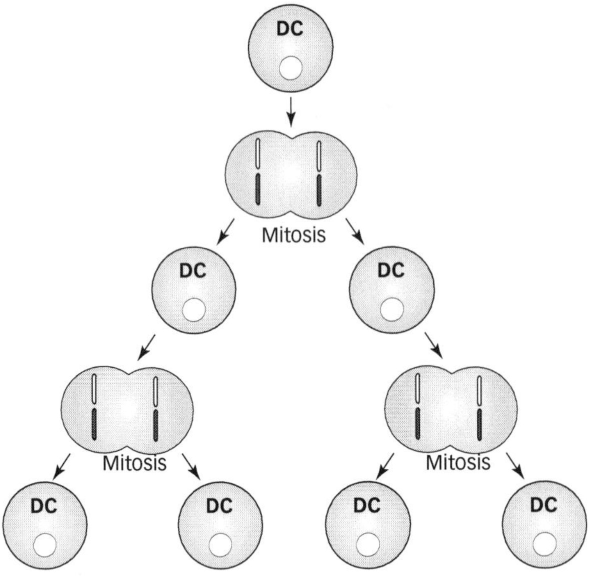

Figure 16.9 Schematic showing proliferative mitosis of a differentiated cell (DC).

Stem cells

- Stem cells are undifferentiated cells that when appropriately stimulated by signalling molecules (Chapter 15) can:
  - undergo an unlimited number of mitotic divisions
  - also give rise to differentiated cells.
- Stem cells maintain their permanent state of proliferation by secreting signals such as **Wnt** (pronounced *wint* and named because their genes are homologous to the *Drosophila wingless* genes and the insertional activation mouse mammary tumour virus gene) **glycoproteins**, together with expressing receptors for these same proteins on their surfaces.
- The binding of a Wnt protein to its receptor initiates an intracellular signalling pathway that stimulates mitosis; hence proliferation is stimulated by a positive feedback mechanism.
- A stem cell is said to be **determined** when it is irreversibly committed to a particular differentiation pathway(s).
- Asymmetric mitosis allows stem cells to simultaneously engage in self-renewal and the formation of differentiated tissue cells: one daughter cell remains a stem cell, like its parent, but the second is a **progenitor cell** that can commit towards becoming a single or one of several kinds of differentiated cells (Figure 16.10).
- **Unipotent** stem cells produce a single type of differentiated cell but divisions of **pluripotent** stem cells result in a small number of different but related differentiated types of cells.
- As differentiation progresses, the individual cells formed have reduced capacities to differentiate into different types of cells.

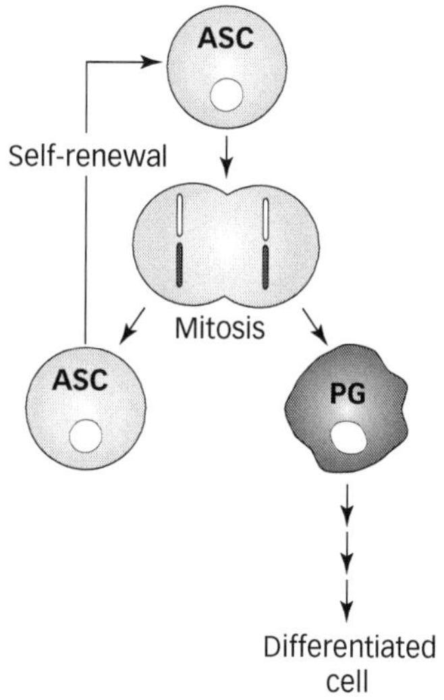

Figure 16.10 Schematic asymmetric mitotic division of an adult stem cell (ASC) to allow its self-renewal and produce a daughter progenitor cell (PG).

Types of stem cells

- Different types of stem cells are found at all stages of development. The two major groups are:
  1. embryonic stem cells
  2. adult stem cells.
- Embryonic stem cells are produced from the fertilized egg during embryogenesis. Following fertilization, a mammalian zygote divides and initially forms a solid mass of cells, the **morula**, which develops to a sphere of cells called a **blastula** or **blastocyst** whose cavity is partially filled with a number of embryonic stem cells (Figure 16.11).
- Embryonic stem cells are pluripotent and capable of developing into all the differentiated cells of the adult (Box 16.1); they are not totipotent because they cannot form the placenta.

Looking for extra marks?

Cancer stem cells are thought to occur in some types of tumours and be responsible for the proliferation of the cancerous cells and tumour growth. Therapies that appear to eliminate the tumour but leave some of these cells alive will allow the tumour to regenerate and patient relapse.

Meiosis

Meiosis or **reduction division** occurs prior to reproduction and results in the formation of daughter cells called gametes (eggs or ova, and spermatozoa).

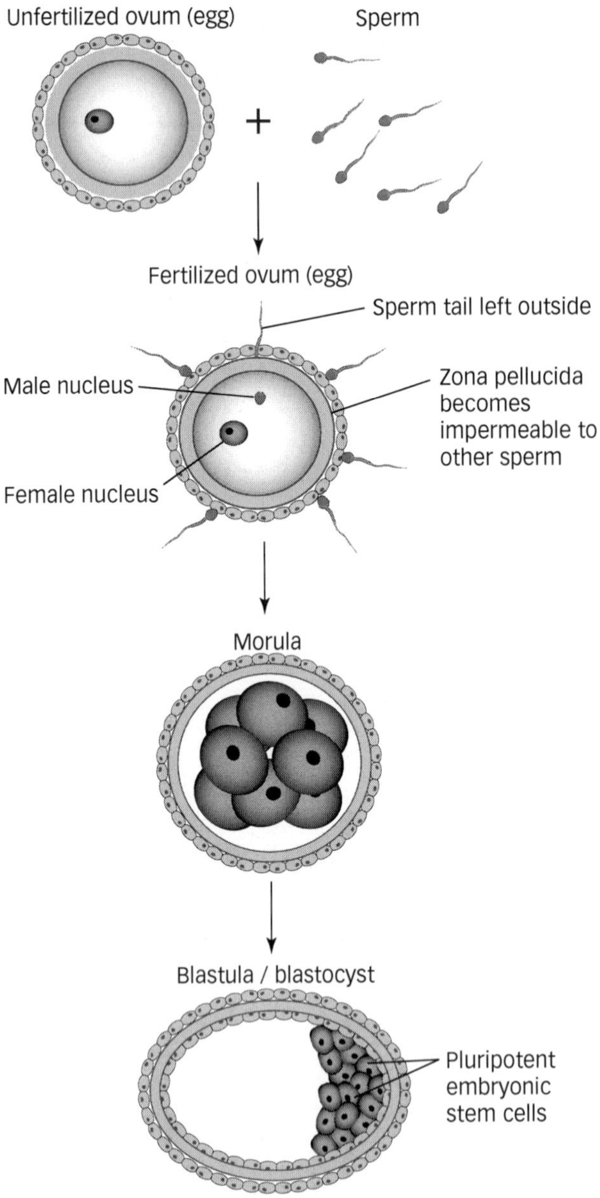

Figure 16.11 Formation of a blastula or blastocyst.

- Meiosis halves the number of chromosomes, resulting in gametes with a haploid number of chromosomes.
- Meiosis is often thought of as two separate but linked cell divisions called **meiosis I** and **II** respectively.

Box 16.1 Embryonic stem cells

There is great clinical interest in embryonic stem cells.

- They can be cultured and propagated in cell culture (Chapter 2) and if they are grown in appropriate conditions they do not differentiate but remain pluripotent.
- If cultured embryonic stem cells are introduced into another early embryo they can develop and populate all tissues of the developing animal including the germline.
- Stem cells therefore have the long-term potential to:
  - correct a range of genetic disorders such as lysosomal storage diseases (Chapter 9)
  - correct a range of neurodegenerative disorders; for example, Alzheimer's and Parkinson's diseases
  - replace defective tissues such as non-insulin producing β cells in the islets of Langerhans of some diabetics
  - provide tissues and organs for use in transplant therapies.
- Adult stem cells are present in many of the tissues of young or adult animals; they are able to divide and provide replacements for a limited range of cell types and are therefore **multipotent** rather than pluripotent.
- There is a limited set of different types of adult stem cells. They include, for example:
  - cells that replace the epidermal tissues of the skin
  - epithelial stem cells that renew the lining of GIT
  - haematopoietic stem cells in bone marrow that provide blood cells.
- The epidermis of the skin is a stratified epithelium (Figure 16.12).
- Unipotent stem cells in the basal germinative layer divide to renew themselves and produce precursor cells that differentiate to keratinocytes.
- Keratinocytes progress through several stages to eventually terminally differentiate into the flat keratin-filled squamae that flake off from the outermost layer of the skin.
- Stem cells found deep in the crypts of Lieberkühn of the GIT are multipotent.
- These stem cells divide and self-renew but also provide precursor cells that form secretory Paneth cells deep in the crypt and others that differentiate into mucus-secreting goblet cells and absorptive cells that move up the crypt toward the lumen.
- Blood cells are constantly replaced by pluripotent **haemopoietic** stem cells in the bone marrow that produce:
  - **lymphoid stem cells** whose progeny form B and T lymphocytes and natural killer cells
  - **myeloid stem cells** capable of giving rise to a range of progenitor cells that differentiate to produce cells that include neutrophils, monocytes, eosinophils, erythrocytes, basophils, mast cells and blood platelets.

continued

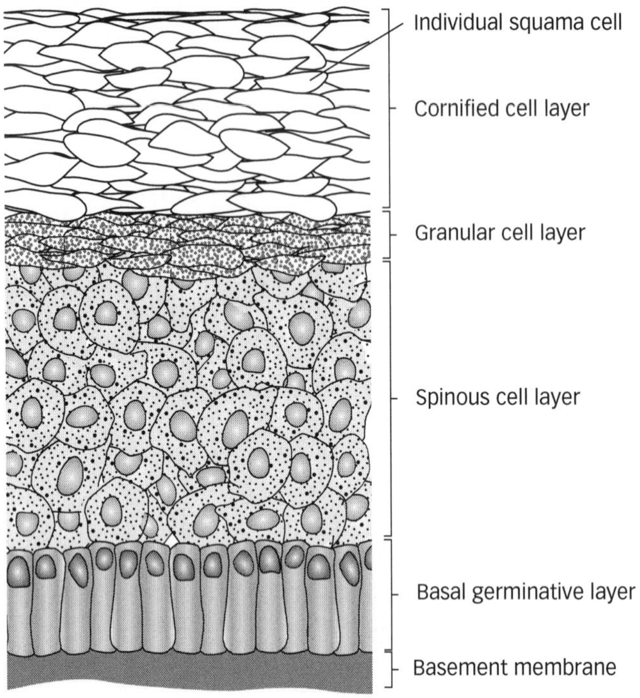

Figure 16.12 Schematic showing the structure of the skin epidermis.

- Division of the parent cell in the first meiotic division forms two daughter cells each with a single copy of each pair of homologous chromosomes.
- Each chromosome in the daughter cell consists of two **chromatids**.
- The second meiotic division separates the chromatids, which then form chromosomes in the daughter cells.
- The net result is *four* daughter cells, each containing a haploid complement of chromosomes.

First meiotic division

- The phases of the first meiotic division are referred to as prophase I to telophase I.
- During prophase I, the chromosomes contract and the nucleolus shrinks in size.
- Homologous chromosomes lie side by side in pairs, a situation called **synapsis**.
- Each member of the pair is called a **bivalent**.
- **Genetic recombination** or **crossing over** occurs (see 'Recombination' section) at synapsis.
- In metaphase I, homologous chromosomes or bivalents align along the equator of the spindle. Sister chromatids are orientated towards the same pole whereas the homologous chromosomes face opposite poles.

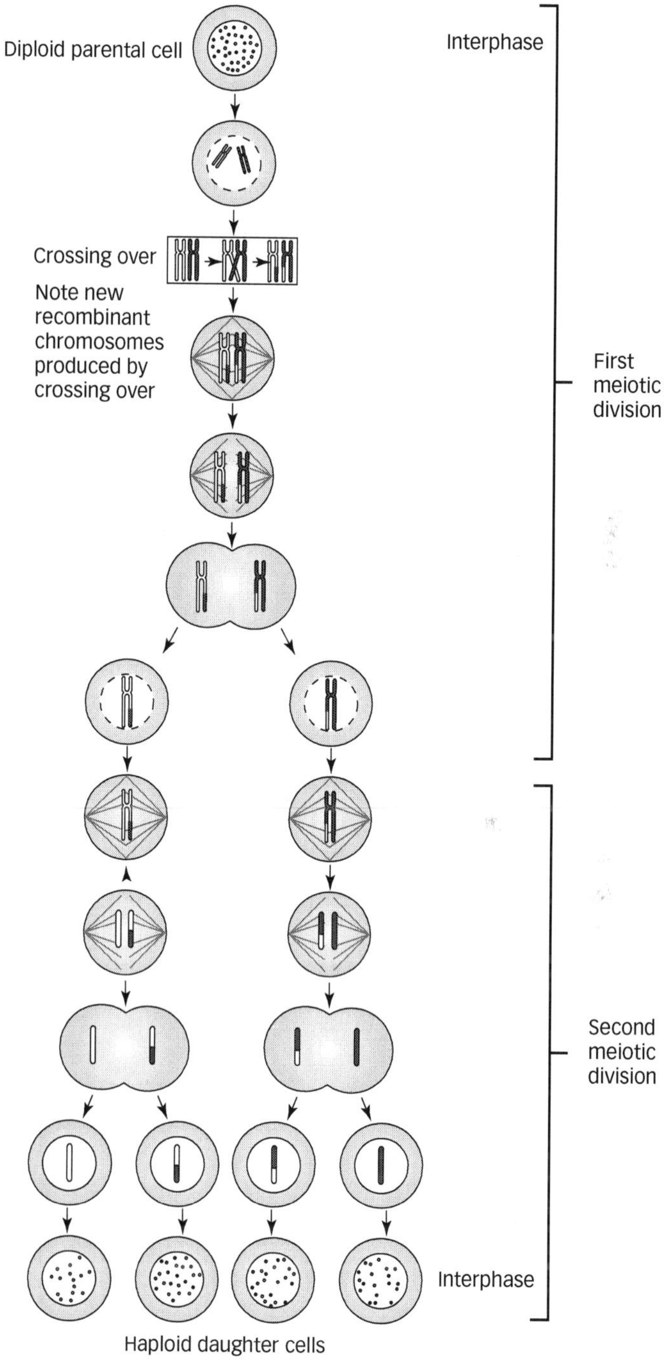

Figure 16.13 Schematic showing the stages of meiosis as outlined in the main text. For simplicity only one pair of homologous chromosomes is shown. Each member consists of two chromatids. Compare with Figure 16.3.

- At anaphase I, the homologous chromosomes (each a pair of chromatids) are pulled by the spindle to opposite poles of the cell.
- During telophase I, the cell divides as in mitosis to give two daughter cells whose chromosome complement consists of paired chromatids.
- Following a brief 'interphase', during which the chromosomes do not usually decondense and which lacks an S phase, the cells enter the second meiotic division.

Second meiotic division

- The phases of the second meiotic division are referred to as prophase II to telophase II.
- During prophase II, the two daughter cells prepare for the second division and form a new spindle.
- At metaphase II, the chromosomes move to the equator of the spindle and the chromatids are oriented to face opposite poles.
- In anaphase II, the chromatids separate from each other and move to opposite poles of the cell.
- Finally, at telophase II, each cell divides to form two daughter cells that have only haploid numbers of chromosomes.
- Thus the parental cell has produced four haploid daughter cells.

Recombination

- While paired together, non-sister chromatids of a homologous pair, one maternal and one paternal (that is *not* the newly synthesized strands), are broken at equivalent positions and exchange homologous pieces of material (Figure 16.13).
- The crossed chromatid structures formed during recombination are called **chiasmata** (singular **chiasma**).
- Recombination produces chromatids with new combinations of genes that differ from the parental types, so the gametes eventually formed also differ from the paternal cells in their gene content.
- Hence, crossing over promotes genetic variation.

Looking for extra marks?

Recombinations in eukaryotic cells do not occur randomly along chromosomes but are concentrated in localized positions commonly called *hot spots*. Typically, hot spots are 1000–2000 bp long and contain a short G–C-rich motif. Hot spots are distributed differently on the chromosomes of humans and chimpanzees despite their 99% identical genome sequences, and it has been suggested this may explain why they have diverged from a recent common ancestor.

Check your understanding

16.3 Which of the following statements is/are true or is/are false?

a. Germline cells such as gametes (eggs or ova, and spermatozoa) are totipotent.

b. Embryonic stem cells are totipotent and capable of developing into all the differentiated cells of the adult.

c. The binding of Wnt glycoproteins to their receptors autostimulates adult stem cells to continuously proliferate.

d. Embryonic stem cells are multipotent but adult stem cells are only pluripotent.

e. Meiosis in oocytes forms four equally sized haploid gametes.

16.3 CONTROL OF THE CELL CYCLE

It is essential that the different phases of the cell cycle are precisely coordinated: phases must follow in the correct sequence and one phase must be completed before the next begins.

- The cell cycle control system is regulated by both external and internal controls.

External regulators of the cell cycle

To move through the cycle and divide, a cell must receive a **mitogenic** or **mitosis-stimulating** signal.

- Typical signals are proteins or large peptides called **growth factors**. For example, platelet-derived growth factor (PDGF) stimulates human fibroblast cells in cell culture to divide.
- Growth factors bind to specific receptors on the cell surface and produce their effects through cell signalling pathways.

➔ *Cell signalling is revised in Chapter 15.*

- *In vivo*, growth factors that affect a cell are usually produced by neighbouring cells and stimulate its division to replace cells that have died. This maintains a constant cell number.
- Other external signals include the amounts of nutrients available to the cell, which must also have grown to an appropriate size.
- Those cells that show density-dependent growth inhibition receive signals in crowded conditions telling them to stop dividing.
- Many animal cell types also exhibit anchorage dependence; they can only divide when appropriate signals tell the cell it is attached to a substratum.

Internal control of the cell cycle

Progression through the cycle is driven by specific chemical signals present in the cytoplasm.

- The cytoplasmic proteins that control the cell cycle belong to three key groups:
  - **cyclin-dependent kinases** (CDKs)
  - **cyclins**
  - **CDK inhibitory proteins**.

CDKs, cyclins, and CDK inhibitory proteins

If a cell receives a mitogenic signal, then specific CDKs are activated appropriately at different stages of the cell cycle.

- However, CDKs are only active when combined with a protein cyclin (Figure 16.14); for example, a cyclin–CDK complex, sometimes called **maturation-promoting complex** (MPC), allows a cell to enter the M phase of the cycle.
- CDKs are a family of proteins, whose different members are identified using numbers (CDK2, 3, …). Cyclins are also a family of proteins and are identified using capital letters (cyclin A, B, …).
- Different cyclins can bind to different CDKs and activate them to phosphorylate a variety of key target enzymes and other proteins, which are appropriately activated or inactivated to drive the cell through different stages of the cycle.
- The control system of the eukaryotic cell involving cyclins and CDKs has been remarkably conserved throughout the evolution of eukaryotes.
- Many of the genes encoding them are universal. For example, the CDK4–cyclin D complex initiates transfer from G_1 to S in all eukaryotes.
- The term *cyclin* was coined because the concentration of these regulatory proteins rises and falls in a regular pattern through each cell cycle (Figure 16.15).

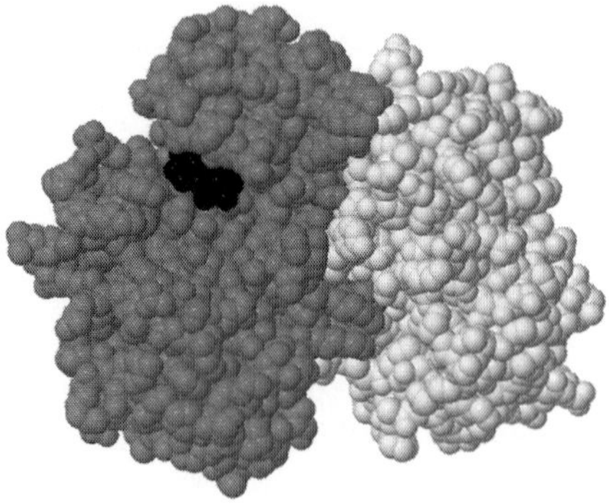

Figure 16.14 Molecular model of a cyclin-dependent kinase–cyclin complex (CDK2–cyclin A). The CDK2 is shown in the darker tone with its bound ATP molecule highlighted in black. The cyclin A portion of the complex is depicted in the lightest tone. Model constructed using PDB file 1FIN.

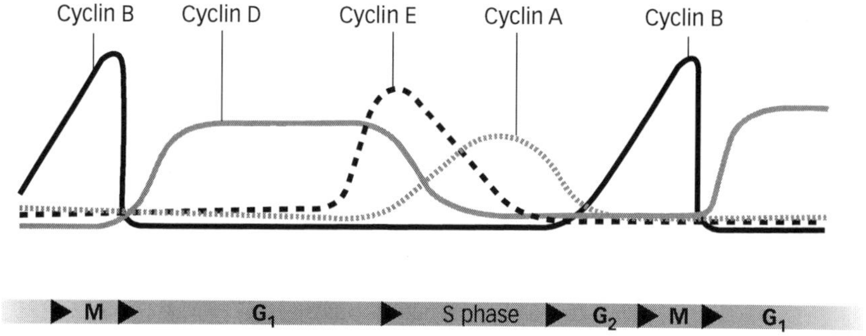

Figure 16.15 Cyclic fluctuations in the concentrations of some cyclins during the phases of the cell cycle.

- Cyclin synthesis is under tight regulation whereby relevant genes are only turned on when cyclins are needed.
- When no longer required they are rapidly degraded by the **ubiquitin-proteasome system**.
- Ubiquitin is a 76-amino acid residue, acidic protein. It is attached to proteins destined for degradation in proteasomes by ubiquitin ligases.
- Proteasomes are large protein complexes (M_r 2×10^6) that hydrolyse proteins earmarked by ubiquitination to form short inactive peptides.
- The rapid synthesis and degradation of cyclins ensures their orderly appearance during the cell cycle.
- CDK inhibitory molecules also comprise a family of molecules, each specific for one type of kinase.
- Internal controls monitor the progress of the cycle at a number of **checkpoints**; a cell can only proceed to the next stage of the cycle when appropriately instructed.

Looking for extra marks?

Loss of control of the cell cycle can lead to uncontrolled cell proliferation and the formation of tumours.

Checkpoints in cell cycle control

The major checkpoints which occur in the cell cycle are:

- at the G_1–S boundary
- at the G_2–M boundary
- multiple DNA damage checkpoints
- M-phase checkpoints.

- Figure 16.16 indicates the positions of some of these checkpoints and a number of the CDK–cyclin complexes associated with each stage.

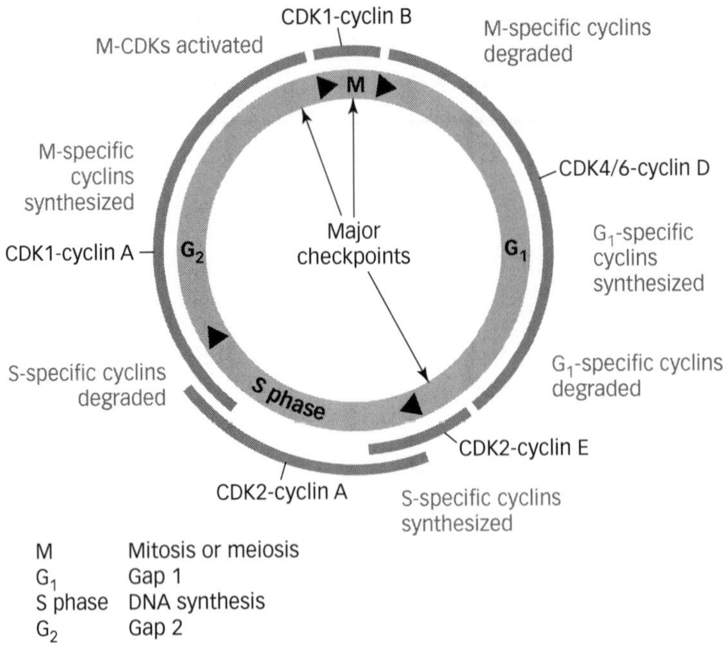

Figure 16.16 Positions of major checkpoints in the cell cycle and a number of the cyclin-dependent kinases (CDK)–cyclin complexes associated with each phase.

- Those CDKs and cyclins required for the cell to traverse the G_1 to S phase transition are collectively called **G_1 CDKs** and **G_1 cyclins** respectively.
- Cyclins associated with the replication of DNA in the S phase are **S cyclins**.
- CDKs and cyclins needed for the cell to traverse the G_2–M transition are called **mitotic CDKs** and **mitotic cyclins** respectively.

G_1–S boundary checkpoint

- For many cells, the G_1–S checkpoint seems to be the most important: mammalian cells that stop dividing nearly always do so in late G_1.
- This late control point is called the **restriction point** in animal cells (and **start** in yeast).
- If a cell receives a signal to pass through this point, it will usually complete the S, G_2, and M phases and divide.
- If the cell does not receive a suitable mitogenic signal, it will exit the cycle and enter the G_0 phase, where it remains until stimulated to re-enter G_1 (see later).
- The G_0–G_1 transition (the reentry into G_1) is regulated by the kinase activities of CDK3-cyclin C.
- Mitogenic signals activate G_1 CDK-cyclins, which catalyse the phosphorylations of target proteins and allows progression through the restriction point.

- A key target of G_1 CDK–cyclins is the **retinoblastoma (RB) protein**, which is an inhibitor of **E2F transcription factor**. Phosphorylation of RB prevents it binding to E2F, which is now able to activate the transcription of genes required for replicating DNA.

➔ *Transcription is revised in Chapters 5 and 6.*

- Given its role in cell cycle progression, mutations in the gene for RB protein have serious clinical consequences.
- Once the G_1–S checkpoint has been passed, G_1 cyclins are degraded by the ubiquitin-proteasome system, S cyclins synthesized and the cell progresses to the S phase.

G_2–M boundary

- Once a cell has committed to the S phase it moves through to the next control point at the end of G_2 called the **DNA replication checkpoint**. This checkpoint ensures that only cells with completely replicated DNA are allowed to enter the M phase.
- S phase cyclins are degraded at the end of S phase and mitotic cyclins produced during G_2.
- The MPF mentioned earlier is a mitotic CDK–cyclin complex that initiates the entry of the cell into mitosis, hence its alternative name of **M-phase-promoting factor**.
- The CDK of MPF is present at a relatively constant concentration throughout the cycle, but the concentration of its cyclin component is gradually increased throughout G_1 to the end of G_2.
- At the end of G_2, MPF is itself activated by a series of phosphorylations and dephosphorylations as shown in Figure 16.17.

DNA damage checkpoints

- DNA damage checkpoints are found at the end of G_1, during S, and in late G_2 phases.
- If these detect damaged DNA, the relevant CDK–cyclin complexes are inhibited and the cell is prevented from proceeding to the next stage of the cycle until the damage is repaired. Such inhibition is necessary because replication of damaged DNA could result in mutated daughter cells.
- Double-stranded breaks in damaged DNA activate a complex called the **ataxia telangiectasia mutated (ATM) protein kinase** (named from the clinical condition, ataxia telangiectasia).
- Single-stranded breaks in DNA activate ATR (ATM-related) protein kinases, which catalyse the phosphorylation of a key cell protein, **protein 53 (p53)**.
- Protein 53 is a transcription factor but is normally rendered inactive by MDM2 (mouse/murine double minute2), a ubiquitin ligase that targets it for degradation by proteasomes.

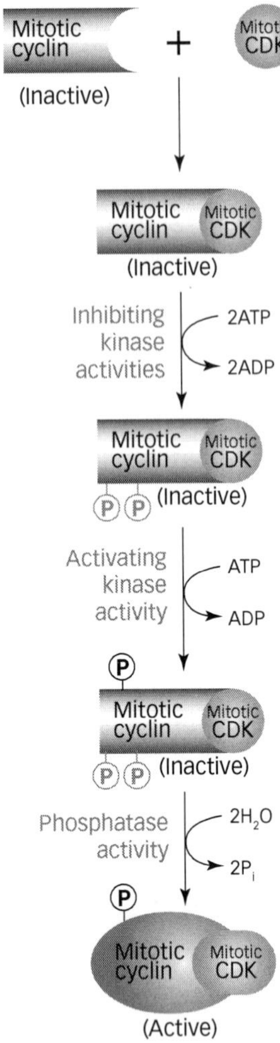

Figure 16.17 Activation of mitotic cyclin-dependent kinase (CDK)–cyclin (maturation-promoting factor, MPF) by its phosphorylation and dephosphorylation.

- Phosphorylation of p53 prevents its destruction and allows it to bind to DNA (Figure 16.18) and activate a variety of genes.
- One of the eventual products of p53-induced transcription is **protein 21 (p21)**, which is able to inhibit the activities of a number of CDK–cyclins and so stop progression through the cell cycle at several different points.
- If the damaged DNA cannot be repaired, p53 activates a group of genes that code for proteins which promote the death of the cell by a process called **apoptosis**.

➔ *Apoptosis is revised in Chapter 17.*

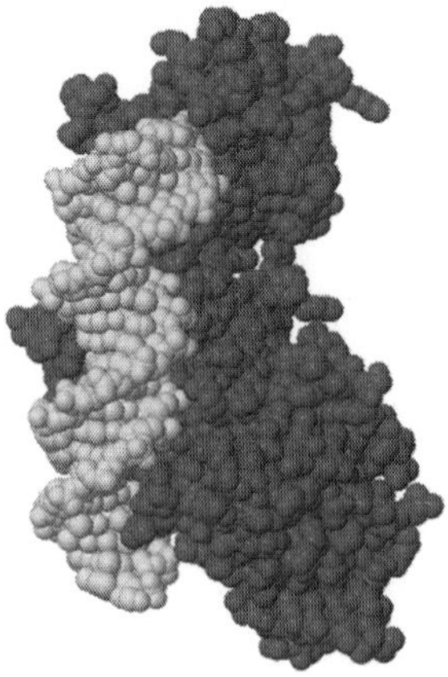

Figure 16.18 Molecular model of a p53-DNA complex. The p53 is shown in the darker tone. Model constructed using PDB file 1TUP.

Looking for extra marks?

The properties of p53 of arresting the cell cycle or initiating apoptosis means it has a key role in protecting the genome by preventing the replication of mutated DNA and its inheritance by daughter cells. For this reason it has been called the 'guardian of the genome'.

Check your understanding

16.4 Which of the following statements is/are true?

a. E2F and p53 are transcription factors.

b. A major checkpoint in the cell cycle occurs at the G_1–M boundary.

c. Retinoblastoma (RB) protein is an inhibitor of p53 transcription activities.

d. Ataxia telangiectasia mutated (ATM) protein kinase is a CDK.

e. One of the products of p53-induced transcription is p21, which promotes cell death by apoptosis.

➔ *DNA repair mechanisms can be revised using the companion volume,* Thrive in Biochemistry and Molecular Biology.

M phase checkpoints

- One M checkpoint is the **mitotic spindle checkpoint** which ensures that mitosis cannot progress until the spindle is fully organized and all the chromosomes are attached to it.

- Those chromosomes whose kinetochores are not attached to spindle MTs send a signal that inhibits a multiprotein complex called the **anaphase-promoting complex**, which functions as an ubiquitin ligase when active.
- Chromosome separation cannot occur until the cohesins holding the pairs of chromosomes together are degraded by the protease, **separase**.
- Separase activity is inhibited by another protein, **securin**.
- Once all the kinetochores are appropriately attached to the spindle, the anaphase-promoting complex inhibitory signal is no longer sent and it is able to target securin for proteasome destruction.
- The removal of its inhibitor allows separase to attack the cohesins and facilitate their separation during anaphase.
- Geminin is also degraded by the anaphase-promoting complex during the metaphase–anaphase transition, which allows the licensing system to become active and promote the succeeding cell cycle.
- Approximately halfway through mitosis, mitotic cyclin is rapidly degraded, which prevents a second mitotic phase until further molecules of it are formed in the next cell cycle.
- Once the M phase (mitosis and cytokinesis) is complete, all cyclins are degraded by the ubiquitin-proteasome complex, which abolishes all CDK activities.
- For the cell(s) to enter a new round of the cell cycle it must now receive a mitogenic signal to activate cyclin biosynthesis.

Looking for extra marks?

The Nobel Prize in Physiology or Medicine in 2001 was awarded jointly to Hartwell, Hunt, and Nurse for their studies on key regulators of the cell cycle.

Check your understanding

16.5 Which of the following statements is/are false?

a. If a cell receives a mitogenic signal, then specific cyclin-dependent kinases are appropriately activated at different stages of the cell cycle.

b. Cyclin-dependent kinases are rapidly degraded by the ubiquitin-proteasome system during control of the cell cycle.

c. Anaphase-promoting complex and MDM2 are both ubiquitin ligases.

d. The late G_1–S checkpoint in animal cells is called start.

e. Maturation promoting factor is an S CDK-cyclin.

16.4 EVOLUTION OF MITOSIS

Mitosis is a universal occurrence in eukaryotic organisms suggesting that it arose at the base of the eukaryotic evolutionary stem (see Figure 1.4).

- It is generally assumed that the earliest eukaryotes were single-celled, haploid, and propagated by mitosis.
- Thus, diploid chromosome complements must have evolved by cell fusion or endomitosis mutations.
- Since bacteria evolved before eukaryotes, and given the involvement of tubulin-like proteins in binary division (Chapter 5), it is probable that mitosis evolved from binary fission.
- Indeed, some single-celled organisms show interesting intermediate stages between binary division and eukaryotic cell division.
- For example, in dinoflagellates, the condensed chromosomes are attached to the nuclear envelope, which remains intact during division.
- The nucleus subsequently divides in a manner resembling binary division.
- A further evolutionary development is seen in diatoms and some yeasts where although the nuclear envelope still remains intact, a spindle forms *within* it, and which separates the chromosomes. The nucleus then divides to give two daughter nuclei.
- The evolution of meiosis has yet to be satisfactorily explained, however, it would seem a reasonable premise that it evolved by modifications to mitosis.

Revision tip

Do browse through the further reading references given for this chapter online. Go to http://www.oxfordtextbooks.co.uk/orc/thrive/

Check your understanding

16.6 Which of the following statements is/are true or is/are false?

a. Eukaryotic chromosomes consist of a complex of DNA and proteins called chromatin and condense during cell division.

b. The cell cycle in somatic cells is the time between successive cell divisions and consists of interphase followed by meiosis and cytokinesis.

c. Mitosis consists of the S and M phases.

d. Wnt glycoproteins form receptors on the surfaces of adult stem cells.

e. To move through the cycle and divide a cell must receive a mitogenic signal.

f. Maturation-promoting factor (MPF) is a cyclin that is able to induce mitosis.

g. Cyclin-dependent kinases only ever require the binding of a cyclin to become active enzymes.

h. Retinoblastoma (RB) protein is a transcription factor.

i. Once all the kinetochores are appropriately attached to the spindle, the anaphase-promoting complex is able to target separase for proteasome destruction.

j. Variations in mitotic separations observed in different organisms suggest mitosis evolved from binary division.

17 Cell death

Key features of cell death

- Cell numbers in multicellular organisms are tightly regulated by:
  - the control of mitosis by mitogens and growth factors that stimulate cell division and growth (Chapter 16)
  - the genetically programmed death of damaged, diseased, or superfluous cells.
- In animals, the commonest form of genetically programmed cell death is apoptosis.
- Necrosis is the death of damaged cells due to tissue lesions.

17.1 APOPTOSIS AND NECROSIS

Apoptosis is a highly regulated, genetically programmed process that eliminates unwanted, dysfunctional, or diseased cells by allowing them to self-degrade and die. **Necrosis**, also called **accidental cell death**, can be caused, for example, by a chemical or physical assault to the cell or tissue.

- Necrosis leads to swelling of the cell and, ultimately, the disruption of the plasma membrane and death.

- Organelles such as the mitochondria or the nucleus remain intact throughout and are released into the surrounding tissue.
- Necrosis usually involves large areas of tissue and is often associated with inflammation.

Apoptosis

Apoptosis, unlike necrosis, usually occurs in single, selected cells.

- During its early stages, DNA becomes segregated near the inner face of the nuclear envelope (Chapter 6).
- The cell shrinks and the fusion of cytoplasmic vesicles with the plasma membrane forms small bubble-like extensions or blebs on the cell surface (Figure 17.1).
- Proteases called **caspases** (see 'Caspases and apoptosis' section) degrade the cell into smaller fragments called **apoptotic bodies** (Figure 17.2).
- DNA condensation and its degradation by an apoptotic-specific endonuclease is one of the last steps before the membrane starts to bleb.
- Activation of a flippase transfers phosphatidylserine from the inner to the outer leaflet of the plasma membrane (Chapter 3) where its accumulation stimulates neighbouring cells, especially macrophages and dendritic cells of the immune system, to ingest and degrade the remnants of the apoptotic cell.
- Thus, the contents of apoptotic cells are retained within the plasma membrane and are not released to stimulate inflammatory responses.
- Additionally, the secretion of cytokines such as interleukin 10 (IL-10) and transforming growth factor β (TGF-β) by the phagocytic cells inhibit inflammation.

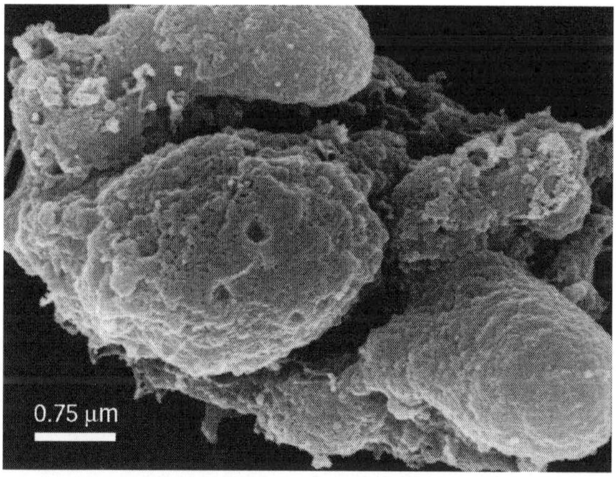

Figure 17.1 Scanning electron micrograph showing the extensions or blebs on the surface of the plasma membrane that are a characteristic of apoptotic cells. Courtesy of Professors S.M. Bowen and I.D. Bowen.

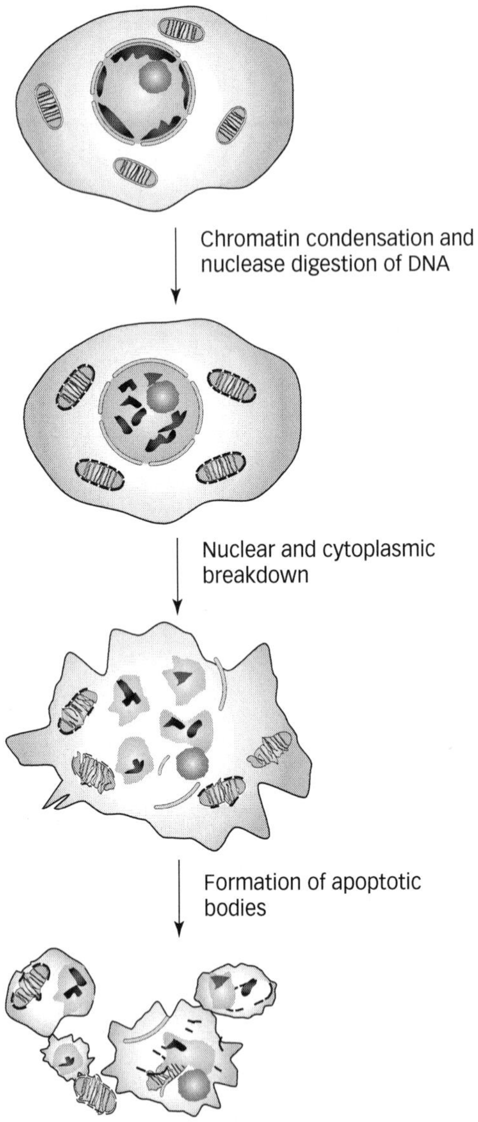

Figure 17.2 Schematic illustration of the major stages of apoptosis.

> ***Looking for extra marks?***
>
> **Terminal differentiation** can also be a form of genetically programmed death. It occurs when specific genes are activated that lead to a loss of mitotic capacity and the synthesis of new types of macromolecules; the cell eventually dies without any of the signs of apoptosis. Examples include the formation of dead cornified keratinocytes of skin epidermis and water-conducting xylem vessels of plants.

Caspases and apoptosis

Caspases are proteases containing a cysteine residue in their active sites, which hydrolyse peptide bonds at specific aspartate residues in their substrates.

- They are produced as inactive **procaspases** that are activated only when needed.
- Two groups of caspases are involved in apoptosis:
  - **initiators** such as caspase-2, -8, -9, -10
  - *effectors* or *executioners* such as caspase-3, -6, -7.
- Initiator procaspases are activated when they are recruited into multiprotein complexes in apoptotic pathways.
- Executioner procaspases are typically activated by their proteolysis at specific aspartate residues by an already active caspase (Figure 17.3).
- The procaspases are hydrolysed at two positions to release unwanted prodomains and form large and small subunits of a heterodimer, which combine to form an active tetrameric caspase.

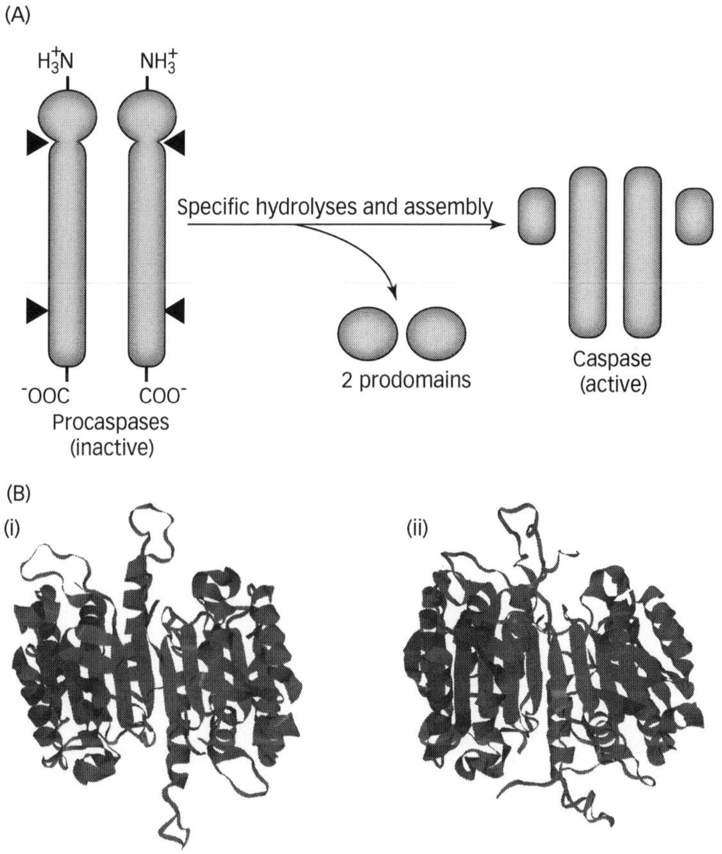

Figure 17.3 (A) Activation of a procaspase. See text for details. (B) Molecular models of (i) procaspase-7 and (ii) caspase-7. Models constructed using PDB files 1K88 and 1K86 respectively.

- Following activation, the caspase can hydrolyse and activate other procaspases in a proteolytic **cascade** that amplifies the proteolytic response to the original apoptotic signal.

Looking for extra marks?

The use of cascade mechanisms to amplify a response is a common feature in biology. Examples include signal transduction mechanisms, which are revised in Chapter 15.

Revision tip

A number of excellent videos showing different aspects of apoptosis can be viewed at: http://www.dnatube.com/search/?search_type=search_videos&search_id=apoptosis&image.x=33&image.y=17

Check your understanding

17.1 Which of the following statements is/are false?

a. Necrosis is associated with the death of single selected cells.
b. DNA condensation, cell shrinkage, and blebs of the plasma membrane are characteristics of an apoptotic cell.
c. During apoptosis, a flippase transfers phosphatidylserine from the outer to the inner leaflet of the plasma membrane.
d. Initiator caspases activate *effector* or *executioner* procaspases.
e. Apoptosis is an unnatural form of cell death.

17.2 ROLES OF APOPTOSIS

Apoptosis is a natural part of the biology of multicellular organisms.

- For example, it is involved in:
  - growth and development
  - maintaining homeostatic cell numbers
  - eliminating cells that are diseased or have been damaged beyond repair.

Apoptosis and development

Apoptosis is involved in embryonic development in multicellular organisms. For example:

- It can greatly affect the morphology of the body during development.
- It eliminates many cells of the immune system during its maturation.

- It is essential in ensuring appropriate neuronal connections are formed during tissue and organ formation.

Morphological development

- During fetal and embryonic development, programmed cell death eliminates tissues that are no longer required.
- Human embryos initially have webbed fingers and toes. As the embryo develops, cells in the webbing die and the fingers and toes become fully separated.
- If the cells in the interdigital spaces fail to undergo apoptosis, the fetus will be born with webbed hands and/or webbed feet, a condition called **syndactyly**.
- **Endochondral ossification** is the replacement of cartilage, formed by chondrocytes, with bone. The process begins inside the cartilage and spreads outwards and occurs in two stages:
  - Firstly, the hypertrophy and death of the chondrocytes by apoptosis
  - Secondly, blood capillaries, and undifferentiated mesenchymal cells penetrate the spaces left by the chondrocytes where they differentiate into osteoblasts that deposit bone tissue that replaces the cartilage.
- In frog development, metamorphosis of a tadpole to an adult frog involves the progressive diminution of the tail by apoptosis until it finally disappears.
- Cells that are eliminated by apoptosis, like those of a tadpole's tail, are typically only required at one stage of development.

Immune system

- Apoptosis is responsible for eliminating many cells of the immune system during their development.
- During their maturation, the immune system eliminates B and T lymphocytes that could produce autoimmune responses and selects those most suitable for immune defence.
- Only 10% of potential B lymphocytes developing in bone marrow reach full maturity; the other 90% are eliminated by apoptosis at various stages in their development.
  - Defects in the gene rearrangements that occur in pre-B cells precipitate the apoptotic death of the cell.
  - Immature B cells that express immunoglobulin M (IgM) and react with self-antigens in the bone marrow are stimulated to undergo apoptosis.
  - Mature B lymphocytes released into the blood only mature into plasma or memory cells if they bind to an appropriate antigen, otherwise they die by apoptosis within a week.
- Mature T lymphocytes are produced with receptors capable of binding to only one antigen but most immature T lymphocytes or thymocytes do not mature: 90–95% of them die by apoptosis.

Roles of apoptosis

Apoptosis and neuron connections

- During the development of chordate embryos many more neurons than will eventually be needed grow out from the central nervous system to innervate peripheral tissues and organs.
- Neurons that reach their destinations receive appropriate signals, such as nerve growth factor (NGF), from their target cells that allows them to survive.
- Over 50% of neurons do not make suitable connections and fail to receive such signals and soon after their formation are cleared from the brain by apoptosis.
- This seemingly wasteful system ensures tissues that require the most nervous connections receive them because they secrete the most NGF.

Maintaining cell numbers

Apoptotic death is not restricted to embryonic development. Mature organisms of constant weight maintain a fixed (homeostatic) number of cells by balancing the rates of mitosis and apoptosis.

- The number of cells dying daily by apoptosis in the human body is approximately 10^{10}–10^{11}, which is appropriately compensated by mitosis.
- The balance between the two can be illustrated by considering liver cell numbers.
    - Hepatocytes normally need divide only once or twice a year to maintain homeostatic numbers.
    - If the size of the liver is reduced by injury or surgery, the remaining cells secrete hepatocyte growth factor that stimulates regeneration of the organ.
    - Conversely, treatment with phenobarbitone stimulates mitosis and excessive growth in a normal liver.
    - Cessation of treatment allows an increased rate of apoptosis to reduce the liver to its normal size.

Check your understanding

17.2 What range of cells (as a percentage) is killed by apoptosis in a human body containing 10^{13}–10^{14} cells if 10^{10}–10^{11} are eliminated daily?

Eliminating damaged or diseased cells

Badly damaged or diseased cells can be eliminated by apoptosis.

- Cells whose DNA is damaged beyond repair cannot be allowed to enter mitosis.
- In such cases, the transcription factor, p53 (Chapter 16) initiates the synthesis of proteins such as the **p53 upregulated modulator of apoptosis** (**Puma**) that promotes apoptosis by binding to anti-apoptotic proteins (see Section 17.3).

- Diseased cells, such as those infected with a virus are induced to die by apoptosis and so prevent the spread of the virus to other cells.

17.3 INDUCING APOPTOSIS

The balance in favour of apoptosis as opposed to cell survival is generally induced in one of three major ways:

1. Certain types of physical or physiological stresses.
2. The receipt of death signals from other cells that induce apoptosis.
3. Lapses in the receipt of **survival** or **trophic** factors from other cells that prevent apoptosis.

Apoptosis following physical and physiological stresses

Stresses that trigger apoptosis by initiating an **intrinsic apoptotic pathway** include:

- irreparable DNA damage
- antibiotics
- heat shock
- short-term anoxia
- excessive production of reactive oxygen species (Chapters 10 to 12)
- high concentrations of cytosolic Ca^{2+}
- viral infections.

- Such factors can stimulate a cell to enter apoptosis by altering the balance between members of a family of intracellular proteins called **B-cell lymphoma gene 2** or **Bcl2 proteins**.

Apoptosis and Bcl2 proteins

- The Bcl2 proteins collectively contain up to four **Bcl2 homology domains** (**BH1** to **BH4**) that allow them to be divided into three groups which prevent or promote apoptosis (Figure 17.4).
- *Anti-apoptotic* proteins such as Bcl2 (Figure 17.5), which gives the group its name, and Bcl-X_L possess all four domains.
- *Pro-apoptotic* proteins consist of BH123 proteins such as Bax and Bak, and BH3-only proteins (for example, Puma, Bad, and Bid (Figure 17.4)).
- Pro- and anti-apoptotic Bcl2 proteins are able to combine as heterodimers, in which each member inhibits the other's functions.

The intrinsic apoptotic pathway

- In healthy cells, pro-apoptotic Bcl2 proteins are absent or inhibited but if the cell receives a suitable stress signal, the balance is shifted in their favour and the intrinsic apoptotic pathway is initiated.

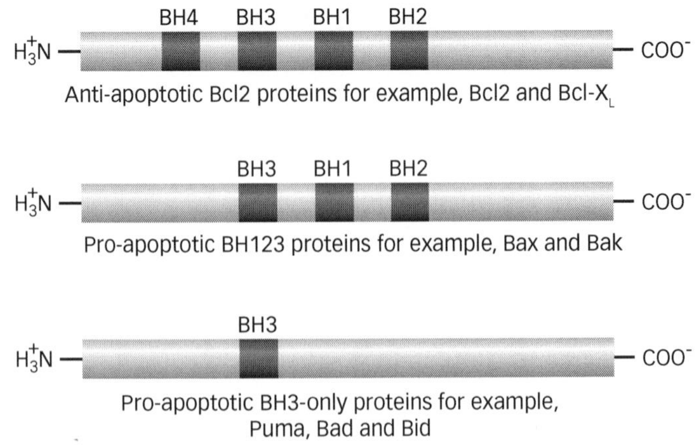

Figure 17.4 Schematic structure of Bcl2 proteins.

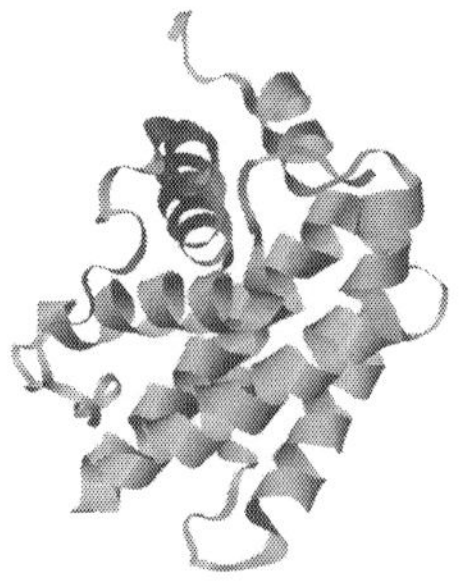

Figure 17.5 Molecular model of Bcl2. Model constructed using PDB file 1GJH.

- Different types of stresses act through different types of BH3-only proteins, which then inhibit the actions of anti-apoptotic proteins such as Bcl2 and Bcl-X_L and stimulate those of pro-apoptotic ones such as Bax and Bak.
- Active Bax undergoes conformational changes that are thought to convert its extended carboxyl terminal region into transmembrane domains, which become inserted into the outer mitochondrial membrane; Bak is associated with this membrane permanently.
- The proteins assemble to form multisubunit channels through the membrane, which facilitates the release of cytochrome *c* into the cytosol.
- Cytochrome *c* is normally found only in the intermembrane space of the mitochondrial envelope and is transiently associated with the outer surface of the inner mitochondrial membrane, where it normally plays a key role in electron transport (Chapter 11).
- Its release irreversibly commits the cell to apoptosis (Figure 17.6).
- In the cytosol, cytochrome *c* combines with factors such as **apoptotic protease activating factors** (**Apaf-1**) and procaspase-9 to form a complex called an **apoptosome** (Figures 17.6 and 17.7).

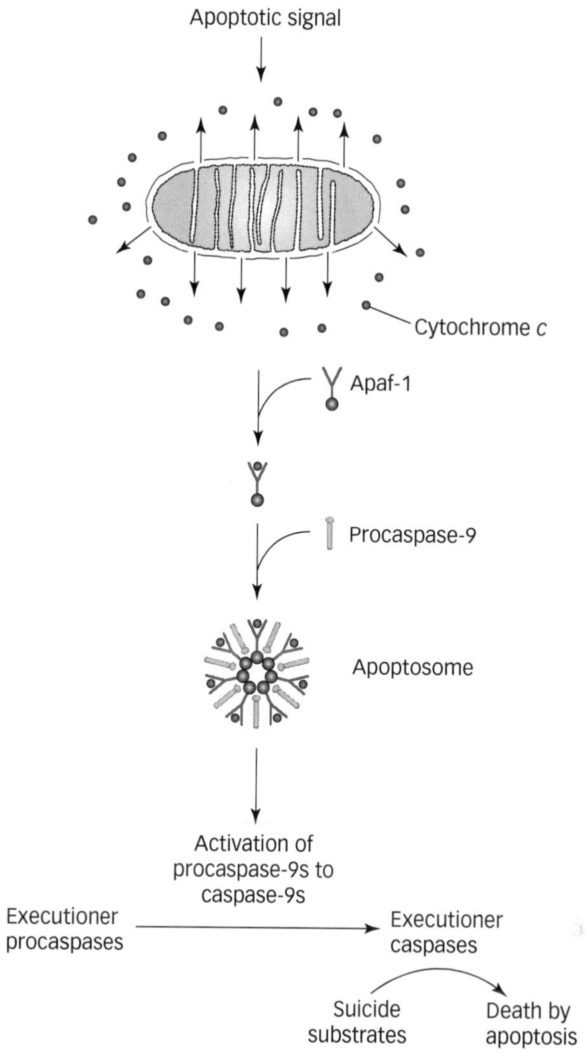

Figure 17.6 Schematic of the intrinsic apoptotic pathway. See text for details. Apaf-1, apoptotic protease activating factor.

- Procaspase-9 is an initiator caspase. It is activated upon joining the apoptosome and catalyses the proteolytic activation of procaspases, such as procaspase-3, -6, and -7, to effector or executioner caspases that initiates cell degradation through the cascade effect described earlier.
- Substrate targets for effector caspases include:
    - structural proteins such as those of the cytoskeleton and nuclear lamins leading to changes in the shape of and break up of the nucleus and cytoplasm
    - proteins concerned with regulating the cell cycle and repairing DNA

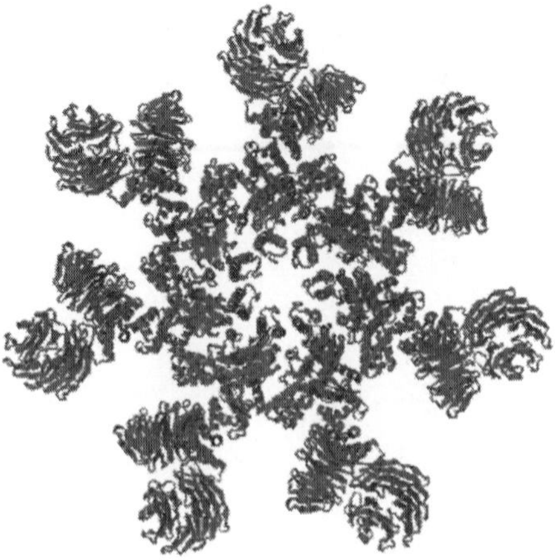

Figure 17.7 Molecular model of an apoptosome. Model constructed using PDB file 3IZA.

- a protein inhibitor of the caspase activated DNase (CAD), which hydrolyses DNA into regularly sized fragments (Box 17.1)
- integrins with loss of cell–cell interactions
- proinflammatory cytokines.

Box 17.1 Apoptotic DNA fragmentation

When apoptosis is induced, the normally high M_r nuclear DNA is catalytically hydrolysed by caspase activated DNase (CAD) into fragments of low M_r. Electrophoresis of the DNA then forms a so-called DNA ladder that in laboratory studies is used to detect apoptosis (Figure 17.8).

Figure 17.8 An electrophoresis agarose gel showing the separation of DNA from a normal cell (lane 1) and the characteristic DNA ladder of an apoptotic cell (lane 2).

Apoptosis following death signals

Apoptosis can be triggered in a cell by instructions transmitted to it by extracellular signal molecules and executed by internal responses that constitute an **extrinsic apoptotic pathway**.

- The extrinsic pathway is initiated in cells possessing **death receptors**; these are integral transmembrane proteins containing:
  - an extracellular ligand binding site
  - a single transmembrane domain
  - a cytosolic death domain.
- Target cells are stimulated to enter apoptosis by the binding of another cell with ligands on *its* surface specific to its death receptors.
- Cells possessing the death receptor **Fas** are induced to enter apoptosis when killer cells, such as cytotoxic T lymphocytes and natural killer cells (CTLs and NKCs), possessing **Fas ligands** (**FasLs**) on their surfaces bind to them (Figure 17.9).
- Binding of FasLs to three Fas receptors on the target cell induces conformational changes in their death domains, which allows the adaptor protein, **Fas-associated death domain protein** (**FADD**) to be recruited to the receptors (Figure 17.9).
- FADD is able to recruit two procaspase-8/-10 molecules by the mutual association of death effector domains to form a **death-inducing signal complex** (**DISC**).
- The low latent protease activities of the procaspases can now mutually activate each other forming an active initiator caspase-8/-10 molecule (Figure 17.9).
- Caspase-8/10 catalyses the conversion of executioner procaspases to active executioners that begins cell degradation.

Tumour necrosis factor

- **Tumour necrosis factor** (**TNF**) or **cachectin** is a soluble signalling molecule that can initiate the extrinsic apoptotic pathway.
- It is produced by cells of the immune system as an immune response and binds to the death receptor, TNF receptor 1 (TNFR1).
- Binding induces the formation of DISC, followed by the activation of the initiator procaspase-8/10, and then executioner caspase cascade.

Natural killer cells and apoptosis

- Natural killer cells can induce apoptosis in virus-infected and cancer cells (Figure 17.10).
- Healthy cells express **killer inhibitory receptors** (**KIRs**) on their surfaces. If an NKC binds to these, the negative response protects the cell.
- However, unhealthy cells express **killer activation receptors** (**KARs**). Binding of an NKC to KARs induces it to kill the cell by two mechanisms.
  1. Natural KCs express FasLs and so can induce apoptosis through the DISC pathway outlined earlier.

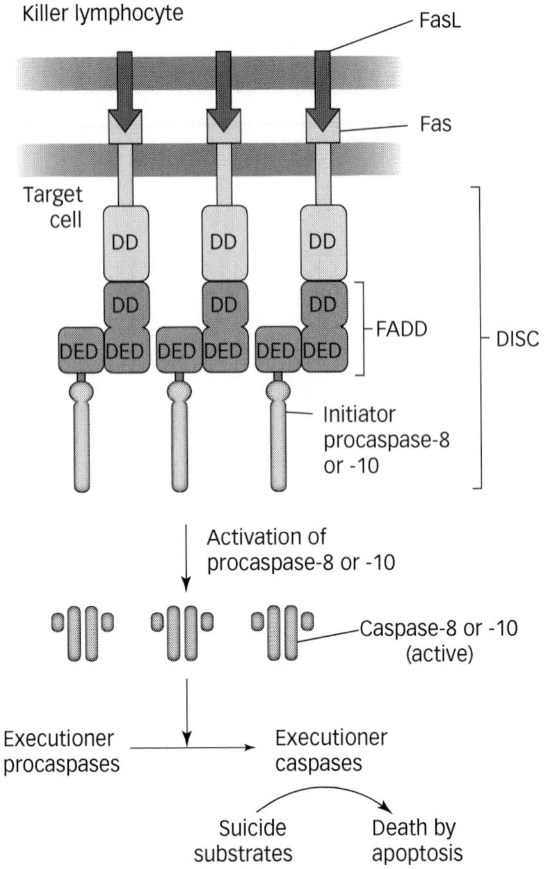

Figure 17.9 Schematic of the extrinsic apoptotic pathway. See text for details. DD, death domain; DED, death effector domains; DISC, death-inducing signal complex; FADD, Fas-associated death domain protein; FasL, Fas ligand.

2. They also secrete **perforins** and **granzymes**. Perforins form pores of approximately 5 nm in diameter in the plasma membrane of the target cell, through which the proteolytic granzymes can enter the cell.

- Granzymes can directly hydrolyse and so activate caspases, leading to apoptosis.
- Given cytoplasm can also leak from the target cell through the perforin pores, death may well be by a combination of necrosis and apoptosis!

The extrinsic and intrinsic apoptotic pathways are linked

The activities of the extrinsic and intrinsic apoptotic pathways need not operate independently but can be linked and amplify the killing response.

- In some cells the extrinsic pathway stimulates the intrinsic one via the BH3-only protein, Bid.

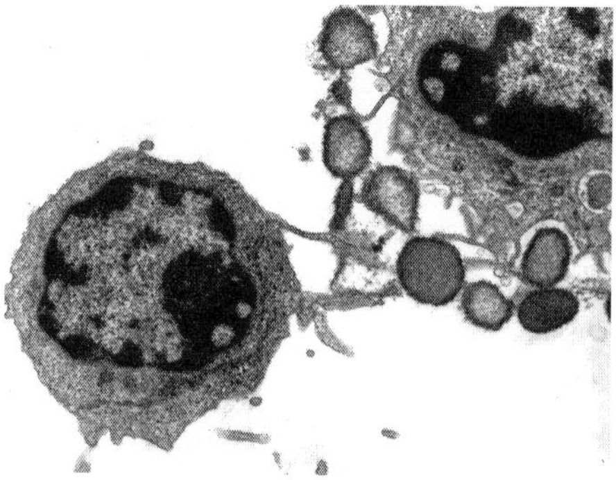

Figure 17.10 Electron micrograph showing a natural killer cell (bottom left) destroying a target tumour cell. Courtesy of Dr J.T. Thornthwaite of the Cancer Research Institute of West Tennessee, USA (http://cancerfoundation.com).

- The initiator caspase-8 hydrolyses Bid to a truncated form, tBid, which is able to inhibit anti-apoptotic Bcl2 proteins, allowing pro-apoptotic BH123 proteins to form membrane pores and release cytochrome *c* and initiate the intrinsic response.

Survival or trophic factors and apoptosis

Apoptosis is prevented if a cell receives survival or trophic factors from other cells.

- These factors bind to cell surface receptors and suppress apoptosis by regulating the activities of Bcl2 proteins.
- For example, members of the insulin-like growth factor (IGF) family (Chapter 15) are survival factors (Figure 17.11).
- Binding of an IGF to a tyrosine kinase receptor (RTK) as described in Chapter 15 activates phosphoinositide 3-kinase (PI-3K) as shown in Figure 17.12.
- Phosphoinositide 3-kinase is now able to catalyse the conversion of phosphatidylinositol 4,5-bisphosphate (PIP_2) to phosphatidylinositol 3,4,5-trisphosphate (PIP_3).
- Phosphatidylinositol 3,4,5-trisphosphate functions as a membrane anchor, which recruits an inactive protein kinase first isolated from the mouse retrovirus AKT8 and therefore called Akt (it is also called protein kinase B (PKB)).
- Akt is activated upon binding and can now prevent apoptosis by phosphorylating the pro-apoptotic protein Bad, which is normally bound in a dimer with Bcl2 (Figure 17.12).
- Thus, the anti-apoptotic protein, Bcl2 and an inactive form of the pro-apoptotic protein, Bad, have been released, which promotes cell survival.
- The corollary of this mechanism is that the non-receipt of cell survival factors allows a cell to enter apoptosis.

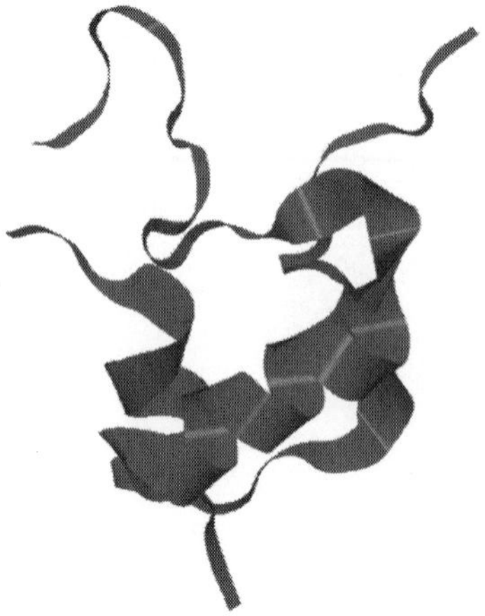

Figure 17.11 Molecular model of insulin-like growth factor. Compare with Figure 15.16. Model constructed using PDB file 1GZR.

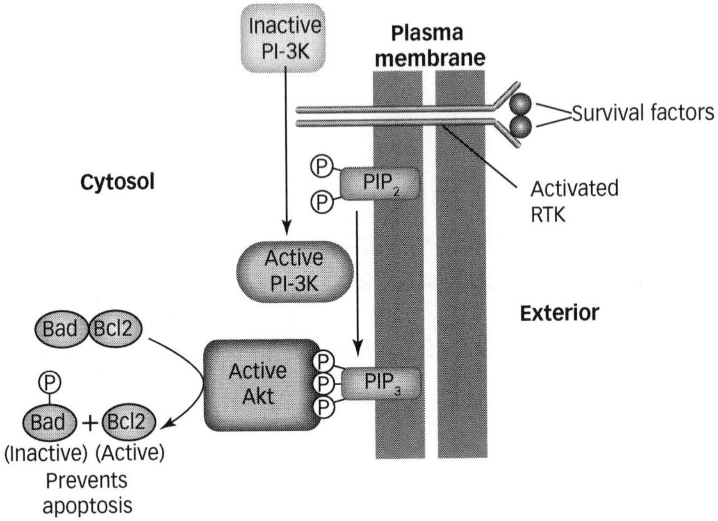

Figure 17.12 Activation of the phosphoinositide 3-kinase (PI-3K)-Akt/ protein kinase B (PKB) pathway by a survival factor. See text for details. PIP_2, phosphatidylinositol 4,5-bisphosphate; PIP_3, phosphatidylinositol 3,4,5-trisphosphate and RTK, tyrosine kinase receptor. Compare with Figure 15.19.

Looking for extra marks?

Defects in apoptosis are involved in many clinical conditions. These include neurodegenerative diseases such as Alzheimer's and Parkinson's diseases, and autoimmune disorders, for example, rheumatoid arthritis, where healthy cells die, and in neoplasias or cancer development, where cells that should be eliminated are protected from apoptosis.

Revision tip

Do browse through the further reading references given for this chapter online. Go to http://www.oxfordtextbooks.co.uk/orc/thrive/

Check your understanding

17.3 Which of the following statements is/are true or is/are false?

a. Apoptotic death is essential for embryonic development and homeostasis in multicellular organisms.

b. Syndactyly is the process by which apoptosis removes the interdigital spaces in a fetus.

c. Apoptosis can be triggered by the excessive production of reactive oxygen species or long-term anoxia.

d. All Bcl2 proteins contain four Bcl2 homology domains.

e. All pro-apoptotic proteins such as Bax and Bid are BH123 proteins.

f. Stress signals can initiate the intrinsic apoptotic pathway through the activities of BH3-only proteins.

g. Initiator caspases include caspase-8, -7 and -9.

h. Fas, Fas ligands, and the Fas-associated death domain protein all possess death domains.

i. Caspase-8 is able to catalyse the hydrolysis of Bid and so links the activities of the extrinsic and intrinsic apoptotic pathways.

j. Survival factors bind to tyrosine kinase receptors leading to the inactivation of phosphoinositide 3-kinase.

18 Viruses

Key features of viruses

- **Viruses** are infectious, obligate, intracellular parasites that infect organisms from all three domains: Bacteria, Archaea, and Eucarya.
- All organisms are infected by a range of viruses, which are largely specific for their host and do not infect other species.
- Viruses are the causative organisms of many human, animal, and plant diseases, although most do not cause disease and do little harm to their hosts.
- Virus particles can be examined by electron microscopy, although they can also be studied by X-ray diffraction methods because many can be crystallized. Molecular modelling is also increasingly being employed to visualize their structures (Figure 18.1).
- Viral particles or virions are not cellular and consist of an RNA or DNA genome enclosed in a protective protein coat called a capsid.
- Double- and single-stranded RNA and DNA molecules are found in different types of viruses.
- Some viral particles also have a lipid bilayer–protein membrane or envelope derived from the plasma membrane of the host cell, which surrounds the nucleocapsid.
- Outside a cell, virus particles are inert: they can only be replicated within a host cell.

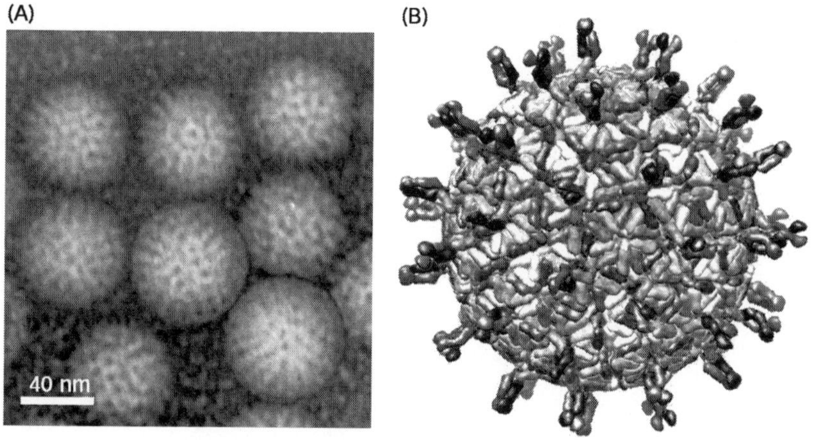

Figure 18.1 (A) Electron micrograph and (B) atomic model of a rotavirus. (A) Courtesy of H. Cotterill, Manchester Royal Infirmary, UK. (B) Atomic model constructed using PDB 3IYU.

- In general, a virus is specific for its host species although some can infect a closely related group of species.
- Within its host, a particular virus normally displays a preference for certain types of cells, a phenomenon called tissue tropism.
- Once infected, a host cell provides some of the enzymes and the biological energy required for the manufacture of new viral particles.
- Release of new virus particles by the host completes the cycle of virus replication and allows the virus to spread and infect new cells.
- Viruses may have originated from degenerate cells or from vagrant pieces of nucleic acids that escaped destruction by host cells.
- Mutations and recombination can quickly alter virus genomes and facilitate their rapid evolution.
- Viruses that use RNA genomes evolve much more quickly than those with DNA.

Looking for extra marks?

Extrapolating from the number of known human viruses (the best studied virus host species) it is estimated that there are at least ten times as many types of viruses as organisms.

18.1 DISCOVERY OF VIRUSES

Viruses were only discovered approximately 120 years ago.

- In 1892, Iwanowski showed that the causative agent of tobacco mosaic disease was not retained by filters that removed bacterial cells from liquid suspensions.

- He surmised he had discovered a new type of infectious agent, which he called *contagium vivum fluidum*.
- In 1892, Loeffler and Frosch found foot and mouth disease of cattle could also be passed from animal to animal in cell-free filtrates.
- They further showed that diluting the filtrate did not decrease its infectivity and therefore the infectious agent must be multiplied (replicated) in the host animals.
- Bacterial viruses were discovered by Twort in 1915 during his studies on micrococci.
- In 1917, d'Hérelle coined the term **bacteriophage** (usually shortened to **phage**) for viruses that infect bacteria.
- In 1933, Schlessinger used differential centrifugation (Chapter 2) to produce a purified suspension of a bacteriophage.
- Chemical analysis of the isolated phage particles showed them to consist of protein and DNA.
- Tobacco mosaic virus (TMV) particles were isolated in 1935 by Stanley.
- Two years later, Bawden and Pirie showed that they consisted of protein and RNA.
- The discovery of phages was a significant observation, since at the time animal and plant viruses could only be studied by infecting whole organisms.
- Since bacteria can be relatively easily grown in defined media this provided a ready means of studying the growth of bacteriophages.
- However, it was realized that viruses only require cells (not organisms) for growth and in 1928 viruses were propagated in minced kidney tissue, and during the 1930s in developing chicken embryos.
- The study of viruses that cause diseases of humans, animals, and plants only started to advance in the 1950s when the development of antibiotics that suppressed the growth of bacterial contaminants allowed the relatively routine culture of eukaryotic cells.
- Modern techniques in molecular biology such as inserting genes into vectors for cloning and studying their expression, and sequencing whole genomes have revolutionized the study of virology.

➔ *Tissue culture and molecular biology techniques are described in Chapter 2 and in the companion book,* Thrive in Biochemistry and Molecular Biology.

18.2 SIZES AND STRUCTURES OF VIRUS PARTICLES

Virus particles or virions range in size from approximately 30 to 300 nm in length/diameter, making them considerably smaller than cells (Figure 18.2).

- They are best observed using electron microscopy or X-ray diffraction techniques (Chapter 2) although computer-based molecular modelling is increasingly employed to visualize their structures (Figure 18.1).
- Virus particles have a generalized structure consisting of a protein coat called a capsid, which encloses a nucleic acid core (Figure 18.3).

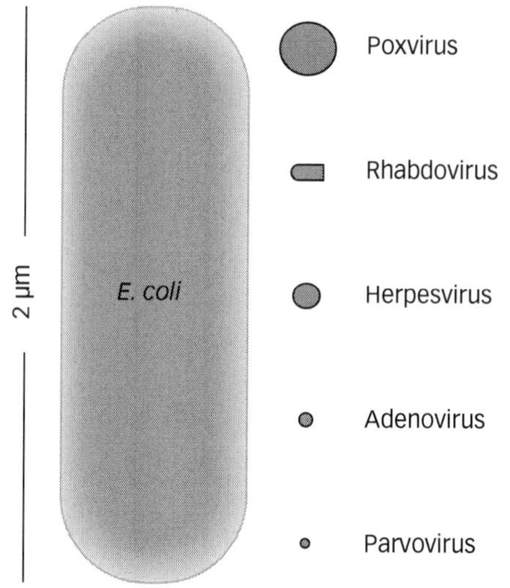

Figure 18.2 Comparative sizes of representative virus particles compared to that of an *Escherichia coli* cell. Redrawn from Collier, L., Kellam, P., and Oxford, J. (2011) *Human Virology*, 4th edition, Oxford: Oxford University Press.

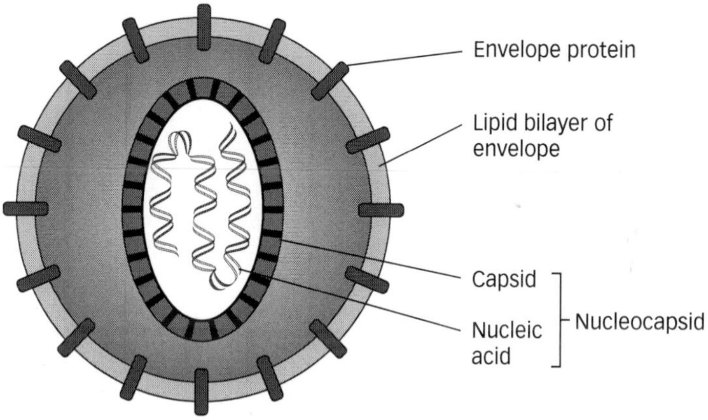

Figure 18.3 Generalized structure of a virus particle. Not all viruses are enveloped.

Capsids

The capsid is constructed of units called **capsomeres**; each capsomere, in turn, is formed of one or more proteins called **protomers**.

- The protomers of a single capsomere may be different or identical.
- The capsid protects the nucleic acid core and recognizes receptors on the surfaces of host cells in non-enveloped viruses (see later).
- Also, in non-enveloped viruses, capsids possess antigenic sites and are able to stimulate a specific immune response.

Nucleic acid

The nucleic acid encodes the genome of the virus, which is considerably smaller than those of cells.

- The nucleic acid present is DNA *or* RNA, never both.
- Depending upon the type of virus, the DNA or RNA may be single or double stranded (ss or ds).
- The lengths of viral nucleic acids are generally only a few thousand nucleotides or base pairs (depending whether they are single or double stranded).
- Generally virus genomes contain fewer than about 20 genes, although larger viruses, such as *Vaccinia*, have more than 100.
- The nucleic acid and protein coat together constitute the **nucleocapsid**.

➔ *The structures of proteins and nucleic acids are revised in the companion text,* Thrive in Biochemistry and Molecular Biology.

Envelope

In the case of **enveloped viruses**, the nucleocapsid is enclosed with a lipoprotein membrane, which is acquired when the viral particle leaves its host cell.

- In enveloped viruses, specific glycoproteins in the envelope, often referred to as 'spikes', bind to receptors on host cells and allow the envelope and plasma membrane of the cell to fuse.
- Spike proteins are also antigenic and able to elicit a specific immune response.
- The envelope is often lined with a layer of matrix proteins.

18.3 DIFFERENT TYPES OF VIRUS SYMMETRIES

Virus particles show a range of different shapes. However, they can be divided into three basic types depending upon the symmetry displayed.

- These include:
  - **helical symmetry** giving a rod or filamentous structure
  - **cuboid symmetry** giving an **icosahedral** structure
  - **complex symmetry**.
- The vast majority of virus particles have a helical or icosahedral symmetry.

Helical viruses

Helical viruses appear rod-shaped when examined by electron microscopy.

- Among the best studied helical viruses are tobacco mosaic, influenza, and vesicular stomatitis viruses (Figure 18.4).

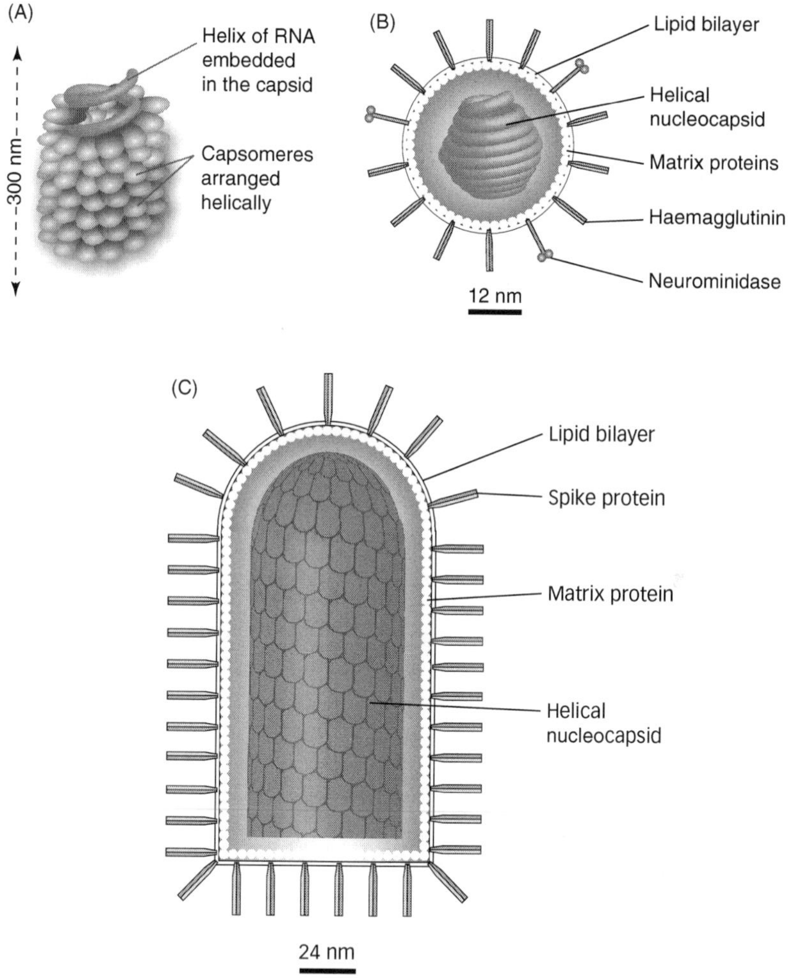

Figure 18.4 Schematic diagrams showing the helical structures of (A) the tobacco mosaic virus, and the nucleocapsids of the enveloped influenza (B) and vesicular stomatitis (C) virus particles.

Tobacco mosaic virus

- TMV particles exist as rigid rods 300 nm long and 18 nm in diameter (Figures 18.4a and 18.5).
- It consists of 2130 single capsid proteins (M_r approximately 17,500) enclosing an ss RNA molecule 6.4 kb in size.
- Each turn of the helix corresponds to 16.3 proteins.
- The RNA molecule winds coaxially with the protein, with each protein binding three nucleotides.
- The negative charges of the nucleotide phosphate groups are neutralized by basic amino acid residues of the capsomeres.

Different types of virus symmetries

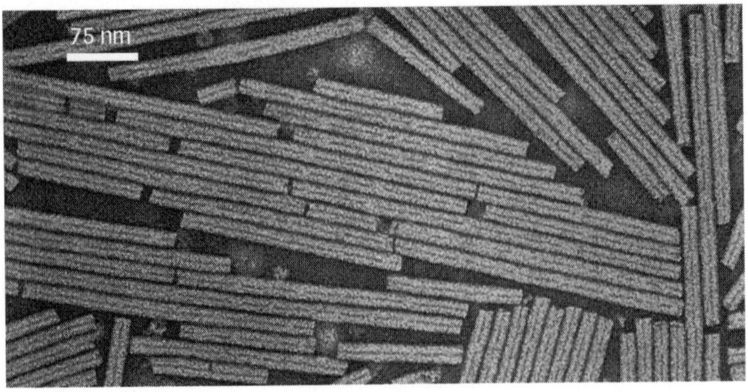

Figure 18.5 Electron micrograph of tobacco mosaic virus. Courtesy of C.M. Clay, Institute for Horticultural Research, Wellesbourne, UK.

Icosahedral viruses

When viewed by electron microscopy, icosahedral viruses usually appear roughly spherical (Figure 18.6).

- Icosahedral viruses are regular solid structures with 20 faces, each one forming an equilateral triangle, and 12 vertices (Figure 18.7).
- The number of possible protein subunits in an icosahedral shell is $60T$, where T is called the **triangulation number**.
- Possible values of T are 1, 3, 4, 7, 9, 12, 13, 16 . . .
- The smallest number of protomers that can form a regular icosahedron is 60, i.e. three in each of the 20 faces.

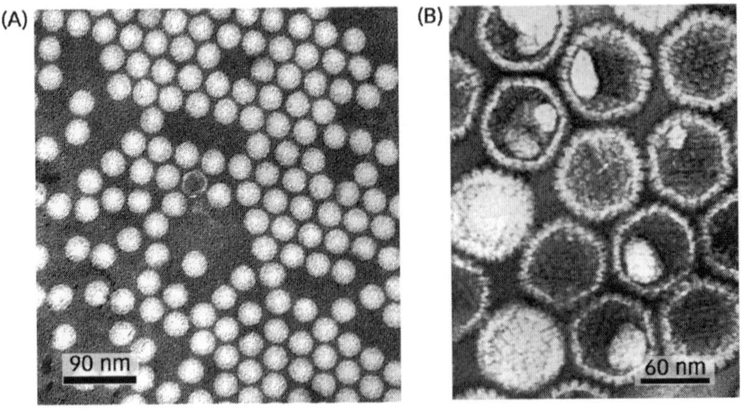

Figure 18.6 Electron micrograph of icosahedral viruses. (A) Polio and (B) herpes viruses. Courtesy of Dr F.A. Murphy and S. Whitfield and the Public Health Image Library, Centers for Disease Control and Prevention, USA.

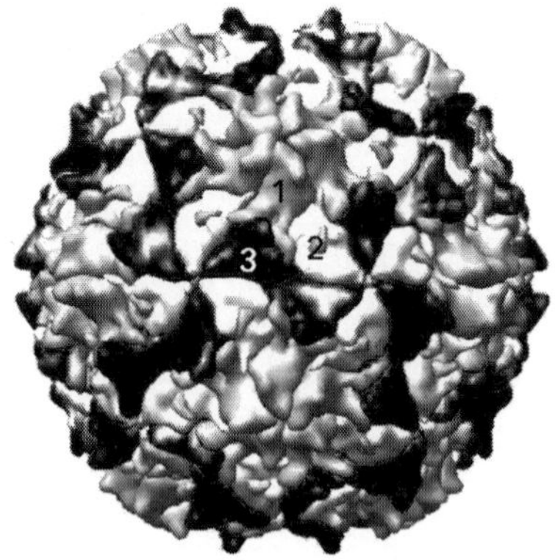

Figure 18.7 Atomic model of a polio virus particle. Note the regular icosahedron structure formed by three different subunits (1, 2, and 3) forming each capsomere. Model built using PDB file 1AR8.

- Each protein 'subunit' in a capsomere can be formed of several polypeptides (polymeric) or a combination of proteins.
- Three protomers are found in some small viruses but most use larger numbers to generate a bigger shell capable of holding a larger nucleic acid.
- Some bacteriophages, such as T4 or λ (lambda), can be regarded as icosahedrons with tails (Figure 18.8), an arrangement sometimes called head-tail morphology.

Check your understanding

18.1 (a) What is the minimum number of protomers in the shell of an icosahedral virus with a triangulation number of 4? (b) If each protein is tetrameric, how many subunits are present?

Revision tip

See a superb animation showing the assembly of a T4 bacteriophage at: http://www.youtube.com/watch?v=Ofd_IgEymto&feature=related

Complex virus symmetry

Many large viruses, such as *Vaccinia* and other pox viruses, do not show helical or icosahedral symmetries (Figure 18.9).

- These viruses have lipids in both their envelopes and an external coat, which encloses tubular or globular protein structures.
- They are often referred to as complex viruses due to their lack of apparent symmetry.

(A)

(B)

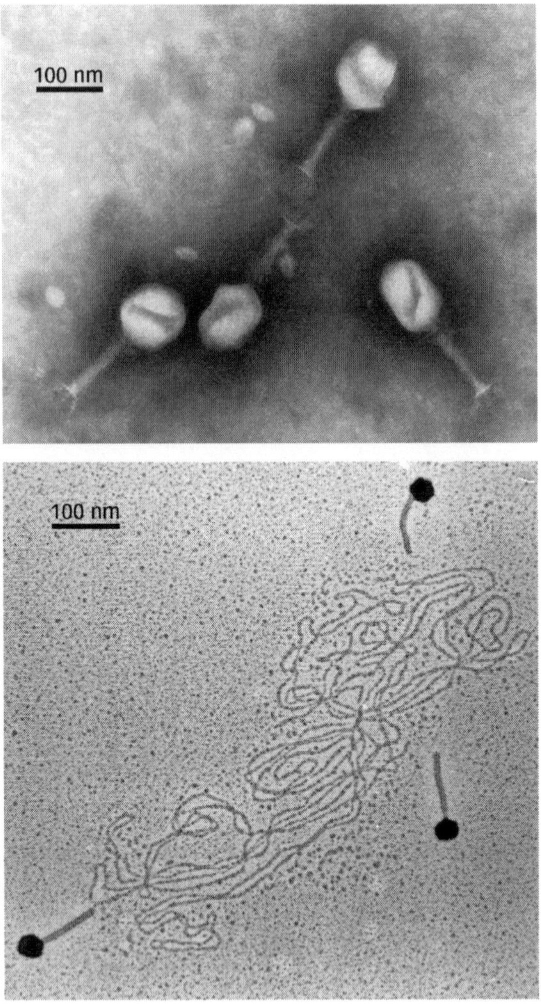

Figure 18.8 Electron micrographs of (A) T4 and (B) λ (lambda) bacteriophages. Courtesy of North West Regional Virus Laboratory, Booth Hall Hospital, Manchester, UK.

Revision tip

Try the keyword search 'complete virus particles' to explore the structure of viruses using the Jmol molecular modeller at the Protein Data Bank (PDB): http://www.rcsb.org/pdb/explore/explore.do?structureId=3IYM

18.4 NOMENCLATURE AND CLASSIFICATION OF VIRUSES

The naming and classification of viruses is formally overseen by the International Committee on Taxonomy of Viruses (ICTV).

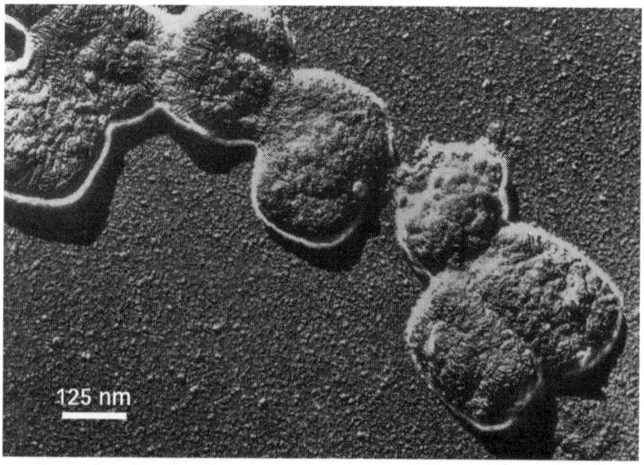

Figure 18.9 Electron micrograph of a *Vaccinia* virus. Courtesy of North West Regional Virus Laboratory, Booth Hall Hospital, Manchester, UK.

Nomenclature

The methods of naming of viruses are diverse.

- Systems used include naming according to:
  - the disease caused
  - the place or organ/tissue from where they were first isolated
  - morphological feature of the virus particle
  - acronyms summarizing a number of features of the virus
  - the name of the first person(s) to isolate the virus
  - idiosyncratic terms such as T2, T4, M13, and φX174.
- Examples of these methods of naming viruses are given in Table 18.1.

Classification of viruses

A number of different systems are used to classify viruses, including those based on:

- diseases caused
- host organism
- virion morphology
- nucleic acid present
- taxonomic classification.

- Classifying viruses on the basis of the diseases they cause, identity of their hosts, and virion morphology have proved problematical because:
  - Most viruses do not cause disease, while some can lead to different diseases at different stages of their replication cycle (see 'Replication cycle of viruses' section).

Nomenclature and classification of viruses

Method of naming	*Examples*	*Comment*
Disease caused	Human immunodeficiency viruses Influenza viruses Tobacco mosaic virus	
Place/organ first isolated	Adenoviruses Coxsackie virus Marburg virus	First isolated from adenoids Small town in New York State, USA German town where first human cases recorded (it normally affects the green monkey)
Morphological feature	Aleutian mink disease virus Coronaviruses	Named because its blue-grey coat is similar in colour to the Aleutian blue fox Virions have a corona of spike proteins
Acronyms	ECHO viruses Papovaviruses Picornaviruses	*E*nteric *c*ytopathic *h*uman *o*rphan *Pa*pilloma *po*lyoma *va*cuolating *Pico* (small) *RNA* (ribonucleic acid)
Name of first person(s) to isolate the virus	Epstein–Barr Rous	Named after two British pathologists USA scientist who received the Nobel Prize in Physiology or Medicine in 1966, 55 years after discovering this virus

Table 18.1 Examples of virus nomenclature

- Host-based classification has problems because some viruses infect a restricted range of hosts; all primates can catch influenza.
- The range of morphologies shown by virus particles is quite restricted (see Sections 18.2 and 18.3) and therefore says little about their biology.

- Classifications based on the types of nucleic acid (Baltimore system) or combinations of taxonomic and molecular biological features (ICTV formal taxonomy) have proved much more useful.

Baltimore classification of viruses

- In 1971, the USA biologist Baltimore suggested using the nature of the nucleic acid in the virus as a basis for classifying viruses.
- His classification grouped viruses according to:
    - the type of nucleic acid present (DNA or RNA)
    - whether the nucleic acid present is double stranded (ds) or single stranded (ss)
    - the polarity of ss RNA molecules when present; RNA strands that can be directly translated are termed positive or plus (+), negative strands (–) require the synthesis of their complementary plus strand for translation to proceed
    - whether a DNA intermediate need be synthesized in the replication of an ss RNA genome
    - whether an RNA intermediate need be synthesized in the replication of a ds DNA genome.
- The Baltimore classification of viruses divides them into seven types as listed in Table 18.2.

Type of virus	*Nucleic acid present*	*Examples*
I	[a]ds DNA	Adenoviruses Herpesviruses Poxviruses Phages T4 and λ
II	[a]ss DNA	Parvoviruses Phage φX174
III	ds RNA	Reoviruses
IV	ss RNA [b]plus (+) configuration	Coronaviruses Picornaviruses Togaviruses
V	ss RNA [b]negative (–) configuration	Influenza virus Rabies virus Vesicular stomatitis virus
VI	ss RNA replicates with DNA intermediate	Retroviruses
VII	ds DNA replicates with RNA intermediate	Hepatitis B virus

Table 18.2 Baltimore system of classifying viruses

[a] ds, double stranded; ss, single stranded
[b] Positive or plus (+) RNA strands can be directly translated; negative strands (–) require the synthesis of their complementary (+) strand before translation can proceed

Looking for extra marks?

A key advantage of the Baltimore classification is that the general mechanisms used to replicate virus nucleic acid and transcribe protein-encoding genes can be predicted knowing the type of nucleic acid present (see 'Virus nucleic acids' section).

International Committee on Taxonomy of Viruses

- The ICTV formally classifies viruses using a hierarchical system of four taxons (formal classification groups).
- Viruses are formally classified using order, family, genus, and species name.
- Higher taxons (class, phyla, and so on) are not used.
- Features used by the ICTV to assign a particular virus to any one taxon include:
  - type of host
  - morphological features of the virion, including the length of tail in bacteriophages
  - nature of the nucleic acid
  - the presence/absence of specific genes in apparently similar viruses
  - the similarity of nucleotide sequences in genes or, for an increasing number, sequences of complete virus genomes.
- The ICTV reassesses the use of these characters and updates its classification as new viruses are discovered and issues updated reports every few years.
- The 2011 release of the ICTV recognizes six orders of viruses; each order is then subdivided into a number of families as illustrated in Table 18.3.

Order	*Number of families*
Caudovirales	3
Herpesvirales	3
Mononegavirales	4
Nidovirales	3
Picornavirales	5
Tymovirales	4
Virus families not assigned to an order	72

Table 18.3 Six orders of viruses recognized by the International Committee on Taxonomy of Viruses in Executive Committee 2011

- In addition, 65 families of viruses had yet to be assigned to a particular order.
- Each family is then further divided into genera, which contain individual species.
- Thus, hepatitis A virus is classified as:
  - Order: Picornavirales
  - Family: *Picornaviridae*
  - Genus: *Hepatovirus*
  - Species: hepatitis A virus

Check your understanding

18.2 Is it true that viruses are formally classified into the same taxa used for organisms?

18.5 REPLICATION OF VIRUSES

Virus particles outside host cells are biologically inert: they can only replicate within host cells. New virions are assembled from virus nucleic acid and virus proteins synthesized by the host.

- Viruses generally subvert the normal cell machinery for their own production.
- Different types of viruses are replicated by the host cell in differing ways.
- The genomes of many viruses encode their own RNA or DNA polymerases, while others use replication enzymes and other proteins of the cell.
- In the latter cases, the virus genome always encodes at least one regulatory protein required for replication.
- Viral genes that are expressed before the formation of virus nucleic acid encode **early proteins**.
- These proteins are normally involved in replicating the virus nucleic acid.
- **Late proteins** are produced following replication of the genome and are mainly structural proteins.

- Virus replication may include an inactive or latent phase during which the virus remains dormant.
  - For example, the virus, *Varicella zoster* (the chicken pox virus) can enter nerve cells and remain inactive.
  - The virus can later be reactivated if, for example, the host immune system becomes weakened, and causes painful attacks of shingles in areas of the body served by the nerve.
- '**Satellite viruses**' can only replicate within their host with the assistance of proteins encoded by other viruses.
- Newly formed virions then exit the host cell and enter new cells, so beginning a new **replication cycle**, which is also called the **multiplication** or **infectious cycle**.

Replication cycle of viruses

Despite the diversity of types of nucleic acids found in viruses (Table 18.2) and the differing replication strategies they show, the replication cycle can be divided into a number of common stages.

- These include:
  - recognition and adsorption
  - entry
  - uncoating
  - synthesis of virus components
  - assembly of new virions
  - release.
- These steps are summarized in Figure 18.10.
- In some cases, the separate steps are combined.

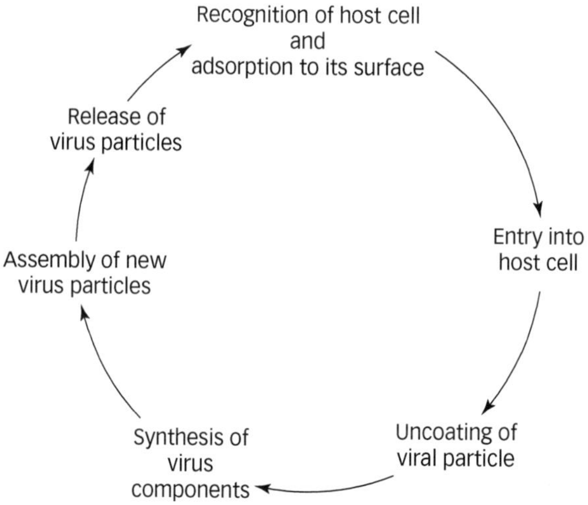

Figure 18.10 Outline of the replication cycle of a virus.

Replication of viruses

Recognition and adsorption

- Virus particles recognize and bind to their host cell by attaching to specific molecules or receptors on its surface (a process sometimes called **docking**).
- Examples of viruses and the cellular receptors they use are listed in Table 18.4.

Entry

- In most cases, a virus infects a cell enclosed within a cell wall. Only after the virus has circumvented or penetrated the cell wall, can it attempt to enter the cell.

Crossing cell walls

- Bacteriophage particles possess endopeptidases and lysozymes that are able to degrade bacterial cell walls and allow the virus to inject its nucleic acid into the bacterial cytoplasm.
- The capsid of the bacteriophage does not enter the bacterial cell but remains outside it (Figure 18.11).
- Viruses that infect archaea are thought to penetrate their cell walls.
- Viruses that infect fungi generally do so persistently and never leave the host cell. This eliminates the need to cross the fungal cell wall.
- These viruses are spread during fungal sexual reproduction when cell–cell fusion occurs.
- Viruses of algae produce $\beta,1\rightarrow$ glucanases and chitinases to help degrade their host cell walls and effect entry.

➔ *Prokaryotic and eukaryotic cell walls are revised in Chapters 4 and 13 respectively.*

- Many plant viruses avoid the problem of crossing cell walls by using arthropod vectors to introduce them directly into the cytoplasm of their hosts or by persistently infecting their hosts and relying on seeds for their transmission.
- Once a plant cell is infected, virus particles can spread to other cells through plasmodesmata.

Virus	*Receptor*
Epstein–Barr virus	Complement receptor CR2
Foot and mouth disease	Specific integrins (Chapter 14)
Herpes simplex virus	Heparin sulphate proteoglycans (Chapter 14)
Human immunodeficiency virus (HIV)	[a]CD4
Influenza virus	Sialic acid residues of glycoproteins and glycolipids (Chapter 3)
Measles virus	CD150
Polio virus	CD155
Rabies virus	Acetylcholine receptor (Chapter 15)
Semliki forest virus	Human leukocyte antigens (HLAs)

Table 18.4 Examples of host cell receptors

[a] CD, cluster of differentiation. See the companion volume, *Thrive in Immunology*.

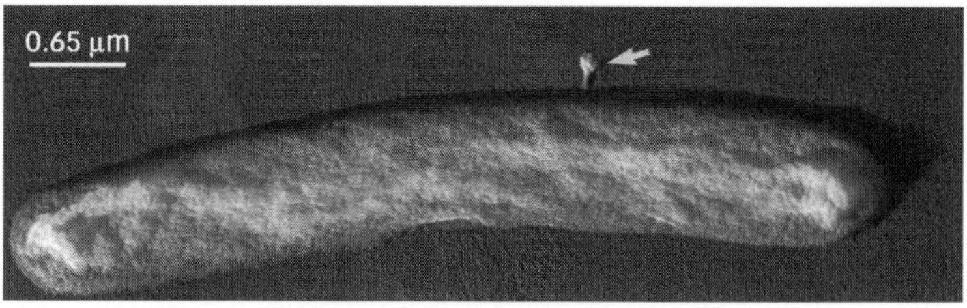

Figure 18.11 Electron micrograph showing a single bacteriophage docked on the surface of an *Escherichia coli* cell and injecting its nucleic acid into the bacterium.

Entering the cell

- Viruses enter cells using one of a number of different routes, including phagocytosis or receptor-mediated endocytosis (Figure 18.12).

➔ *Phagocytosis or receptor-mediated endocytosis are discussed in Chapters 3 and 9.*

- Following internalization, the virus nucleocapsid must still cross a phagocytic vesicle membrane, its own envelope, or both, to be released into the cytosol.
- Fusion proteins on the surface of the virus particle catalyse fusion of the virus envelope and host membrane and allow release of the nucleocapsid.
- In some cases, direct fusion of the virus envelope with the host plasma membrane can release the nucleocapsid directly into the cytoplasm.

Uncoating

- Uncoating releases or makes the viral nucleic acid accessible to the metabolism of the cell.
- Different uncoating mechanisms are employed depending upon the nature of the virus.
  - Acidification in the endosomal-lysosmal system (Chapter 9) is essential for uncoating influenza viruses.
  - Exposure of adenoviruses to the proteolysis in the reducing cytosol (Chapter 7) results in the capsid binding to nucleopore complexes (NPCs) and transport of the virus nucleic acid into the nucleus.
  - Phosphorylation of the hepatitis B virus (HBV) capsid proteins exposes their nuclear locating signals (NLSs) to the nuclear transport machinery and the small capsid (diameter 32–36 nm) can be directly transported through the NPC.
- Following their transport through the central pore of the NPC, capsid protein and DNA from mature viruses are released into the nucleoplasm.
- The 125-nm diameter herpes simplex virus (HSV) particles are too large to traverse NPCs.
- These virus particles dock with the cytoplasmic side of the NPC in an importin-dependent manner and release their DNA through the NPC into the nucleoplasm.

➔ *Nucleocytoplasmic transport is revised in Chapter 6.*

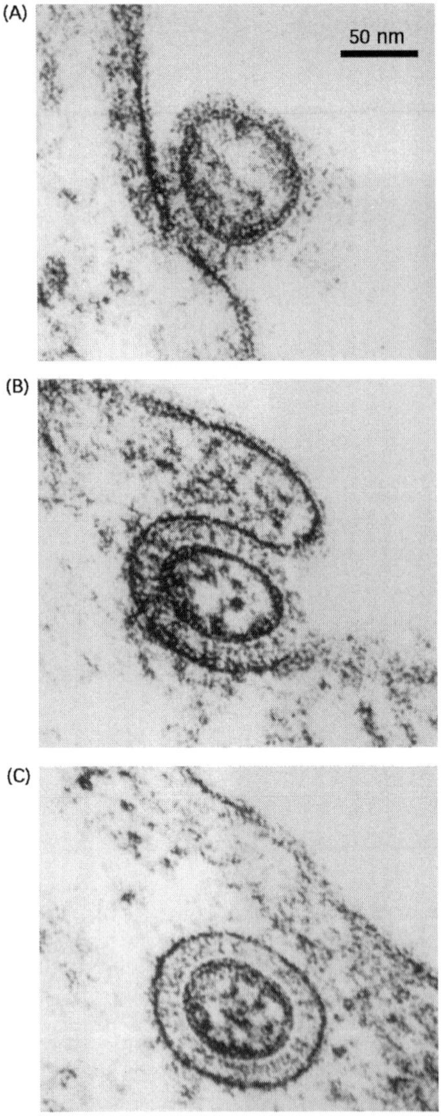

Figure 18.12 Electron micrographs showing the entry of influenza virus by receptor-mediated endocytosis. (A) Binding, (B) envelopment, and (C) internalization of the virus. Courtesy of Dr J.J. Shekel, National Institute for Medical Research, London, UK.

Synthesis of virus components

- Once it has been uncoated and available to enzymes and other proteins, the virus genome directs the metabolism of the cell to produce new virus proteins and nucleic acids.
- Given the diversity of virus genomes (Baltimore Classes I to VII), different routes are used to achieve these objectives.

Virus nucleic acids

- **Class I** viruses have ds DNA genomes that are replicated by the usual semiconservative mechanism described in Chapters 5 and 6.
- Their proteins can be produced by conventional transcription and translation (Chapters 5 and 7).
- **Class II** ss DNA viruses direct the synthesis of their complementary strand to give a ds DNA intermediate, which can be used to produce new genomes and proteins.
- **Class III** viruses transcribe the (–) strand of their ds RNA to form new (+) ss RNA.
- Transcription is catalysed by a virus-specific **RNA-dependent RNA polymerase** or **RNA replicase** that entered the host as part of the virion.
- The newly formed (+) RNA can be translated to give virus proteins and transcribed to form new (–) RNA genomes.
- **Class IV** viruses have a single (+) RNA strand that can be used directly as mRNA.
- One of the proteins formed from its transcription is a virus-specific, RNA-dependent RNA polymerase, which synthesizes complementary (–) RNA molecules that are used as templates to produce more (+) strands.
- **Class V** viruses transcribe their single (–) RNA to form a (+) RNA complement using a virus RNA-dependent RNA polymerase that, as with Class III viruses, is a component of the virus particle.
- The (+) copy can be directly translated to produce virus proteins and transcribed to give new genomes.
- **Class VI** viruses have (+) ss RNA genomes and virions possessing **reverse transcriptase**, which allows the ss RNA genome to be converted to a ds RNA–DNA heteroduplex and finally to a ds DNA one.
- Normal cell transcription and translation can then produce new virus genomes and proteins.
- **Class VII** viruses have ds DNA genomes but replicate using an RNA intermediate.

Check your understanding

18.3 Suggest plausible reasons why the genes of bacteriophages would not be expected to possess introns and exons but those of viruses that infect eukaryotic hosts might.

Virus proteins

- In general, virus proteins are produced initially as a single polypeptide or **polyprotein** that is hydrolysed at specific positions to generate the individual specific virus proteins.
- This situation is illustrated for the polio virus, which has a (+) ss RNA genome, in Figure 18.13.

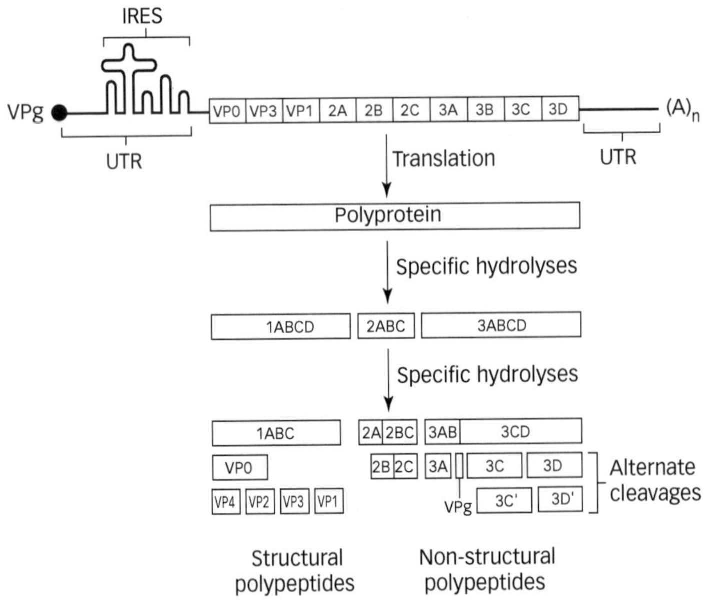

Figure 18.13 Schematic illustrating the production of the polio virus polyprotein and its subsequent hydrolysis to form individual virus proteins. Alternative hydrolyses of the protein 3CD forms 3C and 3D or 3C′ and 3D′ respectively. IRES, internal ribosome entry site; UTR, untranslated regions; VPg, virion-protein, genome-linked.

- The 5′ end of the RNA genome lacks a cap but has a 22-amino acid residue long peptide called VPg (virion-protein, genome-linked) covalently attached to it. A 3′ poly(A) tail is present.
- The lack of cap and presence of VPg means initiation of translation cannot be by the usual eukaryotic method. Indeed, the usual cap-based initiation system is prevented by virus-induced proteolytic damage to some initiation factors.

➔ *Eukaryotic translation is revised in Chapter 7.*

- Rather, the RNA has an **internal ribosome entry site** (**IRES**) on the 5′ side of the AUG initiator codon, which adopts a specific conformation (Figure 18.13).
- Ribosomes are able to bind to the IRES and translation is initiated.
- Elongation of translation produces a polyprotein approximately 2200 amino acid residues long, which is hydrolysed at specific positions to give a series of virus proteins (Figure 18.13).
- The hydrolytic events are catalysed by virus-encoded proteases that also autocatalyse their release from the polyprotein.

Assembly of new virions

- New virus particles are spontaneously assembled from the newly formed virus nucleic acid and proteins.

- Assembly may occur in the nucleus, cytoplasm, or a combination of the two, and may occur in stages, with an initial formation of subassemblies.
- Enveloped viruses acquire their envelope as they bud from the plasma membrane (Figure 18.14); the envelope forming from a portion of the host lipid bilayer enriched in virus membrane (spike) glycoproteins.

Looking for extra marks?

Emphasize that the replication of viruses involves their exploitation of the 'normal' biochemistry and molecular biology mechanisms of the cell.

Release

- Release of virus particles can be by lysis of the cells, by gradual extrusion, or by the budding of viruses at the plasma membrane (Figure 18.13).
- Particles of the bacteriophage T2 causes lysis of its *Escherichia coli* host cell, but the filamentous bacteriophage, M13, uses the same host and is released by a budding mechanism that does not kill the cell (Figure 18.15).

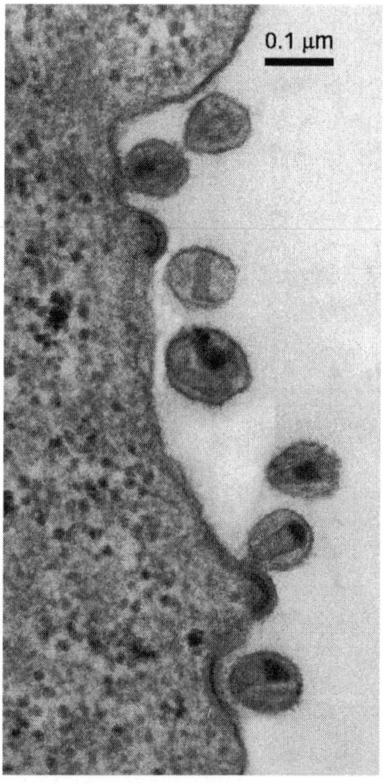

Figure 18.14 Electron micrograph showing the formation of the envelopes of human immunodeficiency viruses type 1 (HIV-1) as they are released by budding from the surface of a human leukaemia host cell. Courtesy of Dr D. Robertson and Professor R.A. Weiss, Institute for Cancer Research, Royal Cancer Hospital, UK.

Revision tip

Look at the excellent animation of the influenza virus replication cycle at: http://www.xvivo.net/zirus-antivirotics-condensed/

Check your understanding

18.4 Which of the following statements is/are true?

a. Early proteins produced before the formation of virus nucleic acid are generally structural proteins.

b. Baltimore Class III, IV, and V viruses all depend upon RNA-dependent RNA polymerase activity when their genomes are replicated.

c. Baltimore Class VI viruses rely on reverse transcriptase or RNA replicase activity when their genomes are replicated.

d. Despite having double-stranded DNA genomes, Baltimore Class VII viruses replicate their genomes using an RNA intermediate.

e. In general, transcription of virus RNA produces a single polypeptide called a polyprotein.

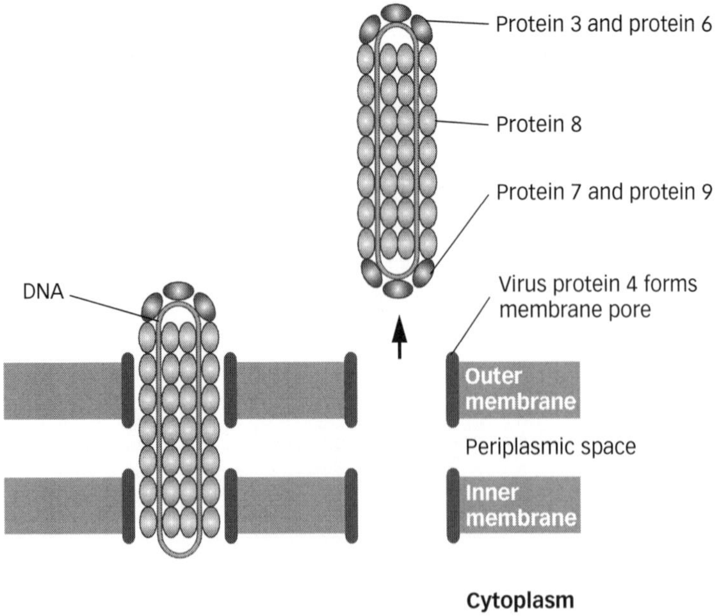

Figure 18.15 Schematic showing release of the filamentous bacteriophage M13 by budding from its *Escherichia coli* host. Budding involves the single-stranded virus DNA molecule leaving the cell through a cell formed of virus genome-encoded proteins. During budding the DNA is coated with the virus proteins P3 and P6 at the front end, the major portion with P8, and the rear with P7 and P9 as it exits the host.

18.6 ORIGIN AND EVOLUTION OF VIRUSES

Since all viruses are obligate intracellular parasites, they must have evolved after or with their host cells.

- Two common theories to explain their origins are:
  1. viruses are derived from degenerate cells
  2. viruses originated from vagrant pieces of nucleic acids.
- The first theory suggests that viruses evolved from prokaryotic or eukaryotic cells that parasitized other cells and came to depend on their host for replication.
- Eventually, the parasites lost their nucleus and cytoplasm, a process referred to as degeneracy.
- The second theory suggests that at some stage nucleic acid was accidentally transferred into a foreign cell, escaped degradation, and was replicated.
- Whatever the origin, the number and variety of types of viruses suggest they had multiple origins rather than having evolved from a single ancestor.

Recombination and mutation in viruses

Viruses, like all biological entities, are subject to evolution.

- The two major processes driving virus evolution are **recombination** and **mutation**, which produce variations that are acted upon by natural selection.
- Recombination can occur when two related viruses infect the same cell, and can generate different combinations of genes leading to virus particles with different phenotypes to either of the 'parental' types.
- Mutations in virus nucleic acids are a major source of new phenotypes.
- The replication of DNA is relatively error-free because of the proofreading capacities of DNA polymerases described in Chapters 5 and 6. Thus mutations in copying viral DNA are of the order of one nucleotide in 10^9 to 10^{10} per replication cycle.
- In contrast, RNA polymerases lack proofreading properties (Chapters 5 and 7). Thus, errors in copying virus RNA genomes occur at a rate of one nucleotide in 3×10^4 per replication cycle.
- Thus, RNA viruses evolve much more rapidly than DNA types.
- It has been estimated that the RNA-containing influenza A virus evolves sufficiently quickly that it takes only 4 years to produce a new phenotype that can no longer be recognized by the immune system of a previously protected individual.
- Like all parasites, viruses tend to change over the long term into less virulent forms during their evolution.

Check your understanding

18.5 Which of the following statements is/are false?

a. Virus particles cannot be observed using any form of light microscopy.

b. Viruses can only replicate within host cells.

c. Viruses can be separated from host cells by centrifugation and filtration.
d. Bacteriophages always induce bacteriolysis on their release.
e. The replication cycle of a virus may include a latent phase.

Revision tip

Do browse through the further reading references given for this chapter online. Go to http://www.oxfordtextbooks.co.uk/orc/thrive/

Check your understanding

18.6 Indicate whether the following statements is/are true or is/are false.

a. Both double- and single-stranded RNA and DNA molecules are found in different types of viruses.
b. Viruses always cause diseases in their host organisms.
c. Viruses were only discovered in 1892 by Iwanowski using electron microscopy.
d. Virus particles or virions range in size from approximately 10 to 300 μm in length/diameter.
e. The number of capsomeres and protomers in a capsid are always identical.
f. Both tobacco mosaic and vesicular stomatitis viruses have helical symmetries.
g. The nucleic acid, protein coat, and envelope of a virus particle form the nucleocapsid.
h. The three major types of symmetries exhibited by virus particles are helical or filamentous, icosahedral, and complex symmetries.
i. Viruses are classified in terms of order, class, genus, and species name.
j. Viruses that employ DNA genomes mutate more slowly than those with RNA genomes.

Index

D

P

Printed and bound by CPI Group (UK) Ltd, Croydon, CR0 4YY